Biodegradable Polymers and Textiles

Biodegradable Polymers and Textiles

Guest Editors

**Sandra Varnaitė-Žuravliova
Jolanta Sereikaitė
Julija Baltušnikaitė-Guzaitienė**

Basel • Beijing • Wuhan • Barcelona • Belgrade • Novi Sad • Cluj • Manchester

Guest Editors

Sandra Varnaitė-Žuravliova
Department of Textile
Technologies
Center for Physical Sciences
and Technology
Kaunas
Lithuania

Jolanta Sereikaitė
Department of Chemistry
and Bioengineering
Faculty of Fundamental
Sciences
Vilnius Gediminas Technical
University
Vilnius
Lithuania

Julija Baltušnikaitė-Guzaitienė
Department of Textile
Technologies
Center for Physical Sciences and
Technology
Kaunas
Lithuania

Editorial Office
MDPI AG
Grosspeteranlage 5
4052 Basel, Switzerland

This is a reprint of the Special Issue, published open access by the journal *Journal of Functional Biomaterials* (ISSN 2079-4983), freely accessible at: https://www.mdpi.com/journal/jfb/special_issues/Biodegradable_Polymers_Textiles.

For citation purposes, cite each article independently as indicated on the article page online and as indicated below:

Lastname, A.A.; Lastname, B.B. Article Title. *Journal Name* **Year**, *Volume Number*, Page Range.

ISBN 978-3-7258-3227-9 (Hbk)
ISBN 978-3-7258-3228-6 (PDF)
https://doi.org/10.3390/books978-3-7258-3228-6

© 2025 by the authors. Articles in this book are Open Access and distributed under the Creative Commons Attribution (CC BY) license. The book as a whole is distributed by MDPI under the terms and conditions of the Creative Commons Attribution-NonCommercial-NoDerivs (CC BY-NC-ND) license (https://creativecommons.org/licenses/by-nc-nd/4.0/).

Contents

About the Editors vii

Sandra Varnaitė-Žuravliova and Julija Baltušnikaitė-Guzaitienė
Biodegradable Polymers and Textiles
Reprinted from: *J. Funct. Biomater.* **2025**, *16*, 26, https://doi.org/10.3390/jfb16010026 1

Esmail M. El-Fakharany, Marwa M. Abu-Serie, Noha H. Habashy, Nehal M. El-Deeb, Gadallah M. Abu-Elreesh, Sahar Zaki and Desouky Abd-EL-Haleem
Inhibitory Effects of Bacterial Silk-like Biopolymer on Herpes Simplex Virus Type 1, Adenovirus Type 7 and Hepatitis C Virus Infection
Reprinted from: *J. Funct. Biomater.* **2022**, *13*, 17, https://doi.org/10.3390/jfb13010017 8

Polina Tyubaeva, Ivetta Varyan, Alexey Krivandin, Olga Shatalova, Svetlana Karpova, Anton Lobanov, et al.
The Comparison of Advanced Electrospun Materials Based on Poly(-3-hydroxybutyrate) with Natural and Synthetic Additives
Reprinted from: *J. Funct. Biomater.* **2022**, *13*, 23, https://doi.org/10.3390/jfb13010023 27

Claudio José Galdino da Silva Junior, Julia Didier Pedrosa de Amorim, Alexandre D'Lamare Maia de Medeiros, Anantcha Karla Lafaiete de Holanda Cavalcanti, Helenise Almeida do Nascimento, Mariana Alves Henrique, et al.
Design of a Naturally Dyed and Waterproof Biotechnological Leather from Reconstituted Cellulose
Reprinted from: *J. Funct. Biomater.* **2022**, *13*, 49, https://doi.org/10.3390/jfb13020049 45

Anjana S. Desai, Akanksha Singh, Zehra Edis, Samir Haj Bloukh, Prasanna Shah, Brajesh Pandey, et al.
An In Vitro and In Vivo Study of the Efficacy and Toxicity of Plant-Extract-Derived Silver Nanoparticles
Reprinted from: *J. Funct. Biomater.* **2022**, *13*, 54, https://doi.org/10.3390/jfb13020054 59

Joseph Merillyn Vonnie, Kobun Rovina, Rasnarisa Awatif Azhar, Nurul Huda, Kana Husna Erna, Wen Xia Ling Felicia, et al.
Development and Characterization of the Biodegradable Film Derived from Eggshell and Cornstarch
Reprinted from: *J. Funct. Biomater.* **2022**, *13*, 67, https://doi.org/10.3390/jfb13020067 80

Khmais Zdiri, Aurélie Cayla, Adel Elamri, Annaëlle Erard and Fabien Salaun
Alginate-Based Bio-Composites and Their Potential Applications
Reprinted from: *J. Funct. Biomater.* **2022**, *13*, 117, https://doi.org/10.3390/jfb13030117 95

Artem Egorov, Bianca Riedel, Johannes Vinke, Hagen Schmal, Ralf Thomann, Yi Thomann and Michael Seidenstuecker
The Mineralization of Various 3D-Printed PCL Composites
Reprinted from: *J. Funct. Biomater.* **2022**, *13*, 238, https://doi.org/10.3390/jfb13040238 126

María José Lovato, Luis J. del Valle, Jordi Puiggalí and Lourdes Franco
Performance-Enhancing Materials in Medical Gloves
Reprinted from: *J. Funct. Biomater.* **2023**, *14*, 349, https://doi.org/10.3390/jfb14070349 144

Matheus F. Celestino, Lais R. Lima, Marina Fontes, Igor T. S. Batista, Daniella R. Mulinari, Alessandra Dametto, et al.
3D Filaments Based on Polyhydroxy Butyrate—Micronized Bacterial Cellulose for Tissue Engineering Applications
Reprinted from: *J. Funct. Biomater.* **2023**, *14*, 464, https://doi.org/10.3390/jfb14090464 . . . . . . **178**

Sandra Varnaitė-Žuravliova and Julija Baltušnikaitė-Guzaitienė
Properties, Production, and Recycling of Regenerated Cellulose Fibers: Special Medical Applications
Reprinted from: *J. Funct. Biomater.* **2024**, *15*, 348, https://doi.org/10.3390/jfb15110348 . . . . . . **196**

About the Editors

Sandra Varnaitė-Žuravliova

Sandra Varnaitė-Žuravliova is a Senior Researcher at Department of Textile Technologies of the State Research Institute Center for Physical Sciences and Technology (FTMC). Her projects, scientific interests and publications cover the following topics: electrospinning; biodegradable materials; microcapsules; phase change materials; textiles; conductive materials; antiradar coatings.

Jolanta Sereikaitė

Jolanta Sereikaitė obtained her PhD degree from Vilnius University and now is a principal investigator in the Laboratory of Molecular Biotechnology at Vilnius Gediminas Technical University. Her current research mainly focuses on the encapsulation of bioactive compounds and the application of biopolymers for the development of delivery systems for carotenoids and antimicrobials.

Julija Baltušnikaitė-Guzaitienė

Julija Baltušnikaitė-Guzaitienė is the senior scientific researcher at Department of Textile Technologies of the Center for Physical Sciences and Technology. Her projects, scientific interests and publications cover the following topics: biodegradable materials; functional textile materials; sustainability; adaptive materials; conductive textiles.

Editorial

Biodegradable Polymers and Textiles

Sandra Varnaitė-Žuravliova * and Julija Baltušnikaitė-Guzaitienė

Department of Textile Technologies, Center for Physical Sciences and Technology, Demokratu Str. 53, LT-48485 Kaunas, Lithuania; julija.baltusnikaite@ftmc.lt
* Correspondence: sandra.varnaite.zuravliova@ftmc.lt

The increasing interest in developing biodegradable polymers through chemical treatments, microorganisms, and enzymes highlights a commitment to environmentally friendly disposal methods. These polymers are engineered to decompose within a specific period, transforming into products that can be easily discarded. They are sourced from a variety of waste materials and bioresources, including food and agricultural waste, starch, and cellulose. Typically, bioplastics made from renewable resources are more cost-effective than those derived from microbial sources, encouraging manufacturers to prioritize these options [1,2].

In general, polymers are classified into natural (polysaccharides and proteins) and synthetic (amides, esters, ethers, urethanes) biodegradable polymers, and synthetic-biopolymers (or hybrid systems) [1]. Biodegradable polymers (also called biopolymers) are materials designed to function for a defined period before breaking down into easily disposable products through a controlled process [2].

The environmental benefits of biodegradable polymers are significant. They contribute to the regeneration of raw materials, promote biodegradation, and help reduce carbon dioxide emissions, which are major contributors to global warming. Key factors influencing the biodegradation process include the material's composition, polymer morphology, structural characteristics, and molecular weight [1,2].

Both natural and synthetic biodegradable polymers are being developed to meet specific physical and chemical requirements across various applications, such as cosmetics, coatings, wound dressings, and gene delivery. Furthermore, modified biopolymers are gaining traction in advanced fields like nanotechnology, medical implants, prosthetics, cryopreservation, and sanitation products, including surgical sutures, indicating their versatile potential for addressing environmental challenges [1,2].

The study by El-Fakharany, E. M. et al. [3] highlights the promising antiviral properties of bacterial polymeric silk produced by Bacillus sp. strain NE (a petroleum-originated bacteria), specifically composed of fibroin and sericin proteins. These proteins demonstrated significant inhibitory effects against various viruses, including herpes simplex virus type 1 (HSV1), adenovirus type 7 (ADV7), and hepatitis C virus (HCV). The research found that these proteins effectively block viral entry and replication, with specific IC50 values (representing the concentration at which 50% inhibition of cell lysis occurs) indicating their potency. Additionally, the bacterial silk proteins showed a reduction in reactive oxygen species (ROS) generation in HCV-infected cells, further supporting their antiviral efficacy.

The research [3] underscores the potential of bacterial silk proteins as a biopharmaceutical candidate due to their ability to be produced in large quantities at low cost, offering a sustainable option for antiviral drug development. These findings suggest that bacterial silk proteins could play a crucial role in managing viral infections, including potential applications in controlling COVID-19, either alone or in combination with other antiviral drugs. The high safety profile on normal cells also supports their consideration for medicinal applications.

The investigation by Tyubaeva, P. et al. [4] focuses on the impact of natural and synthetic porphyrins, specifically those containing the same metal atom, on the structure and properties of poly(-3-hydroxybutyrate) (PHB) in electrospun fibrous materials. The research highlights the potential of incorporating hemin (Hmi) and tetraphenylporphyrin with iron (Fe(TPP)Cl) into PHB to enhance its properties for biomedical applications. It was concluded that even small concentrations of porphyrins can significantly enhance the antimicrobial properties (by 12 times) and improve the physical and mechanical properties (by at least 3.5 times) of the PHB fibers. The additives affect the hydrophobicity of the materials, altering it by at least 5%. The study used various analytical techniques such as microscopy, X-ray diffraction, and differential scanning calorimetry to investigate the structural changes in the materials. The porphyrin complexes influence the supramolecular structure differently, affecting the crystalline and amorphous phases of PHB. The presence of trivalent iron in the tetrapyrrole ring helps achieve an optimal balance of electrical conductivity and viscosity for producing defect-free fibers.

These findings in the study [4] provide insights into designing PHB-based fibrous materials with tailored properties, including enhanced mechanical strength, antibacterial activity, and controlled wettability, through the strategic use of porphyrin additives.

The fashion industry, driven by consumerism and the rapid pace of trend cycles, has become a significant contributor to excessive consumption and textile waste. The demand for new products and materials is fueled by globalization and the increasing involvement of new designers. This has led to a need for more sustainable alternatives, especially in the face of growing concerns over environmental impacts. The study by da Silva Junior, C. J. G. et al. [5] presented highlights the potential of bacterial cellulose (BC)-based vegan leather as a sustainable option for fashion. The biopolymer, produced using plant-based natural dyes extracted from Allium cepa L., Punica granatum, and Eucalyptus globulus L and waterproofing agents from Melaleuca alternifolia essential oil and Copernicia prunifera wax, demonstrates impressive mechanical properties, including high tensile strength and flexibility, which are essential for durable fashion products. The material also exhibits promising water resistance, with minimal water interaction, making it a viable option for waterproof fashion. The research in [5] underscores the potential of BC as an alternative to traditional animal leather, offering both environmental and ethical benefits. By utilizing non-toxic, plant-based dyes, the process also reduces potential harm to both consumers and the environment. While the waterproofing process needs further refinement, the results of the investigations presented in [5] indicate a promising future for biotechnological materials in sustainable fashion. As consumer demand for eco-friendly options continues to grow, these innovations could lead to more efficient and cost-effective production methods, contributing to a shift towards sustainability in the fashion industry. Further research in this area will likely result in the continuous improvement of these materials, making them increasingly viable for large-scale production and adoption by both manufacturers and conscientious consumers.

Silver nanoparticles (AgNPs) possess remarkable plasmonic and antimicrobial properties, making them valuable for a wide range of industrial and consumer applications. However, concerns about their toxicity, even at low doses, necessitate the development of safer synthesis methods. Desai, A. S. et al. [6] explored the synthesis of AgNPs using plant extracts, focusing on turmeric and aloe vera, and assessed their physico-chemical properties, cytotoxicity, and wound-healing potential. Researchers in [6] revealed that AgNPs derived from turmeric extract exhibited superior wound-healing properties and the least toxicity in both in vitro and in vivo studies. The in vitro analysis, using HEK-293 cells and a dose-dependent Drosophila model, demonstrated that turmeric-extract-derived AgNPs promoted sustained cell growth without cytotoxic effects. Additionally, the in vivo study using Drosophila showed no adverse effects, such as climbing disability, even at high doses, highlighting the biocompatibility of these nanoparticles. The research also showed that the choice of plant extract plays a crucial role in the synthesis process and the resulting properties of the AgNPs. While aloe vera accelerated nanoparticle growth, it also led to the formation of various silver oxides, which could affect the overall efficacy. Thus, selecting the right plant extract is essential for optimizing the properties and biological interactions of AgNPs. The study in [6] suggested that turmeric-extract-derived AgNPs are a promising, environmentally friendly, and low-cost alternative to chemically synthesized silver nanoparticles. The biocompatibility and minimal cytotoxicity of turmeric-derived AgNPs make them an excellent candidate for further development and potential commercialization. With additional research into toxicity analysis and synthesis standardization, these nanoparticles could provide a safer, more sustainable option for various applications, particularly in the biomedical field.

Vonnie, J. M. et al. [7] successfully developed a biodegradable composite film made from cornstarch (CS) and eggshell powder (ESP) through a casting technique. The morphological and physicochemical properties of the ESP/CS film were thoroughly characterized, revealing a smooth, crack-free surface with a spherical and porous irregular shape, indicative of a large surface area. SEM and TEM analysis confirmed the uniform dispersion of ESP particles within the CS matrix and highlighted the phase separation, contributing to the film's enhanced characteristics. The addition of ESP significantly improved several properties of the film compared to the pure CS film. The ESP/CS film exhibited enhanced moisture content, swelling power, water absorption, and water solubility. These improvements were attributed to the interactions between ESP and CS molecules, which strengthened the film structure. Fourier Transform Infrared Spectroscopy (FTIR) analysis confirmed the presence of carbonate minerals in the film and the formation of hydrogen bonds, further enhancing the film's compactness and potential for adsorption applications. X-ray diffraction (XRD) analysis showed the semi-crystalline structure of the composite film, with an increased peak intensity, demonstrating the strong interactions between ESP and CS. The study in [7] also highlighted that ESP improved the mechanical strength, hardness, and barrier properties of the film without the need for cross-linking agents. The ESP/CS composite film, with its promising physicochemical and mechanical properties, can serve as an effective adsorbent for various target analytes, such as contaminants in food products, making it a potential candidate for sustainable and environmentally friendly applications in the food industry and other sectors.

Bio-polymer fibers, particularly alginate, have garnered significant attention over the past two decades due to their versatile applications in gene therapy, tissue engineering, wound healing, and controlled drug delivery. Alginate's unique properties, such as its ease of sol–gel transformation, ion exchange capabilities, and acid stability, make it a highly valuable material in various fields. The study of Zdiri, K. et al. [8] has provided a comprehensive review of the modifications made to alginate fibers to enhance their properties, including the incorporation of nanoparticles, adhesive peptides, and both natural and synthetic polymers.

The authors in [8] highlighted the structure, source, and specific properties of alginate, as well as the various spinning techniques and the impact of solution parameters on the thermo-mechanical and physico-chemical properties of alginate fibers. The incorporation of fillers, such as nanoparticles, has proven to be an effective strategy for improving the physical, thermal, mechanical, and wound-healing properties of calcium alginate fibers, which hold substantial promise for specialized applications, particularly in the medical field. While the potential applications of alginate composite fibers are vast, including uses in cosmetics, sensors, drug delivery, tissue engineering, and water treatment, there remains considerable room for further research. Future studies could focus on the mixed spinning of alginate and other bio-polymers, such as chitosan, to explore new properties and applications. The continued development of innovative, sustainable materials made from alginate and other bio-polymers is crucial for expanding their use and enhancing their performance in medical and environmental applications. Overall, alginate fibers offer a promising, eco-friendly option for numerous advanced technologies, and their continued modification and optimization will likely lead to even more impactful applications in the future [8].

Egorov, A. et al. in their research [9] investigated various calcification methods for collagen and collagen coatings, assessing their suitability for 3D printing and the production of collagen-coated scaffolds. PCL scaffolds were printed and subsequently coated with collagen using four different methods: direct addition of hydroxyapatite (HA) powder to the collagen solution, incubation in Simulated Body Fluid (SBF), coating with alkaline phosphatase (ALP), and coating with poly-L-aspartic acid (poly ASP). The results of these methods were analyzed using ESEM, CT, TEM, and EDX.

The authors of [9] also found that the direct addition of HA powder to the collagen solution caused a pH shift and an increase in viscosity, leading to clumping on the scaffold surface. In contrast, incubation in SBF and with ALP resulted in an increase in HA layer thickness, but no coating was apparent on the collagen layer with poly-L-aspartic acid. Interestingly, only the poly-L-aspartic acid treatment led to the formation of nanocrystalline HA within the collagen layer, as detected by ultrathin sections and TEM with SuperEDX. Overall, the project demonstrated that HA layers could form at the surface of collagen-coated PCL scaffolds with most calcification methods, except for poly-L-aspartic acid. However, poly-L-aspartic acid treatment uniquely facilitated the incorporation of HA crystals within the collagen layer, offering a promising approach for enhancing the properties of collagen-based scaffolds for applications in tissue engineering and regenerative medicine. These findings contribute to a better understanding of the methods for producing collagen-coated scaffolds with controlled HA incorporation, which can be optimized for future biomedical applications [9].

Medical gloves have long been essential in protecting healthcare workers and individuals from infectious microorganisms and hazardous substances, playing a pivotal role in infection prevention. During the COVID-19 pandemic, their importance was further emphasized as they became widespread preventive tools. Lovato M. J et al. in their review [10] highlighted the evolution of medical gloves, beginning with the discovery of vulcanization by Charles Goodyear in 1839, which laid the foundation for the industry's development. The paper [10] compared the properties, benefits, and drawbacks of various commonly used gloves, including those made from natural rubber (NR), polyisoprene (IR), acrylonitrile butadiene rubber (NBR), polychloroprene (CR), polyethylene (PE), and poly(vinyl chloride) (PVC). Additionally, it addressed the environmental impacts of conventional rubber glove production and discussed mitigation strategies such as bioremediation and rubber recycling. To improve glove performance, the review [10] examined the potential of biopolymers, biodegradable fillers, reinforcing fillers, and antimicrobial agents. These materials, such as poly(vinyl alcohol), starch, cellulose, chitin, and silica, offer enhanced properties for medical gloves, including biodegradability, antimicrobial effects, and improved mechanical characteristics. The incorporation of antimicrobial agents, including biguanides, quaternary ammonium salts, and metal nanoparticles (e.g., silver, copper, and zinc oxide), has also shown promise in combating drug-resistant bacteria and preventing cross-contamination. The review [10] further emphasized the importance of sustainability in medical glove production, particularly in terms of environmental, social, and financial impacts. Using performance-enhancing materials such as bio-fillers and biodegradable polymers presents a viable path to creating new gloves with better properties without compromising functionality. The integration of these materials into rubber formulations or through additional dipping processes offers a cost-effective and scalable solution for the industry. It was summarized in a review [10] that the continued innovation in medical glove production, including the optimization of manufacturing processes and the integration of advanced materials, will ensure the development of high-quality, sustainable, and effective gloves. These gloves will not only meet regulatory standards but also cater to the growing demand for performance-enhanced personal protective equipment.

Celestino, M. F. et al. in their study [11] successfully developed scaffolds based on poly(hydroxybutyrate) (PHB) and micronized bacterial cellulose (BC) through 3D printing. The incorporation of varying concentrations of micronized BC (0.25%, 0.50%, 1.00%, and 2.00%) into the PHB matrix allowed for the production of biocomposite filaments that predominantly retained the functional groups of PHB, as confirmed by Fourier Transform Infrared Spectroscopy (FTIR). Thermogravimetric analyses (TG and DTG) indicated that increasing the BC concentration lowered the peak temperature of PHB degradation, with the lowest degradation temperature observed in the filament with the highest BC concentration (PHB/2.0% BC). However, the thermal variation did not hinder the 3D printing process, as the melting temperature of PHB remained above the degradation points. The biological assays showed that the PHB/BC scaffolds were non-cytotoxic and provided suitable cell anchorage sites, suggesting their potential for use in tissue engineering applications. Additionally, the optimal micronization of BC at 20 Hz was found to offer the best granulometry and homogeneity for producing biocomposite filaments. X-ray diffraction (XRD) analysis revealed that BC micronized at this frequency maintained a balanced crystalline and amorphous structure, which is crucial for tissue regeneration as it allows BC to function as a reinforcing agent while also being biodegradable. The characterization techniques, including TGA, FTIR, and SEM, confirmed the structural integrity of the filaments and their suitability for 3D printing, with no significant thermal stability issues for the intended application. Moreover, the use of BC industrial scraps in the production of these filaments aligns with the principles of circular economy and sustainability, adding value to waste materials.

Overall, the results of the research outlined in Ref. [11] demonstrate the promising potential of PHB/BC-based biocomposite filaments for 3D printing applications in tissue engineering. The successful integration of BC micronization and the resulting biocomposite filaments not only highlight their mechanical and biological advantages but also contribute to sustainable practices in material development. Further studies will be needed to explore the full potential of these materials in clinical applications.

In the review [12] by Varnaitė-Žuravliova, S. and Baltušnikaitė-Guzaitienė, J., the authors summarized that regenerated cellulose fibers (RCFs) offer a versatile and highly valuable biomaterial with a range of medical applications due to their biocompatibility, biodegradability, and mechanical strength. These fibers are widely utilized in wound care, tissue engineering, and absorbable sutures, making them essential for medical textiles. RCFs help create moist environments conducive to healing in wound care, support cell attachment in tissue engineering, and eliminate the need for removal in suturing due to their biodegradability. Additionally, their use in surgical garments and diagnostic materials further highlights their importance in the medical sector. As the field progresses, the future of RCFs in medical textiles will be shaped by advancements aimed at improving production processes, such as closed-loop systems for solvent recovery and reducing environmental impact. Furthermore, the integration of nanotechnology promises to enhance the properties of RCFs, including antimicrobial activity, increased surface area, and improved mechanical strength. However, challenges remain in terms of the long-term clinical performance of RCFs, especially in complex medical applications like tissue scaffolds and implants. Research focusing on their degradation rates, mechanical stability, and tissue regeneration over extended periods will be crucial for advancing their use in these areas. There is also potential in developing reinforced RCF composites that can withstand higher mechanical stresses while maintaining biodegradability, which would enable new applications in fields such as orthopedics and cardiovascular surgery. Additionally, the functionalization of RCFs with antimicrobial agents or drug-loaded fibers holds promise, though cost-effective manufacturing processes for large-scale production need to be established to ensure affordability in global healthcare markets. The development of smart textiles incorporating RCFs and biosensors presents exciting opportunities, but challenges regarding scalability, durability, and cost remain. Furthermore, the regulatory approval process for RCF-based medical products, particularly in innovative areas like tissue engineering and drug delivery systems, requires clearer frameworks to accelerate commercialization [12].

Overall, while regenerated cellulose fibers offer immense potential for medical applications, further research, technological advancements, and regulatory clarity are essential to fully realize their capabilities and to ensure they can be effectively utilized in a wide range of medical contexts [12].

This Special Issue with 10 papers covers reviews and investigations made in the areas of biodegradable polymers, especially natural macromolecules, and textiles. Hopefully, all contributions will bring practical benefits and encourage further research in this field.

We would like to express our sincere thanks to the authors, reviewers, and the MDPI *JFB* editors, along with their teams, for their invaluable contributions.

Conflicts of Interest: The authors declare no conflicts of interest.

References

1. Mukherjee, C.; Varghese, D.; Krishna, J.S.; Boominathan, T.; Rakeshkumar, R.; Dineshkumar, S.; Brahmananda Rao, C.V.S.; Sivaramakrishna, A. Recent advances in biodegradable polymers–properties, applications and future prospects. *Eur. Polym. J.* **2023**, *192*, 112068. [CrossRef]
2. Samir, A.; Ashour, F.H.; Hakim, A.A.; Bassyouni, M. Recent advances in biodegradable polymers for sustainable applications. *Npj Mater. Degrad.* **2022**, *6*, 68. [CrossRef]
3. El-Fakharany, E.M.; Abu-Serie, M.M.; Habashy, N.H.; El-Deeb, N.M.; Abu-Elreesh, G.M.; Zaki, S.; Abd-EL-Haleem, D. Inhibitory Effects of Bacterial Silk-like Biopolymer on Herpes Simplex Virus Type 1, Adenovirus Type 7 and Hepatitis C Virus Infection. *J. Funct. Biomater.* **2022**, *13*, 17. [CrossRef]
4. Tyubaeva, P.; Varyan, I.; Krivandin, A.; Shatalova, O.; Karpova, S.; Lobanov, A.; Olkhov, A.; Popov, A. The Comparison of Advanced Electrospun Materials Based on Poly(-3-hydroxybutyrate) with Natural and Synthetic Additives. *J. Funct. Biomater.* **2022**, *13*, 23. [CrossRef] [PubMed]
5. da Silva Junior, C.J.G.; de Amorim, J.D.P.; de Medeiros, A.D.M.; de Holanda Cavalcanti, A.K.L.; do Nascimento, H.A.; Henrique, M.A.; do Nascimento Maranhão, L.J.C.; Vinhas, G.M.; SOUTO, K.K.O.; de Santana Costa, A.F.; et al. Design of a Naturally Dyed andWaterproof Biotechnological Leather from Reconstituted Cellulose. *J. Funct. Biomater.* **2022**, *13*, 49. [CrossRef] [PubMed]
6. Desai, A.S.; Singh, A.; Edis, Z.; Haj Bloukh, S.; Shah, P.; Pandey, B.; Agrawal, N.; Bhagat, N. An In Vitro and In Vivo Study of the Efficacy and Toxicity of Plant-Extract-Derived Silver Nanoparticles. *J. Funct. Biomater.* **2022**, *13*, 54. [CrossRef] [PubMed]
7. Vonnie, J.M.; Rovina, K.; Azhar, R.A.; Huda, N.; Erna, K.H.; Felicia, W.X.L.; Nur'Aqilah, M.N.; Halid, N.F.A. Development and Characterization of the Biodegradable Film Derived from Eggshell and Cornstarch. *J. Funct. Biomater.* **2022**, *13*, 67. [CrossRef] [PubMed]
8. Zdiri, K.; Cayla, A.; Elamri, A.; Erard, A.; Salaun, F. Alginate-Based Bio-Composites and Their Potential Applications. *J. Funct. Biomater.* **2022**, *13*, 117. [CrossRef] [PubMed]
9. Egorov, A.; Riedel, B.; Vinke, J.; Schmal, H.; Thomann, R.; Thomann, Y.; Seidenstuecker, M. The Mineralization of Various 3D-Printed PCL Composites. *J. Funct. Biomater.* **2022**, *13*, 238. [CrossRef] [PubMed]

10. Lovato, M.J.; del Valle, L.J.; Puiggalí, J.; Franco, L. Performance-Enhancing Materials in Medical Gloves. *J. Funct. Biomater.* **2023**, *14*, 349. [CrossRef] [PubMed]
11. Celestino, M.F.; Lima, L.R.; Fontes, M.; Batista, I.T.S.; Mulinari, D.R.; Dametto, A.; Rattes, R.A.; Amaral, A.C.; Assunção, R.M.N.; Ribeiro, C.A.; et al. 3D Filaments Based on Polyhydroxy Butyrate—Micronized Bacterial Cellulose for Tissue Engineering Applications. *J. Funct. Biomater.* **2023**, *14*, 464. [CrossRef] [PubMed]
12. Varnaitė-Žuravliova, S.; Baltušnikaitė-Guzaitienė, J. Properties, Production, and Recycling of Regenerated Cellulose Fibers: Special Medical Applications. *J. Funct. Biomater.* **2024**, *15*, 348. [CrossRef]

Disclaimer/Publisher's Note: The statements, opinions and data contained in all publications are solely those of the individual author(s) and contributor(s) and not of MDPI and/or the editor(s). MDPI and/or the editor(s) disclaim responsibility for any injury to people or property resulting from any ideas, methods, instructions or products referred to in the content.

Article

Inhibitory Effects of Bacterial Silk-like Biopolymer on Herpes Simplex Virus Type 1, Adenovirus Type 7 and Hepatitis C Virus Infection

Esmail M. El-Fakharany [1,*], Marwa M. Abu-Serie [2], Noha H. Habashy [3], Nehal M. El-Deeb [4], Gadallah M. Abu-Elreesh [5], Sahar Zaki [5] and Desouky Abd-EL-Haleem [5]

[1] Proteins Research Department, Genetic Engineering and Biotechnology Research Institute (GEBRI), City of Scientific Research and Technological Applications, Alexandria 21934, Egypt
[2] Medical Biotechnology Department, Genetic Engineering and Biotechnology Research Institute (GEBRI), City of Scientific Research and Technological Applications, Alexandria 21934, Egypt; marwaelhedaia@gmail.com
[3] Biochemistry Department, Faculty of Science, Alexandria University, Alexandria 21511, Egypt; noha.habashi@alexu.edu.eg
[4] Biopharmacetical Products Research Department, Genetic Engineering and Biotechnology Research Institute, City of Scientific Research and Technological Applications, Alexandria 21934, Egypt; nehalmohammed83@gmail.com
[5] Environmental Biotechnology Department, Genetic Engineering and Biotechnology Research Institute (GEBRI), City of Scientific Research and Technological Applications, Alexandria 21934, Egypt; g_abouelrish@yahoo.com (G.M.A.-E.); saharzaki@yahoo.com (S.Z.); abdelhaleemm@yahoo.de (D.A.-E.-H.)
* Correspondence: esmailelfakharany@yahoo.co.uk

Citation: El-Fakharany, E.M.; Abu-Serie, M.M.; Habashy, N.H.; El-Deeb, N.M.; Abu-Elreesh, G.M.; Zaki, S.; Abd-EL-Haleem, D. Inhibitory Effects of Bacterial Silk-like Biopolymer on Herpes Simplex Virus Type 1, Adenovirus Type 7 and Hepatitis C Virus Infection. *J. Funct. Biomater.* **2022**, *13*, 17. https://doi.org/10.3390/jfb13010017

Academic Editors: Sandra Varnaitė-Žuravliova, Jolanta Sereikaitė, Julija Baltušnikaitė-Guzaitienė and Vijay Kumar Thakur

Received: 15 November 2021
Accepted: 30 December 2021
Published: 2 February 2022

Publisher's Note: MDPI stays neutral with regard to jurisdictional claims in published maps and institutional affiliations.

Copyright: © 2022 by the authors. Licensee MDPI, Basel, Switzerland. This article is an open access article distributed under the terms and conditions of the Creative Commons Attribution (CC BY) license (https://creativecommons.org/licenses/by/4.0/).

Abstract: Bacterial polymeric silk is produced by *Bacillus* sp. strain NE and is composed of two proteins, called fibroin and sericin, with several biomedical and biotechnological applications. In the current study and for the first time, the whole bacterial silk proteins were found capable of exerting antiviral effects against herpes simplex virus type-1 (HSV-1), adenovirus type 7 (AD7), and hepatitis C virus (HCV). The direct interaction between bacterial silk-like proteins and both HSV-1 and AD7 showed potent inhibitory activity against viral entry with IC_{50} values determined to be 4.1 and 46.4 µg/mL of protein, respectively. The adsorption inhibitory activity of the bacterial silk proteins showed a blocking activity against HSV-1 and AD7 with IC_{50} values determined to be 12.5 and 222.4 ± 1.0 µg/mL, respectively. However, the bacterial silk proteins exhibited an inhibitory effect on HSV-1 and AD7 replication inside infected cells with IC_{50} values of 9.8 and 109.3 µg/mL, respectively. All these results were confirmed by the ability of the bacterial silk proteins to inhibit viral polymerases of HSV-1 and AD7 with IC_{50} values of 164.1 and 11.8 µg/mL, respectively. Similarly, the inhibitory effect on HCV replication in peripheral blood monocytes (PBMCs) was determined to be 66.2% at concentrations of 100 µg/mL of the bacterial silk proteins. This antiviral activity against HCV was confirmed by the ability of the bacterial silk proteins to reduce the ROS generation inside the infected cells to be 50.6% instead of 87.9% inside untreated cells. The unique characteristics of the bacterial silk proteins such as production in large quantities via large-scale biofermenters, low costs of production, and sustainability of bacterial source offer insight into its use as a promising agent in fighting viral infection and combating viral outbreaks.

Keywords: bacterial silk; biopolymer proteins; antiviral; HSV-1; adenovirus; HCV

1. Introduction

Natural products have gathered special attention because of the surviving biodiversity of flora worldwide and the luxury of obtaining extracts and crude forms from these sources with the help of technological innovation [1,2]. In addition to collagen, polymeric silk is one of the most plentiful naturally derived macromolecular protein polymers obtained mainly from animal origins [3]. More than 120,000 metric tons of silks are manufactured worldwide annually, and the main manufacturers are located in China, India, and Japan [4]. Polymeric silk is a natural protein of importance in medical applications and commercial life. Increased uses of bacterial polymeric silk-like proteins are expected in the near future for several reasons such as low cost–effect, rapid secretion, and modest product recapture. Silk proteins consist mostly of two types of proteins in a core–shell type shape: fibroin composes the core and is enclosed by the glue-like sericin, which cements itself to the fibroin fibers. Sericin, a major constituent of silk proteins, is selectively removed as waste material from silk fiber throughout the silk manufacturing process to make silk more lustrous. Commonly, the removal of sericin from fiber leads to the formation of industrial byproducts that are disposed of as waste materials, causing environmental pollution and constituting a huge waste of natural resources. Thus, sericin will need to be recovered and recycled in an adequate manner to provide social and significant economic benefits [5]. Sericin is easily hydrolyzed although it is considered an insoluble protein in cold water. Sericin is an useful protein owing to its unique features such as oxidation resistance, antibacterial potential, UV resistance, capacity to absorb and release moisture easily, and tyrosine kinase inhibitory activity [6–12]. Sericin is also a glycoprotein that possesses many biological characteristics, including antioxidation, immunomodulation, and inhibition of elastase and tyrosinase activities. Recently, sericin was found to reduce reactive oxygen species (ROS), protect normal cells from oxidative damage caused by UVB radiation, and inhibit tumor progression, being a strong antibacterial candidate [13,14]. Meanwhile, researchers suggest that viruses are the most abundant biological entities on the planet [15]. Currently, millions of people are infected with different viruses, and many of these people are not receiving treatment or vaccination such as infection with retroviruses [16,17]. Enveloped viruses consist of a protein coat called a capsid that surrounds central genetic material, which is either DNA or RNA and is unable to replicate without a host cell. Therefore, to survive, viruses must infect cells and use these cells to replicate themselves. In the manner of doing this, they can kill these cells and cause damage to the host organism, which is why viral infections can make people ill. Hepatitis C virus (HCV), adenovirus type 7 (ADV7), and herpes simplex virus type 1 (HSV1) infections are universal serious diseases, and they lead to several hepatic sequels, death, and genital herpes [18–20].

The high cost of viral disease treatment and/or vaccinations leads to patients using alternative remedies as a traditional medicine for viral infection control. There are many examples of traditional medicine such as the use of camel milk for HCV treatment [21], as camel milk contains many important proteins that play a crucial role in viral prevention [22–24]. In addition, there are many microbial metabolites are used as effective compounds in viral treatment, including mushroom and cyanobacterial lectins [25–28]. Medical usage of natural bioactive ingredients as a main source of treatment to date is well known, even considering the great contribution of chemotherapeutic drugs to modern therapy. However, there is no information on the potential effects of the silk proteins of animal or bacterial origin on viral diseases such as infection with HCV, HSV-1, and AD7. If either one of these proteins could be shown to be effective against pathogenic viruses, it might be a more affordable source for use in medicinal/pharmacological applications, thus contributing to providing a sustainable production and reduction in medical costs.

In this study, we evaluated the antiviral inhibitory activity of whole bacterial silk proteins in an obtained silk of nonanimal origin, which was produced by *Bacillus* sp. strain NE (a petroleum-originated bacteria) [10,29]. This antiviral effect was measured to investigate the ability of the whole bacterial silk proteins as a nonanimal silk source, produced by *Bacillus* sp. strain NE, to inhibit the infection of HSV-1, ADV7, and HCV, viruses that cause a severe challenge to the worldwide public health system owing to their limited vaccination or symptomatic treatment.

2. Materials and Methods

2.1. Whole Bacterial Polymeric Silk-like Protein Isolation and Silk-like Protein Preparation

The bacterial polymeric silk-like proteins were produced by *Bacillus* sp. strain NE as previously identified according to Kamoun et al. [29]. In brief, *Bacillus* sp. strain NE was inoculated in 100 mL nutrient broth (SNB) medium (containing 500 mL of nutrient broth and 500 mL mineral salt solution per liter). A 500 mL salt solution was prepared by mixing various salts (K_2HPO_4, 5.0 g; KH_2PO_4, 20 g; NaCl, 0.1 g; $(NH_4)_2SO_4$, 30 g; $FeSO_4 \cdot 7H_2O$, 0.01 g; $CaCl_2 \cdot 2H_2O$, 0.01 g; $MgSO_4 \cdot 7H_2O$, 0.2 g; $MnSO_4 \cdot 7H_2O$, 0.2 g; $MnSO4 \cdot H_2O$, 0.002 g) $CaCl_2 \cdot 2H_2O$, 0.01 g; glucose, 0.03% (w/v) (Fluka Chemie, Gillingham, UK); and yeast extract, 0.03% (w/v) [30], and it was incubated at 30 °C overnight in a shaker incubator at 150 rpm. Then, 5 mL of bacterial culture was inoculated in 100 mL TSM (containing per liter: 5 g of KNO3, 30 g trypticase soy broth, and 20 g of L-glutamic acid) production medium and incubated for 5 days under the same conditions. The formed exo-biopolymer was precipitated by centrifugation of the cultured media for 30 min at 8000 rpm. After the crude supernatant was concentrated by a rotary evaporator and dialyzed against distilled water at 4 °C overnight, about 30 mL cold ethanol was added to 10 mL of dialyzed concentrate. Then, the precipitate was mixed with 10% cetylpyridinium chloride (CPC) during gentle stirring. The obtained broth was left under room conditions for several hours to obtain the polymeric precipitate by centrifugation at 5000 rpm for 30 min. The precipitate was dissolved in 0.5 M of sodium chloride to obtain the polymeric broth. The polymeric broth was lyophilized after washing three times with cold ethanol to obtain the purified bacterial polymeric-like silk proteins [29].

2.2. Serum Sample Collection

HCV-4a patient serum samples used in this investigation were obtained from the Institute of Medical Research, Alexandria, Egypt. Serum samples were stored at -80 °C prior to viral inoculation experiments. The patients' written consent and approval for this study were obtained from the institutional ethics committee.

2.3. Human Peripheral Blood Mononuclear Cell Separation

Peripheral blood mononuclear cells (PBMCs) were separated from whole human blood using gradient centrifugation by Ficoll-Paque Plus (MP Biomedicals, Illkirch, France) as reported by Lohr et al. [31]. Briefly, whole blood sample was 5 times diluted using freshly prepared PBS and then overlaid dropwise on Ficoll. The monolayer of PBMCs was separated using gradient centrifugation at 2000 rpm for 30 min. The recovered cells were collected and washed 3 times using PBS.

2.4. Cytotoxicity Assays

African green monkey kidney epithelial cell line (Vero cells) was obtained from American Type Culture Collection (ATCC) via VACSERA (Cairo, Egypt). Vero cells were cultured in DMEM medium and used as adenovirus type 7 (ADV7) and herpes simplex virus type 1 (HSV-1) host cells. While PBMCs (hepatitis C virus (HCV) host cells) were cultured in RPMI medium. Silk protein cytotoxicity was tested on both Vero cells and PBMCs using MTT assay [32,33]. Briefly, both Vero cells and PBMCs were seeded in 96-well cell culture plates at densities of 10^4 and 10^5 cells/well and incubated at 37 °C in 5% CO_2 for 24 h. After incubation, serial dilutions of the tested silk protein were incubated with Vero and PBMCs for 72 h. Serial dilutions of the HSV-1 and ADV7 standard drugs (ribavirin and acyclovir, respectively) were incubated with Vero cells for 72 h. At the end of incubation, 20 µL of 5 mg/mL MTT (Sigma, St. Louis, MO, USA) was added to each well and the plates were incubated at 37 °C for 3 h. After discarding MTT solution, 100 µL DMSO (dimethyl sulfoxide) was added and the dye intensity was quantified using the automated ELISA microplate reader adjusted to 570 nm to quantify the cell viability [26,34]. The effective safe concentration (EC_{100}) doses (at 100% cell viability) of the tested protein were estimated by the GraphPad Prism 9 InStat software.

2.5. The Antiviral Activity of Silk Proteins against ADV7 and HSV-1

Ten-fold dilutions of ADV7 and HSV-1, separately, were incubated with a monolayer of Vero cells in 96-well plates for 2 h in a 5% CO_2 incubator. At the end of incubation, the unabsorbed viruses were aspirated and replaced with DMEM containing 10% FBS. Then, the plates were incubated in a 5% CO_2 incubator for 3 days. MTT assay was used to determine cell viability (%), as described above, in the infected and uninfected cells to calculate $TCID_{50}$ (50% tissue culture infectious dose) using the formula of the Ramakrishnan method [35].

2.5.1. MTT and RT-qPCR Assays for Investigation of Antiviral Activity of Silk Protein

The utilized $100TCID_{50}$ (100 times the $TCID_{50}$) viral inocula (10^{-4} and 10^{-3} for ADV7 and HSV-1, respectively) were used to test the mode of antiviral action of silk protein. For assessment of the direct virucidal effect, different concentrations of silk protein and standard drugs were incubated for 2 h with HDV7 and HSV-1, before being added to the Vero cell monolayer in a 96-well plate. The adsorption inhibition effect was evaluated by pretreating the host cells with serial dilutions of the tested protein and standard drugs for 2 h, and then the viruses were added to Vero cells for another 2 h. For the antireplicative effect, serial doses of silk protein and standard drugs were added after incubating Vero with viruses for 2 h. After 3 days, MTT assay was used as described above to determine the percentage of cell lysis inhibition for each protein concentration to calculate the IC_{50} (using GraphPad InStat software) at which silk protein causes 50% viral inhibition.

Moreover, quantitative real-time PCR (qPCR) was used to confirm the results of silk protein antiviral mode action obtained from the MTT assay. Following the seeding of Vero cells in 6-well culture plates for 24 h, the above-mentioned experiments were repeated using 100 µg/mL and 10 µg/mL of silk protein against ADV7 and HSV-1, respectively, and compared to the standard drugs with the same concentrations. After 72 h incubation in CO2, the untreated and treated infected Vero cells, in each well, were collected for total DNA extraction using Qiagen extraction kit. TaqMan-based real-time PCRs (CFX, BIO RAD) were performed in accordance with [36,37]. The ADV7 qPCR primers were 5'-GAGGAGCCAGATATTGATATGGAATT-3' and 5'-AATTGACATTTTCCGTGTAAAGCA-3' with the probe 5'-6-carboxyfluorescein (FAM)-AAGCTGCTGACGCTTTTTCGCCTGA-6-carboxytetramethylrhodamine (TAMRA)-3'. For HSV-1 qPCR, primers were 5'-CATCACCG ACCCGGAGAGGGAC-3' and 5'-GGGCCAGGCGCTTGTTGGTGTA-3' and the probe was 5'-FAM-CCGCCGAACTGAGCAGACACCCGCGC-6-TAMRA-3'. The reaction mixture contained Taq polymerase (0.05 U/µL) and reaction buffer (0.4 mM of dNTP, 250 nM probe, 400 nM forward/reverse primers, and 4 mM $MgCl_2$). PCR program started with 95 °C for 15 min, followed by 45 cycles of 95 °C for 10 s, 55 °C for 30 s, and 72 °C for 20 s. Viral load was estimated using standard curve.

2.5.2. Evaluation of the Inhibitory Impact of Silk Protein on DNA Polymerase Activity of ADV7 and HSV-1

The inhibitory effect of silk protein on viral DNA polymerase activity was determined using acid-precipitated radioactivity. The reaction mixture of ADV polymerase consisted of 25 mM Tris-HCl pH 7.8, 7 mM $MgCl_2$, 1 µg aphidicolin, 10 mM DTT, activated DNA, and 40 µM deoxynucleotides with 1µ Ci radiolabeled [α-^{32}P]dATP [38]. Meanwhile, the reaction mixture of HSV polymerase contained 50 mM Tris-HCl pH 8, 0.5 µg/mL albumin, 100 mM ammonium sulfate, 8 mM $MgCl_2$, 0.5 mM DTT, and 100 µM deoxynucleotides with 1µ Ci radiolabeled [^{3}H-dTTP] [39,40]. The above-mentioned mixtures of ADV7 and HSV-1 were incubated with serial concentrations of silk protein for 30 and 60 min, respectively. After reaction termination by acid precipitation, radioactivity was measured using a scintillation counter.

2.5.3. The Antiviral Activity of Silk Protein against HCV

The inhibitory effect of silk protein on HCV in the isolated PBMCs of healthy donors was recorded. A nontoxic concentration of silk protein was simultaneously incubated with the virus (sterile filtered infected serum) for 90 min. Then, cells were infected with 100 µL of the cotreated virus for 72 h at 37 °C and 5% CO2. At the end of incubation, the infected cells were washed three times with 1 mL of PBS, and the total RNA was extracted using Qiagen Kit. The positive control of the infected untreated cells and the negative control of uninfected cells were included in the experiment.

2.5.4. Quantification of HCV Genomic and Antigenomic RNA Strands Using RT-PCR

To quantify HCV genomic and antigenomic RNA strands (minus strand), reverse transcription nested PCR was carried out as described previously and with minor changes [31]. Briefly, the reaction was carried out in 25 µL with 20 U of reverse transcriptase (Clontech, Mountain View, CA, USA) with 400 ng of the total extracted PBMC RNA, 40 U of RNAsin (Clontech, Mountain View, CA, USA), dNTP (Promega, Madison, WI, USA, at final concentration of 0.2 mmol/L) and the reverse primer 1CH (for plus strand, 50 pmol) or forward primer 2CH (for minus strand, 50 pmol). The reaction was developed for 60 min at 42 °C and then denatured for 10 min at 98 °C. The amplification of highly conserved 5'-UTR sequences was completed using two PCR rounds (with two pairs of nested primers). The first-round amplification was conducted in 50 µL reaction mixture having 2CH forward primer and P2 reverse primer (50 pmol from each), as well as dNTPs (0.2 mmol/L). RT reaction mixture (10 µL) was used as a template with 2 U of Taq DNA polymerase (Promega, Madison, WI, USA). The thermal cycling protocol was as follows: 94 °C for 1 min, 55 °C for 1 min, and 1 min at 72 °C for 30 cycles. The second amplification cycle was similar to the first one, except for use of the nested reverse primer D2 and forward primer F2 (50 pmol). Primer sequences for the 5' HCV noncoding (NC) region amplification were as follows: 1CH: 5'-ggtgcacggtctacgagacctc-3', 2CH: 5'-aactactgtcttcacgcagaa-3', P2: 5'-tgctcatggtgcacggtcta-3', D2: 5'-actcggctagcagtctcgcg-3', and F2: 5'-gtgcagcctccaggaccc-3'. To overcome the false detection of negative-strand HCV RNA and known variations in PCR competence, specific control assays and rigorous standardization of the reaction were completed: (1) cDNA was synthesized without RNA templates to avoid product contamination, (2) cDNA synthesis was synthesized without RTase to avoid Taq polymerase RTase activity, and (3) cDNA synthesis and PCR step were completed without any reverse or forward primers to avoid the contamination from mixed primers. In addition, cDNA synthesis was conducted using one primer followed by heat inactivation of RTase activity at 95 °C for 1 h, in an effort to reduce the false presence of negative strands before the addition of the second primer. Finally, the RT-PCR was conducted using the final PCR product based on the SYBR Green I dye and LightCycler fluorimeter [41]. An external standard curve was done using 10-fold serial dilutions of a modified synthetic HCV 5' NC RNA [42].

2.5.5. Quantification of the Induced ROS Using Oxidized DCFDA and Flow Cytometry

ROS and oxidative damage are assumed to have a vital role in many human diseases. Using a cell-permeable fluorescent and chemiluminescent probe, 2'-7'-dichlorodihydrofluorescein diacetate (DCFH-DA), we quantified the induced ROS in HCV-infected and treated cells using flow cytometry [43]. Briefly, after cellular treatment with the nontoxic dose of silk protein, all silk-protein-treated, HCV-infected, and control PBMCs were incubated with DCFH-DA at a final concentration of 10 µM for 30 min at 37 °C and 5% CO2. After incubation, cells were washed with prewarmed PBS and suspended in FAC buffer solution. The intensity of fluorescence was quantified by flow cytometry (Partec GmbH, Germany), and the sample redox state was monitored by checking the increase in fluorescence that could be measured at 530 nm when the sample is excited at 485 nm.

2.6. Protein Modeling and Validation

Based on our previous data [10,29], there is a great similarity between the *Bombyx mori* silk protein and the *Bacillus* silk protein. The analysis of the amino acid constituents showed that the block protein is rich in Ala-Pro-Gly, while the molecular weight (Mwt) of fibroin protein H-chain was around 400 kDa, which exceeded that of the natural silk of *Bombyx mori* [44]. The bacterial silk-like nanofibers gave an adequate and uniform ribbon-shaped structure with an average nanofiber diameter of around 110 nm. Meanwhile, the block silk-like protein exhibited high thermal degradation monitored at 140–373 °C, which is close to that of natural silkworms like *Bombyx mori* [45]. In our recently published work [30], the fibroin H-chain exhibited a band at ~400 kDa, while the fibroin L-chain exhibited a band at ~35 kDa and P25 proteins exhibited a band at ~30 kDa. However, sericin proteins exhibited bands at 95, 66, 16, and 6.5 kDa. The *Bombyx mori* silk fibroin P25 protein and sericin 1A' were used in the current study for the docking analysis due to the availability of their full sequences in the NCBI database. In addition, the Mwt of these proteins is similar to those of our bacterial silk proteins. Therefore, the current study used *Bombyx mori* silk fibroin p25 and sericin 1A' for the in silico analysis due to the availability of their full sequences in the NCBI database. The Protein Data Bank (PDB, Long Island, NY, USA, https://www.rcsb.org/ (accessed on 6 October 2021) was used to derive the 3D structures of HCV-NS5B (PDB: 3FQK, 576 res, chain A and B) and HSV1 DNA polymerase (PDB: 2GV9, 1193 res, chain A and B). However, the PDB of the unavailable 3D structures of silk fibroin, silk sericin, and human ADV7 DNA polymerase were generated by the Swiss-Model protein-modeling server (https://swissmodel.expasy.org/ (accessed on 13 October 2021) [46]. The amino acid sequences of silk fibroin (Accession: AAL83649, 262 amino acids), sericin 1A' (Accession: BAD00699, 722 amino acids), and ADV7DNA polymerase (Accession: ASK85767, 1193 amino acids) were retrieved from the NCBI protein database and then submitted to the Swiss-Model for analysis. The validation of the generated structural models was done by the PROCHEK Ramachandran plot [47,48]. The theoretical molecular weight values of the two *Bombyx mori* silk protein subunits were predicted from the protein amino acids by the EXPASY server (https://web.expasy.org/compute_pi/ (accessed on 14 October 2021) [49] to compare them with those obtained from the *Bacillus* strain.

2.7. Molecular Docking Analysis

The predicted mechanism of the silk sericin and silk fibroin inhibitory impact on the activity of HCV-NS5B and the DNA polymerase of HSV1 and ADV7 was assessed using molecular docking. The docking of the highest C-score proposed 3D structure models of silk sericin and silk fibroin, individually with each of the studied viral polymerases, was established by the GRAMM-X Protein-Protein Docking Server (http://vakser.compbio.ku.edu/resources/gramm/grammx/ (accessed on 14 October 2021) [50]. Then, the binding pocket atoms of the created docked structural complexes were visualized and analyzed using the Discovery Studio 2020 Client program (v20.1.0.19295, Dassault Systèmes, Vélizy-Villacoublay, France).

2.8. Assessment of the Binding Affinity in the Docked Complexes

The PDBePISA (Proteins, Interfaces, Structures, and Assemblies) website (https://www.ebi.ac.uk/msd-srv/prot_int/pistart.html (accessed on 15 October 2021) [51] was used to analyze the interface of the docked complexes. The numbers of interface residues and hydrogen bonds, along with the change in Gibbs free energy (ΔG, solvation free energy) gained upon interface formation, are provided by this platform. The ΔG value indicates the docked complex's binding affinity.

2.9. Active Site Prediction

The PDBsum web-based database (http://www.ebi.ac.uk/pdbsum (accessed on 19 October 2021) [48] was used in the current study to retrieve the active site residues of the HCV and HSV1 polymerases. The PDB IDs of these viral enzymes were required by this website; hence, it provides structural data on PDB database entries. However, the active site residues of ADV7 polymerase, which has no PDB ID, were predicted using the GASS-WEB server (https://gass.unifei.edu.br/ (accessed on 23 October 2021) [52].

2.10. Statistical Analysis

All data are expressed as mean ± standard error of the mean (SEM). The unpaired two-tailed Student's *t*-test of SPSS 16.0 was used. Statistical differences were expressed as *p*-value < 0.05 *, <0.001 **, <0.0001 ***.

3. Results

3.1. Cytotoxicity Assay

Cytotoxicity assay was used to quantify both IC_{50} and nontoxic dose of silk protein on PBMCs and Vero cells using MTT assay protocol. The obtained results indicated that the nontoxic dose (EC_{100}) of silk protein on PBMCs and Vero cells reached 100 µg/mL with IC_{50} determined to be 460 and 743.3 µg/mL, respectively. So, we selected a 100 µg/mL dose for completion of the antiviral assay (Figure 1). The recorded safe doses of silk protein, ribavirin, and acyclovir on Vero cells were 1000, 284.3, and 266.1 µg/mL, respectively.

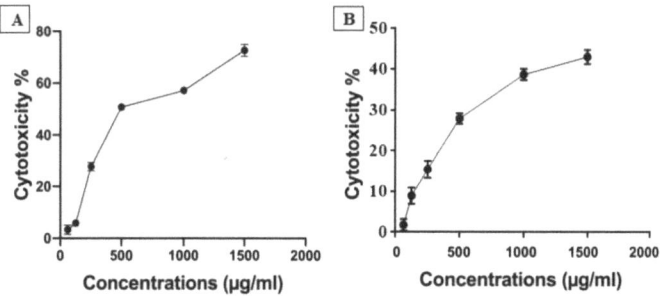

Figure 1. The safety patterns of silk protein on PBMCs (**A**) and Vero cells (**B**); different concentrations of silk protein (from 1500 to 62.5 µg/mL) were tested on PBMCs to detect the nontoxic dose and IC_{50} value of silk protein.

3.2. Antiviral Assays

3.2.1. Antiviral Assays on ADV7 and HSV1

The antiviral mode of silk protein was investigated, at $100TCID_{50}$ (10^{-4} and 10^{-3} of ADV7 and HSV-1, respectively), by quantifying the percentages of cell lysis inhibition and viral elimination using MTT and qPCR, respectively. This was achieved by preincubating the tested proteins with viruses before application to host cells (direct virucidal), pretreatment of host cells before adding viruses (antiadsorption), and addition after infection of host cells (antireplicative). Figure 2 illustrates that silk protein inhibited cell lysis in a dose-dependent manner. From these curves (Figure 2), IC_{50} values, representing the concentration at which 50% inhibition of cell lysis occurs, were calculated for each antiviral mode, and the lowest value refers to the most effective action mode(s) of this protein for combating ADV7 and HSV-1. The estimated IC_{50} values of silk protein for ADV7 inactivation were 46.4 ± 0.5 µg/mL, 222.4 ± 1.0 µg/mL, and 109.2 ± 2.1 µg/mL by direct virucidal, antiadsorption, and antireplicative modes of action, respectively. The values of IC_{50} for HSV-1 inactivation were 4.1 ± 0.7 µg/mL, 12.5 ± 0.1 µg/mL, and 9.8 ± 0.3 µg/mL by direct virucidal, antiadsorption, and antireplicative modes of action, respectively. These IC_{50} values and the highest percentage of cell lysis inhibition indicate that silk protein can inhibit virus-mediated cell lysis mainly via a direct virucidal effect with the lowest antiadsorption effect on both viruses. This protein exhibited a comparable antireplicative effect on both viruses. Standard drugs (ribavirin and acyclovir) only exhibited antireplicative effects with IC_{50} equivalent to 48.9 ± 1.6 µg/mL and 30.9 ± 0.7 µg/mL, respectively, but IC_{50}

values for other effects could not be determined because their maximum safe concentration did not reach 50%. Accordingly, the anti-ADV and anti-HSV efficacy of silk protein is mainly achieved by direct virucidal and antireplicative manner, respectively. Therefore, these modes were selected for the following evaluation of its antiviral activity using more specific parameters.

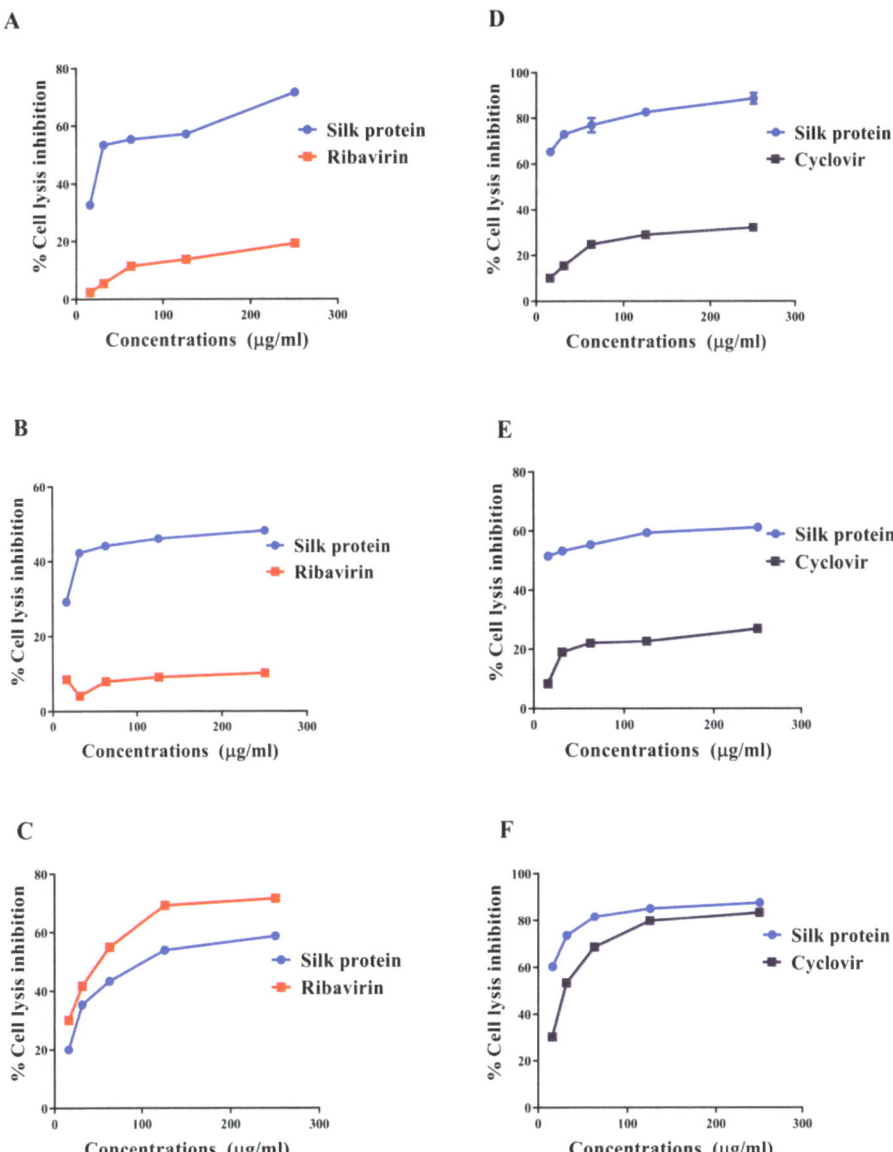

Figure 2. Percentage lysis inhibition of adenovirus (ADV7)- and herpes simplex virus (HSV-1)-infected Vero cells after treatment with silk protein with modes of action illustrated. (**A**) Direct virucidal, (**B**) antiadsorption, and (**C**) antireplication effects of silk protein against ADV7 in comparison with standard drug (ribavirin) in the term of % cell lysis inhibition. (**D**) Direct virucidal, (**E**) antiadsorption, and (**F**) antireplication activity of silk protein against HSV1 in comparison with standard drug (acyclovir) in the term of % cell lysis inhibition. All data are expressed as mean ± SEM.

Data of qPCR for cellular virus contents supported MTT results of cell lysis inhibition. As shown in Figure 3, this protein eliminated 85.7% and 89.9% of ADV7 and HSV1, respectively, by direct virucidal activity, while its antireplication effect caused viral elimination by 63.2% and 77.5%, respectively. On the other hand, standard drugs achieved viral elimination (77.9% and 62.9%, respectively) only via their antireplicative effect. Figure 3 also shows that silk protein had significantly stronger direct ADV7 and HSV1 virucidal effect and anti-HSV1 replicative potential than standard drugs. Meanwhile, no significant difference in anti-ADV7 replicative activity was recorded between silk protein and ribavirin.

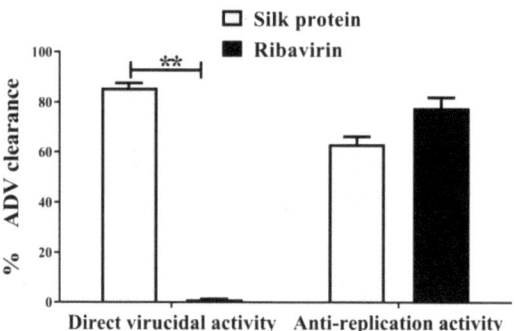

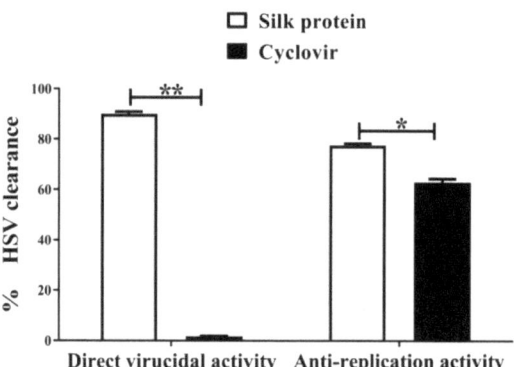

Figure 3. Percentage of adenovirus (ADV7) and herpes simplex virus (HSV-1) eliminated by direct virucidal and antireplication potentials of silk protein in comparison with standard drugs (ribavirin and acyclovir). All data are expressed as mean ± SEM and considered significantly different at $p < 0.05$ *, $p < 0.005$ **.

The antireplicative activity of silk protein was further assessed by estimating the IC_{50} for inhibition of viral DNA polymerases (Figure 4). It was found that silk protein can inhibit these viral polymerases at 164.1 ± 11.1 μg/mL and 11.8 ± 0.6 μg/mL for ADV7 and HSV1, respectively, with no significant difference when compared to standard drugs (Figure 4).

3.2.2. Anti-HCV Activity of Silk Protein

The ability of silk protein to inhibit HCV replication on PBMCs was quantified using qPCR. The obtained results indicated that silk protein at 100 μg/mL showed the ability to inhibit HCV replication on the PBMC model by 66.2% via reducing the viral load from 1.36×10^6 to 0.9×10^5 copies/mL (Figure 5 and Table 1).

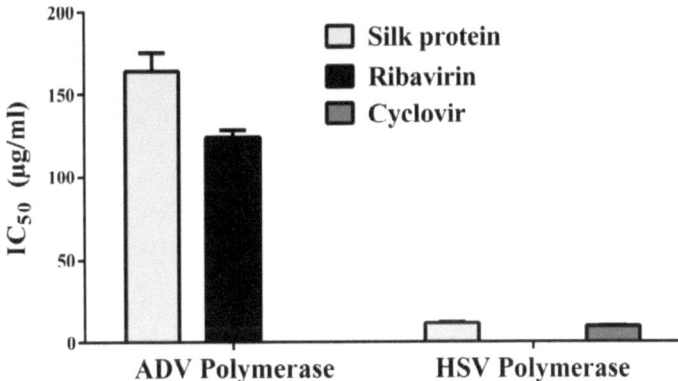

Figure 4. IC$_{50}$ values of silk protein for inhibiting DNA polymerases of ADV and HSV. All data are expressed as mean ± SEM.

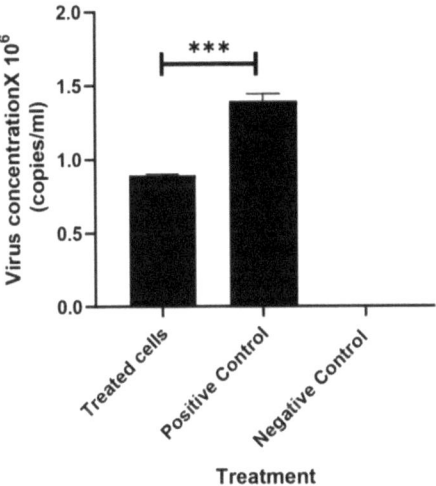

Figure 5. Quantification of HCV viral replication inhibition in PBMCs using RT-qPCR. All data are expressed as mean ± SEM and considered significantly different at $p < 0.0005$ ***.

Table 1. Determination of HCV viral count in PBCs using RT-qPCR.

Samples	Fluor	Cq	Virus Conc. (Copies/mL)	Inhibition%
Treated cells	SYBR	23.89	0.9×10^5	66.2
Positive control	SYBR	28.33	1.36×10^6	0.0
Negative control	SYBR	4.69	-	-

3.2.3. Quantification of the Induced ROS Using Flow Cytometry in HCV-Infected Model

The induced cellular ROS in PBMCs after HCV infection and treatment were quantified using flow cytometry (Figure 6). The obtained results indicated a great induction of cellular ROS in PBMCs after HCV infection (87.9) compared with the uninfected cells (9.5). After treatment, silk protein dramatically reduced the induced ROS from 87.9 to 44.5 with 50.6% inhibition (Figure 7). It is also worth mentioning that silk protein did not induce a significant ROS induction (6.6) in PBMCs compared with the negative untreated cells.

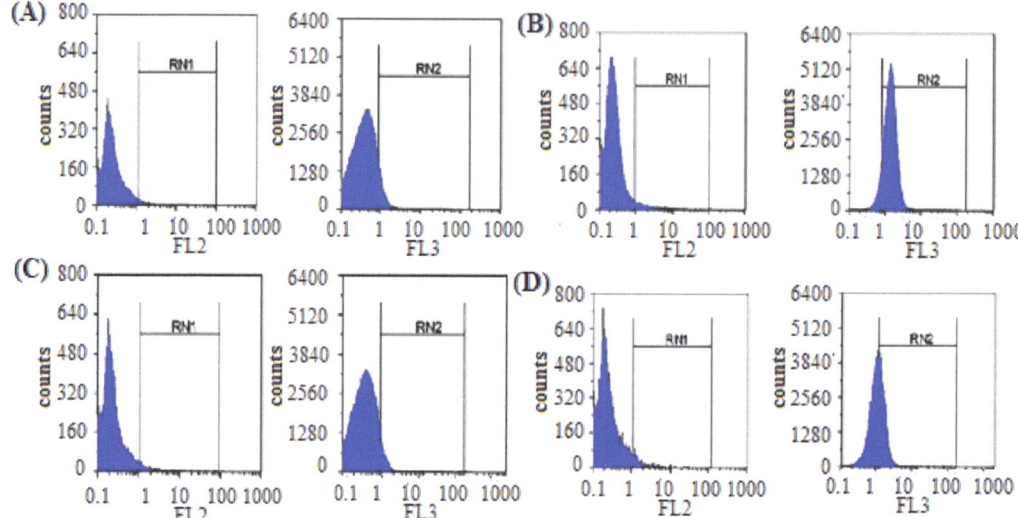

Figure 6. The flow cytometry analysis of the cellular induced ROS in untreated control cells (**A**), positive control HCV infected cells (**B**), silk-protein-treated cells (**C**), and silk-protein-treated HCV-infected cells (**D**). RN1 is the gating region for parameter number 1 using red laser and RN2 is the gating region for parameter number 1 using blue laser.

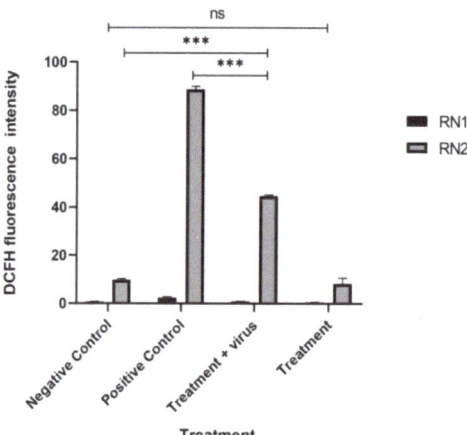

Figure 7. The fluorescence intensity in HCV-infected PBMC models after silk protein treatment. All data are expressed as mean ± SEM and considered significantly different at $p < 0.05$ *, $p < 0.005$ **, $p < 0.0005$ ***.

3.2.4. 3D Predicted Structure Models

The 3D structures of the two silk protein subunits and ADV7 polymerase were modeled by the Swiss-Model server. The protein sequences in Fast Adaptive Shrinkage Threshold Algorithm (FASTA) format were submitted to this online tool to supply the most accurate predictions for their structure. Model 1, with the highest quality and a good C-score value, was established (Figure 8A–C). The molecular weight values of the *Bombyx mori* silk fibroin and sericin 1A′ were computed using the Expasy online server. The results showed a great similarity between them (27.6 and 69.9 kDa, respectively) and those obtained from *Bacillus* sp. (30 and 66 kDa, respectively) in our recently published work [10].

Therefore, these two types of *Bombyx mori* silk protein subunits were chosen here for the computational studies.

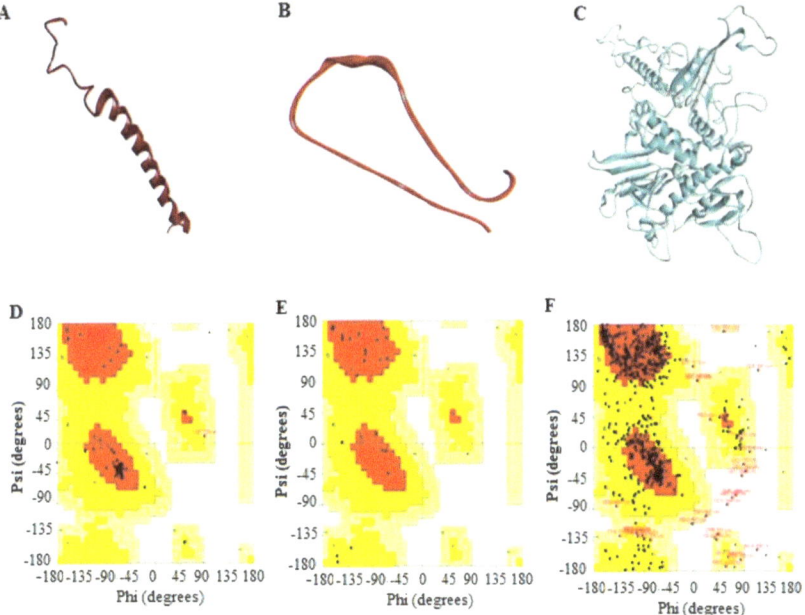

Figure 8. The 3D predicted structures of the silk protein subunits, fibroin and sericin, and ADV7 polymerase and their validation. (**A–C**) Backbone structure of silk fibroin, sericin, and ADV7 polymerase, respectively, as given by Swiss-Model protein-modeling server (https://swissmodel.expasy.org/ (accessed on 13 October 2021). (**D–F**) Ramachandran plots of silk protein fibroin, sericin, and ADV7 polymerase, respectively.

The quality of the predicted structures was confirmed by the Ramachandran plot (Figure 8D–F), which analyzed psi (ψ) and phi (F) torsion angles of the structural backbone. The residues of the ADV7 polymerase 3D structure in the most favored, allowed, and disallowed regions were 74.7%, 24.0%, and 1.3%, respectively. For the 3D structure of silk fibroin, these values were 89.5%, 10.5%, and 0%, respectively. For the 3D structure of silk sericin, these values were 76.2%, 23.8%, and 0%, respectively. The overall G-factors (measurement of unusualness for main-chain dihedral angles and covalent forces) of the predicted models of ADV7 polymerase, fibroin, and sericin were −0.41, −0.27, and −0.31, respectively. These results indicated that the dihedral angles, ψ and F, in the selected model backbone were reasonably accurate.

3.3. The Predicted Inhibitory Mechanism of Silk Fibroin and Sericin on ADV7, HCV, and HSV-1 Polymerases

To predict the inhibitory mechanism of the silk protein on the polymerase activity of the target viruses, the silk fibroin or sericin was docked separately with the viral polymerase (Figures 9 and 10). Then, the docked complex was analyzed by the PDBePISA tool to explore the binding affinity, interface residues, and other details. The outcomes of PDBePISA showed that fibroin interacted with chain A of ADV7 polymerase (47 res, 6 hydrogen bonds) and both chains A and B of HCV (48 res, 4 hydrogen bonds) and HSV-1 (29 res, 3 hydrogen bonds) polymerases. Sericin bound to chain A of ADV7 polymerase (48 res, 7 hydrogen bonds), chain B of HCV polymerase (32 res, 11 hydrogen bonds), and chains A and B of HSV-1 polymerase (27 res, 4 hydrogen bonds). Furthermore, the binding affinity of fibroin and sericin to the studied polymerases was deduced from the predicted

ΔG values. These values were −23.0 (ADV7 polymerase), −11.2 (HCV polymerase), and −14.4 kcal/mol (HSV-1 polymerase) for fibroin and −17.6, −10.5, and −9.0, respectively, for sericin. The ΔG p-value was also provided by the PDBePISA and indicated that the interface surface in the studied fibroin– or sericin–polymerase docked complexes was interaction-specific (p-value < 0.5). The active site residues of ADV7, HCV, and HSV-1 polymerases were compared with the interface residues in the fibroin– or sericin–polymerase docked complexes. The results showed that either fibroin or sericin interacted with the active site residues of the examined viral polymerases, except for HCV polymerase, for which fibroin bound to R200A only (Figure 9B).

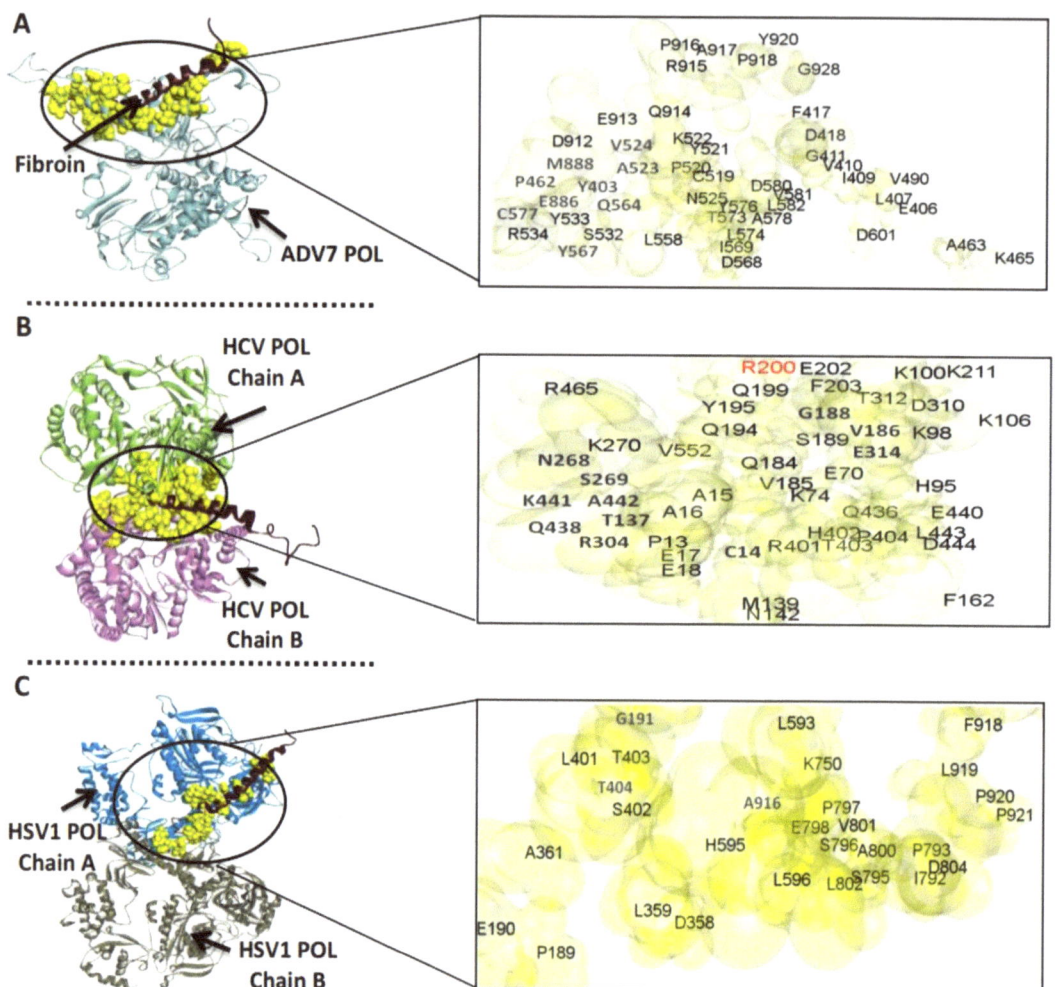

Figure 9. Molecular docking of ADV7, HCV, and HSV-1 polymerase (POL) and silk fibroin. (**A**–**C**) The docked complexes of ADV7 POL (shown in light blue), HCV POL (shown in light green "chain A" and purple "chain B"), and HSV-1 POL (shown in blue "chain A" and gray "chain B") with fibroin (shown in dark red), respectively, as provided by the GRAMM-X Protein-Protein Docking platform and visualized by Discovery Studio software. The interacting pocket residues of the docked complex are indicated by yellow space-filling spheres style, magnification of these regions shows the interface residues on the viral polymerase (shown in pale yellow-gray surface), and the red-colored residue represents the matched residue with the enzyme active site.

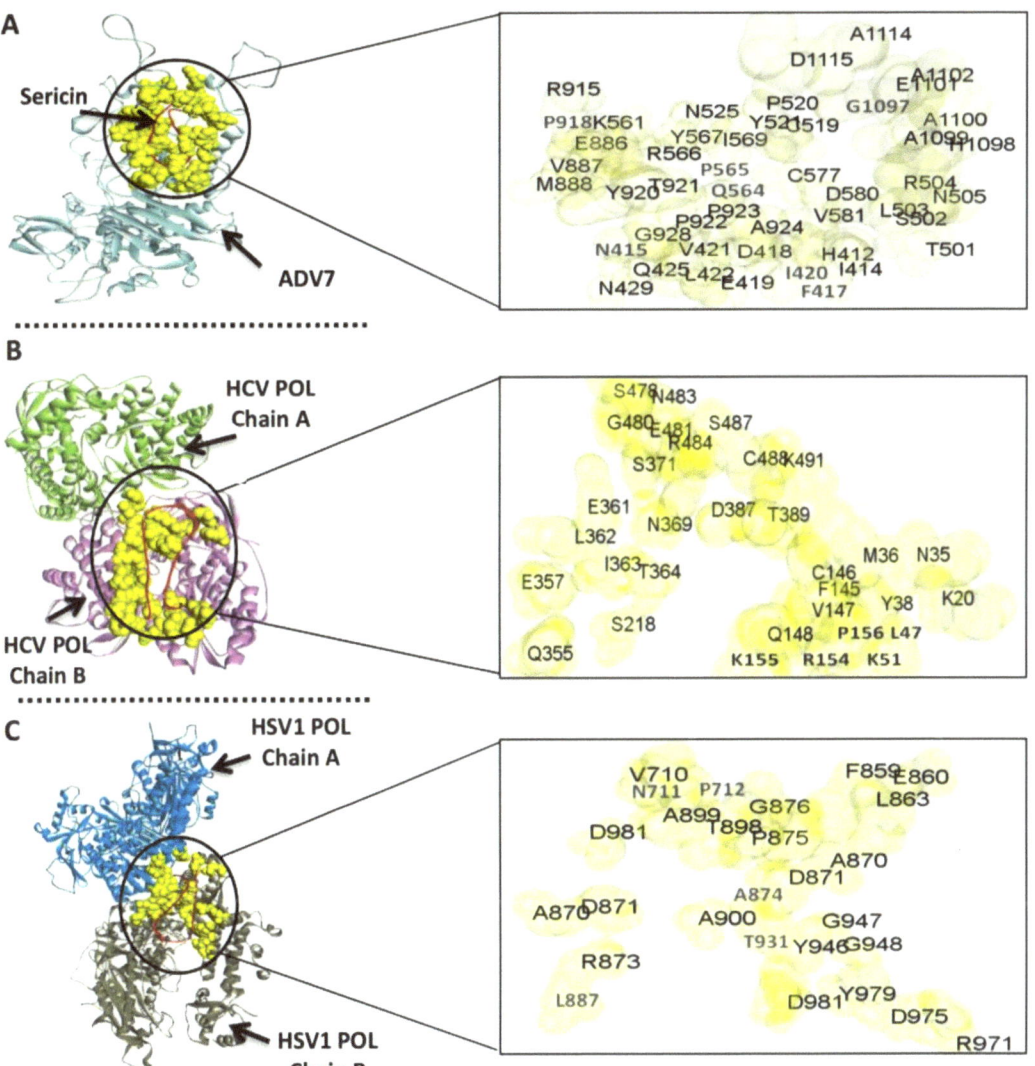

Figure 10. Molecular docking of ADV7, HCV, and HSV-1 polymerase (POL) and silk sericin. (**A–C**) The docked complexes of ADV7 POL (shown in light blue), HCV POL (shown in light green "chain A" and purple "chain B"), and HSV1 POL (shown in blue "chain A" and gray "chain B") with sericin (shown in dark red), respectively, as provided by the GRAMM-X Protein-Protein Docking platform and visualized by Discovery Studio software. The interacting pocket residues of the docked complex are indicated by yellow space-filling spheres style, and magnification of these regions shows the interface residues on the viral polymerase (shown in pale yellow-gray surface).

4. Discussion

There are essential requirements for the discovery and exploration of novel natural agents against both RNA and DNA viruses. A variety of natural compounds derived from microorganisms and plants as medicinal products have been investigated for the management and control of numerous viral diseases. Silk is considered one of the most important natural fibrous proteins and is mainly obtained from animal origins, including silkworms and spiders. Therefore, the most popular studies in this field have focused on the production of polymeric silk proteins from animal origins or using genetically engineered bacteria [53–55]. We identified a previously isolated bacterial strain from petroleum oil origin called *Bacillus sp. strain NE (MK231249)* with the ability to form an exo-biopolymeric extract. Using molecular identification and exo-polymer chemical and physical characterizations, this bacterial exo-polymeric extract was identified as a bacterial silk-like protein [29]. In a previous study, El-Fakharany et al. revealed that the extracted sericin from bacterial silk was found to be like that isolated from animal origin, with potent biological activities, including antioxidant, anticancer, and antibacterial activities [10]. Silk proteins of animal origin have many biological and pharmacological properties, including antioxidant, antitumor, antimicrobial, and anti-inflammatory functions. The available studies on silk proteins show that these proteins may reduce the free radicals in the surrounding media and contamination by microbial pathogens such as bacteria and fungi. Therefore, silk proteins can be used in numerous applications, such as in the treatment of fabrics for medicinal uses and the preparation of wound healing gels. A particular coating of sericin on polyamide or polyester had been used as media for air filters, which were found to assist in sanitizing the contaminated air [9,52,53]. In the literature, there are several studies about the investigation of the antibacterial and antifungal activities of silk proteins, but reports on their antiviral properties are relatively rare. The present study, for the first time, was aimed to investigate the action of bacterial silk-like proteins against different types of viruses. For this purpose, the whole bacterial silk-like proteins were checked for their potential antiviral activity against HSV-1, ADV7, and HCV infectivity in Vero cells and PBMCs. The obtained results revealed that the bacterial silk-like proteins can display a direct virucidal effect on HSV-1 and ADV7, which might be through their amino acid structure and the action of secondary metabolites such as polyphenols and flavonoids [8]. In fact, the bacterial silk-like proteins were able to inhibit and neutralize the infection of HSV-1 and ADV7 upon entry into Vero cells with IC_{50} values of 4.1 ± 0.7 µg/mL and 46.4 ± 0.5 µg/mL, respectively. At similar conditions, the IC_{50} values of adsorption inhibitory effect on HSV-1 and ADV7 were estimated to be 12.5 ± 0.1 µg/mL and 222.4 ± 1.0 µg/mL, respectively. Furthermore, the bacterial silk-like proteins showed potent antireplication activities against HSV-1 and ADV7 with IC_{50} values of 9.8 ± 0.3 µg/mL and 109.3 ± 2.1 µg/mL, respectively. Results of the present study indicated that the bacterial silk proteins had the ability to eliminate ADV7 and HSV-1 by direct virucidal activity determined to be 85.8% and 89.9%, respectively, using qPCR and MTT methods. We showed that the antireplication effect of these proteins exhibited viral elimination of about 63.2% and 77.5%, respectively. However, ribavirin and acyclovir exhibited an antireplicative effect determined to be 77.9% and 62.9%, respectively. On the other hand, the bacterial silk proteins showed inhibitory mechanisms of viral polymerases with IC_{50} estimated to be 164.1 ± 11.1 µg/mL and 11.8 ± 0.6 µg/mL against ADV7 and HSV-1, respectively, similar to the inhibitory effect of the standard drugs.

For the anti-HCV inhibitory effect of bacterial silk proteins, the replication of HCV was inhibited by 66.2% inside infected PBMCs, as determined using RT-qPCR technique, at a concentration of 100 µg/mL. Furthermore, the bacterial silk proteins showed a potent reduction in cellular ROS inside the HCV-infected PBMCs with an inhibition percentage of 50.6%, as compared to untreated HCV-infected cells which showed ROS activity determined to be 87.9%. The docking analysis revealed that silk fibroin and sericin can interact with ADV7, HSV-1, and HCV polymerases by hydrogen bonds and salt bridge interactions, and the interface surface in the obtained docked complexes is interaction-specific ($\Delta^i G$ *p*-values < 0.5). From the $\Delta^i G$ values, we can deduce that the binding affinity of silk fibroin to ADV7 and HSV-1 polymerases was greater than that of silk sericin. Both silk proteins had nearly the same binding affinity to the HCV polymerase. The computational findings also demonstrated the inability of either silk fibroin or sericin to bind to the predicted active site residues of ADV7, HSV-1, or HCV polymerases. So, these viral polymerases can be inhibited by silk fibroin or sericin in an uncompetitive or noncompetitive manner [56]. This can be attributed to the distinctive capability of binding regions to be engaged in multiple interactions with various binding partners of bacterial silk proteins. One of the main possible mechanisms for this antiviral potential could be related to the relatively high cationic nature of the silk polymer, which enables it to bind to the infected host cells and block the viral particles from entry.

The biological activities of silk-like proteins are attributed to their unique composition of amino acids; in particular, their structures contain hydroxyl groups of serine and threonine, which chelate many essential elements like iron. In addition, silk sericin has many aromatic amino acids in its structure, which provide an electron-donating property, besides the action of secondary metabolites such as flavonoids and polyphenols [8,57]. Furthermore, alteration in pH of protein during the extraction process (e.g., using base or acid environment with heating) changes the ionization of amino acid, consequently producing sericin proteins with different lengths which contain large proportions of β-sheets, α-helix, turns, and random coils [58]. Additionally, percentages of random coils and β-sheets reflect the amorphous nature or crystallinity of sericin protein, respectively [59]. Moreover, the bacterial silk-like protein was found to contain unique amino acids such as glycine and proline and contain aliphatic hydrophobic amino acids such as alanine, valine, leucine, and isoleucine with overall composition of 32% [29]. The bacterial silk-like biopolymer can be developed into medicinally useful lead candidates for antiviral therapeutics and control.

5. Conclusions

The results obtained from the present study confirm that the bacterial silk-like proteins have potent antiviral activity against both DNA and RNA viruses. Considering the other biomedical properties and various biotechnological uses of bacterial silk-like proteins, our findings establish the further significance of these proteins as a biopharmaceutic candidate that may be incorporated with other potent drugs for delivery enhancement and achievable treatment. In vitro studies revealed that the bacterial silk-like proteins showed efficient antiviral activities using many molecular mechanisms with high safety on normal cells. Consequently, these results indicate that bacterial silk-like proteins can also be widely applied in controlling and managing viral infection and pandemics (especially in the control of COVID-19) alone or incorporated with other viral drugs. Furthermore, the demand for efficient antiviral drugs with high safety might prompt the consideration of the use of the bacterial silk proteins as a potent candidate in medicinal applications.

Author Contributions: E.M.E.-F.: Conceptualization, formal analysis, methodology, software, data curation, collected literature data, wrote and edited the manuscript. M.M.A.-S.: Conceptualization, formal analysis, methodology, software, data curation, collected literature data, wrote and edited the manuscript. N.H.H.: Conceptualization, formal analysis, methodology, software, data curation, collected literature data, wrote and edited the manuscript. N.M.E.-D.: Conceptualization, formal analysis, methodology, software, data curation, collected literature data, wrote and edited the manuscript. G.M.A.-E.: Conceptualization, formal analysis, methodology, software, data curation, collected literature data, wrote and edited the manuscript. S.Z.: Conceptualization, formal analysis, methodology, software, data curation, collected literature data, wrote and edited the manuscript. D.A.-E.-H.: Conceptualization, formal analysis, methodology, software, data curation, collected literature data, wrote and edited the manuscript. All authors have read and agreed to the published version of the manuscript.

Funding: This study did not receive any specific grant from funding agencies in the public, commercial, or not-for-profit sectors.

Institutional Review Board Statement: Not applicable.

Informed Consent Statement: This research did not include any animal experiments, and the protocols of blood sample collection were approved by the Research Ethical Committee at City of Scientific Research and Technological Applications (SRTA-City), Centre of Excellence for Drug Preclinical Studies (CE-DPS), Alexandria, Egypt, under international, national, and/or institutional guidelines. The blood samples were collected from hepatitis C patient volunteers, and all volunteers provided written informed consent in conformity with our all guidelines.

Data Availability Statement: The data presented in this study are available on request from the corresponding author.

Conflicts of Interest: The authors declare that there is no conflict of interest regarding the publication of this paper.

References

1. Zhang, W.; Jiang, X.; Bao, J.; Wang, Y.; Liu, H.; Tang, L. Exosomes in Pathogen Infections: A Bridge to Deliver Molecules and Link Functions. *Front. Immunol.* **2018**, *9*, 90. [CrossRef]
2. Saad, M.H.; Badierah, R.; Redwan, E.M.; El-Fakharany, E.M. A Comprehensive Insight into the Role of Exosomes in Viral Infection: Dual Faces Bearing Different Functions. *Pharmaceutics* **2021**, *13*, 1405. [CrossRef] [PubMed]
3. Holland, C.; Numata, K.; Rnjak-Kovacina, J.; Seib, F.P. The Biomedical Use of Silk: Past, Present, Future. *Adv. Health Mater.* **2019**, *8*, e1800465. [CrossRef]
4. Pereira, A.M.; Machado, R.; da Costa, A.; Ribeiro, A.; Collins, T.; Gomes, A.; Leonor, I.B.; Kaplan, D.L.; Reis, R.L.; Casal, M. Silk-based biomaterials functionalized with fibronectin type II promotes cell adhesion. *Acta Biomater.* **2017**, *47*, 50–59. [CrossRef]
5. Silva, V.R.; Ribani, M.; Gimenes, M.L.; Scheer, A.P. High Molecular Weight Sericin Obtained by High Temperature and Ultrafiltration Process. *Procedia Eng.* **2012**, *42*, 833–841. [CrossRef]
6. Zhaorigetu, S.; Sasaki, M.; Kato, N. Consumption of Sericin Suppresses Colon Oxidative Stress and Aberrant Crypt Foci in 1,2-Dimethylhydrazine-Treated Rats by Colon Undigested Sericin. *J. Nutr. Sci. Vitaminol.* **2007**, *53*, 297–300. [CrossRef] [PubMed]
7. Nuchadomrong, S.; Senakoon, W.; Sirimungkararat, S.; Senawong, T.; Kitikoon, P. Antibacterial and antioxidant activities of sericin powder from Eri Silkworm Cocoons Correlating to Degumming processes. *Int. J. Wilk Silkmoth Silk* **2008**, *13*, 69–78.
8. Kumar, J.P.; Mandal, B.B. Antioxidant potential of mulberry and non-mulberry silk sericin and its implications in biomedicine. *Free. Radic. Biol. Med.* **2017**, *108*, 803–818. [CrossRef] [PubMed]
9. Chanu, S.B.; Devi, S.K.; Singh, L.R. Silk Protein Sericin: Structure, Secretion, Composition and Antimicrobial Potential. *Front. Anti-Infect. Agents* **2019**, *1*, 183–194. [CrossRef]
10. El-Fakharany, E.M.; Abu-Elreesh, G.M.; Kamoun, E.A.; Zaki, S.; Abd-El-Haleem, D.A. In vitro assessment of the bioactivities of sericin protein extracted from a bacterial silk-like biopolymer. *RSC Adv.* **2020**, *10*, 5098–5107. [CrossRef]
11. Chlapanidas, T.; Faragò, S.; Lucconi, G.; Perteghella, S.; Galuzzi, M.; Mantelli, M.; Avanzini, M.A.; Tosca, M.C.; Marazzi, M.; Vigo, D.; et al. Sericins exhibit ROS-scavenging, anti-tyrosinase, anti-elastase, and in vitro immunomodulatory activities. *Int. J. Biol. Macromol.* **2013**, *58*, 47–56. [CrossRef]
12. Ahamad, M.S.I.; Vootla, S. Extraction and evaluation of antimicrobial potential of antheraeamylitta silk sericin. *Inter J Recent Sci Res* **2018**, *9*, 32019–32022.
13. Kunz, R.I.; Brancalhão, R.M.C.; Ribeiro, L.D.F.C.; Natali, M.R.M. Silkworm Sericin: Properties and Biomedical Applications. *BioMed Res. Int.* **2016**, *2016*, 8175701. [CrossRef]
14. Chithrashree, G.C.; Kumar, M.S.; Sharada, A.C. Sericin, a Versatile Protein from Silkworm—Biomedical Applications. *Shanlax Int. J. Arts, Sci. Humanit.* **2021**, *8*, 6–11. [CrossRef]
15. Griffin, D.W. The Quest for Extraterrestrial Life: What About the Viruses? *Astrobiology* **2013**, *13*, 774–783. [CrossRef] [PubMed]
16. World Health Organisation (WHO). *Progress Report on HIV, Viral Hepatitis and Sexually Transmitted Infections 2019: Accountability for the Global Health Sector Strategies, 2016–2021*; WHO: Geneva, Switzerland, 2019.
17. Forsythe, S.S.; McGreevey, W.; Whiteside, A.; Shah, M.; Cohen, J.; Hecht, R.; Bollinger, L.A.; Kinghorn, A. Twenty Years of Antiretroviral Therapy for People Living With HIV: Global Costs, Health Achievements, Economic Benefits. *Health Aff.* **2019**, *38*, 1163–1172. [CrossRef] [PubMed]
18. Ehwarieme, R.; Agarwal, A.N.; Alkhateb, R.; Bowling, J.E.; Anstead, G.M. A Surprising Cause of Liver Abscesses in a Post-Chemotherapy Patient: Herpes Simplex Virus. *Cureus* **2021**, *13*. [CrossRef]
19. El-Tantawy, W.H.; Temraz, A. Natural products for the management of the hepatitis C virus: A biochemical review. *Arch. Physiol. Biochem.* **2018**, *126*, 116–128. [CrossRef] [PubMed]
20. Tovo, P.-A.; Calitri, C.; Scolfaro, C.; Gabiano, C.; Garazzino, S. Vertically acquired hepatitis C virus infection: Correlates of transmission and disease progression. *World J. Gastroenterol.* **2016**, *22*, 1382–1392. [CrossRef]
21. El Fakharany, E.; El-Baky, N.A.; Linjawi, M.H.; AlJaddawi, A.A.; Saleem, T.H.; Nassar, A.Y.; Osman, A.; Redwan, E.M. Influence of camel milk on the hepatitis C virus burden of infected patients. *Exp. Ther. Med.* **2017**, *13*, 1313–1320. [CrossRef]
22. Redwan, E.M.; Uversky, V.N.; El-Fakharany, E.M.; Al-Mehdar, H. Potential lactoferrin activity against pathogenic viruses. *Comptes Rendus. Biol.* **2014**, *337*, 581–595. [CrossRef] [PubMed]
23. El-Fakharany, E.M. Nanoformulation of lactoferrin potentiates its activity and enhances novel biotechnological applications. *Int. J. Biol. Macromol.* **2020**, *165*, 970–984. [CrossRef]
24. Saad, M.H.; El-Fakharany, E.M.; Salem, M.S.; Sidkey, N.M. The use of cyanobacterial metabolites as natural medical and biotechnological tools: Review article. *J. Biomol. Struct. Dyn.* **2020**, 1–23. [CrossRef] [PubMed]
25. El-Fakharany, E.M. Nanoformulation approach for improved stability and efficiency of lactoperoxidase. *Preparative Biochemistry Biotechnology* **2020**, *51*, 629–641. [CrossRef]

26. El-Fakharany, E.M.; Saad, M.H.; Salem, M.S.; Sidkey, N.M. Biochemical characterization and application of a novel lectin from the cyanobacterium Lyngabya confervoides MK012409 as an antiviral and anticancer agent. *Int. J. Biol. Macromol.* **2020**, *161*, 417–430. [CrossRef]
27. Saad, M.H.; El-Fakharany, E.M.; Salem, M.S.; Sidkey, N.M. In vitro assessment of dual (antiviral and antitumor) activity of a novel lectin produced by the newly cyanobacterium isolate, Oscillatoria acuminate MHM-632 MK014210.1. *J. Biomol. Struct. Dyn.* **2020**, 1–21. [CrossRef]
28. El-Maradny, Y.A.; El-Fakharany, E.M.; Abu-Serie, M.M.; Hashish, M.H.; Selim, H.S. Lectins purified from medicinal and edible mushrooms: Insights into their antiviral activity against pathogenic viruses. *Int. J. Biol. Macromol.* **2021**, *179*, 239–258. [CrossRef] [PubMed]
29. Kamoun, E.A.; Abu-Elreesh, G.M.; El-Fakharany, E.M.; Abd-El-Haleem, D. A Novel Bacterial Polymeric Silk-Like Protein from a Petroleum Origin Bacillus sp. strain NE: Isolation and Characterization. *J. Polym. Environ.* **2019**, *27*, 1629–1641. [CrossRef]
30. Zaki, S.; Farag, S.; Abu Elreesh, G.; Elkady, M.; Nosier, M.; El Abd, D. Characterization of bioflocculants produced by bacteria isolated from crude petroleum oil. *Int. J. Environ. Sci. Technol.* **2011**, *8*, 831–840. [CrossRef]
31. Lohr, H.F.; Goergen, B.; Biischenfelde, K.-H.M.Z.; Gerken, G. HCV replication in mononuclear cells stimulates anti-HCV-secreting B cells and reflects nonresponsiveness to interferon-α. *J. Med. Virol.* **1995**, *46*, 314–320. [CrossRef]
32. Mosmann, T. Rapid colorimetric assay for cellular growth and survival: Application to proliferation and cytotoxicity assays. *J. Immunol. Methods* **1983**, *65*, 55–63. [CrossRef]
33. Redwan, E.M.; Almehdar, H.A.; El-Fakharany, E.M.; Baig, A.-W.K.; Uversky, V.N. Potential antiviral activities of camel, bovine, and human lactoperoxidases against hepatitis C virus genotype 4. *RSC Adv.* **2015**, *5*, 60441–60452. [CrossRef]
34. Abdallah, A.E.; Alesawy, M.S.; Eissa, S.I.; El-Fakharany, E.M.; Kalaba, M.H.; Sharaf, M.H.; Shama, N.M.A.; Mahmoud, S.H.; Mostafa, A.; Al-Karmalawy, A.A.; et al. Design and synthesis of new 4-(2-nitrophenoxy)benzamide derivatives as potential antiviral agents: Molecular modeling and in vitro antiviral screening. *New J. Chem.* **2021**, *45*, 16557–16571. [CrossRef]
35. Ramakrishnan, M.A. Determination of 50% endpoint titer using a simple formula. *World J. Virol.* **2016**, *5*, 85–86. [CrossRef]
36. Kessler, H.H.; Mühlbauer, G.; Rinner, B.; Stelzl, E.; Berger, A.; Dörr, H.-W.; Santner, B.; Marth, E.; Rabenau, H. Detection of Herpes Simplex Virus DNA by Real-Time PCR. *J. Clin. Microbiol.* **2000**, *38*, 2638–2642. [CrossRef]
37. Heim, A.; Ebnet, C.; Harste, G.; Pring-Åkerblom, P. Rapid and quantitative detection of human adenovirus DNA by real-time PCR. *J. Med. Virol.* **2003**, *70*, 228–239. [CrossRef]
38. Mentel, R.; Kurek, S.; Wegner, U.; Janta-Lipinski, M.V.; Gürtler, L.; Matthes, E. Inhibition of adenovirus DNA polymerase by modified nucleoside triphosphate analogs correlate with their antiviral effects on cellular level. *Med. Microbiol. Immunol.* **2000**, *189*, 91–95. [CrossRef] [PubMed]
39. Schnute, M.E.; Anderson, D.J.; Brideau, R.J.; Ciske, F.L.; Collier, S.A.; Cudahy, M.M.; Eggen, M.; Genin, M.J.; Hopkins, T.A.; Judge, T.M.; et al. 2-Aryl-2-hydroxyethylamine substituted 4-oxo-4,7-dihydrothieno[2,3-b]pyridines as broad-spectrum inhibitors of human herpesvirus polymerases. *Bioorganic Med. Chem. Lett.* **2007**, *17*, 3349–3353. [CrossRef]
40. Knopf, K.-W. Properties of Herpes Simplex Virus DNA Polymerase and Characterization of Its Associated Exonuclease Activity. *JBIC J. Biol. Inorg. Chem.* **1979**, *98*, 231–244. [CrossRef]
41. Redwan, E.M.; El-Fakharany, E.M.; Uversky, V.N.; Linjawi, M.H. Screening the anti infectivity potentials of native N- and C-lobes derived from the camel lactoferrin against hepatitis C virus. *BMC Complement. Altern. Med.* **2014**, *14*, 219. [CrossRef]
42. Pradel, F.; Paranhos-Baccalà, G.; Sodoyer, M.; Chevallier, P.; Mandrand, B.; Lotteau, V.; André, P. Quantitation of HCV RNA using real-time PCR and fluorimetry. *J. Virol. Methods* **2001**, *95*, 111–119. [CrossRef]
43. Anticoli, S.; Amatore, D.; Matarrese, P.; De Angelis, M.; Palamara, A.T.; Nencioni, L.; Ruggieri, A. Counteraction of HCV-Induced Oxidative Stress Concurs to Establish Chronic Infection in Liver Cell Cultures. *Oxid. Med. Cell. Longev.* **2019**, *2019*, 6452390. [CrossRef]
44. Jastrzebska, K.; Kucharczyk, K.; Florczak, A.; Dondajewska, E.; Mackiewicz, A.; Dams-Kozlowska, H. Silk as an innovative biomaterial for cancer therapy. *Rep. Pr. Oncol. Radiother.* **2015**, *20*, 87–98. [CrossRef]
45. Agarwal, N.; Hoagland, D.A.; Farris, R.J. Effect of moisture absorption on the thermal properties of Bombyx mori silk fibroin films. *J. Appl. Polym. Sci.* **1997**, *63*, 401–410. [CrossRef]
46. Waterhouse, A.; Bertoni, M.; Bienert, S.; Studer, G.; Tauriello, G.; Gumienny, R.; Heer, F.T.; De Beer, T.A.P.; Rempfer, C.; Bordoli, L.; et al. SWISS-MODEL: Homology modelling of protein structures and complexes. *Nucleic Acids Res.* **2018**, *46*, W296–W303. [CrossRef]
47. Laskowski, R.A.; MacArthur, M.W.; Moss, D.S.; Thornton, J.M. PROCHECK: A program to check the stereochemical quality of protein structures. *J. Appl. Crystallogr.* **1993**, *26*, 283–291. [CrossRef]
48. Laskowski, R.A.; Jabłońska, J.; Pravda, L.; Vařeková, R.S.; Thornton, J. PDBsum: Structural summaries of PDB entries. *Protein Sci.* **2018**, *27*, 129–134. [CrossRef] [PubMed]
49. Wilkins, M.R.; Gasteiger, E.; Bairoch, A.; Sanchez, J.-C.; Williams, K.L.; Appel, R.D.; Hochstrasser, D.F. Protein Identification and Analysis Tools in the ExPASy Server. In *2-D Proteome Analysis Protocols*; Methods in Molecular Biology; Humana Press: Totowa, NJ, USA, 1999; Volume 112, pp. 531–552. [CrossRef]
50. Tovchigrechko, A.; Vakser, I.A. GRAMM-X public web server for protein-protein docking. *Nucleic Acids Res.* **2006**, *34*, W310–W314. [CrossRef]

51. Krissinel, E.; Henrick, K. Inference of Macromolecular Assemblies from Crystalline State. *J. Mol. Biol.* **2007**, *372*, 774–797. [CrossRef] [PubMed]
52. Moraes, J.P.A.; Pappa, G.L.; Pires, D.E.V.; Izidoro, S.C. GASS-WEB: A web server for identifying enzyme active sites based on genetic algorithms. *Nucleic Acids Res.* **2017**, *45*, W315–W319. [CrossRef] [PubMed]
53. Fahnestock, S.R.; Yao, Z.; Bedzyk, L.A. Microbial production of spider silk proteins. *Rev. Mol. Biotechnol.* **2000**, *74*, 105–119. [CrossRef]
54. Yang, Y.J.; Choi, Y.S.; Jung, D.; Park, B.R.; Hwang, W.B.; Kim, H.W.; Cha, H.J. Production of a novel silk-like protein from sea anemone and fabrication of wet-spun and electrospun marine-derived silk fibers. *NPG Asia Mater.* **2013**, *5*, e50. [CrossRef]
55. Antony, V.A.R.; Chinnamal, S.K. Enzymatic degumming of silk using *Bacillus* sp. *Int. J. Sci. Tech. Mang.* **2015**, *4*, 458–465.
56. Sharma, R. Enzyme Inhibition: Mechanisms and Scope. In *Enzyme Inhibition and Bioapplications*; IntechOpen: London, UK, 2012.
57. Fan, J.-B.; Wu, L.-P.; Chen, L.-S.; Mao, X.-Y.; Ren, F.-Z. Antioxidant Activities of Silk Sericin from Silkwormbombyx MorI. *J. Food Biochem.* **2009**, *33*, 74–88. [CrossRef]
58. Da Silva, T.L.; Da Silva, A.C.; Ribani, M.; Vieira, M.G.A.; Gimenes, M.L.; Da Silva, M.G.C. Evaluation of molecular weight distribution of sericin in solutions concentrated via precipitation by ethanol and precipitation by freezing/thawing. *Chem. Eng. Trans.* **2014**, *38*, 103–108.
59. Turbiani, F.R.B.; Tomadon, J.; Seixas, F.L.; Gimenes, M.L. Properties and structure of sericin films: Effect of the crosslinking degree. In Proceedings of the 10th International Conference on Chemical and Process Engineering, Moscow, Russian, 28–30 April 2011.

Article

The Comparison of Advanced Electrospun Materials Based on Poly(-3-hydroxybutyrate) with Natural and Synthetic Additives

Polina Tyubaeva [1,2,*], Ivetta Varyan [1,2], Alexey Krivandin [2], Olga Shatalova [2], Svetlana Karpova [2], Anton Lobanov [1,2], Anatoly Olkhov [1,2] and Anatoly Popov [1,2]

[1] Academic Department of Innovational Materials and Technologies Chemistry, Plekhanov Russian University of Economics, 36 Stremyanny Per., 117997 Moscow, Russia; ivetta.varyan@yandex.ru (I.V.); avlobanov@mail.ru (A.L.); aolkhov72@yandex.ru (A.O.); popov.ana@rea.ru (A.P.)

[2] Department of Biological and Chemical Physics of Polymers, Emanuel Institute of Biochemical Physics, Russian Academy of Sciences, 4 Kosygina Str., 119334 Moscow, Russia; a.krivandin@sky.chph.ras.ru (A.K.); shatalova@sky.chph.ras.ru (O.S.); karpova@sky.chph.ras.ru (S.K.)

* Correspondence: polina-tyubaeva@yandex.ru; Tel.: +79-268-805-508

Abstract: The comparison of the effect of porphyrins of natural and synthetic origin containing the same metal atom on the structure and properties of the semi-crystalline polymer matrix is of current concern. A large number of modifying additives and biodegradable polymers for biomedical purposes, composed of poly(-3-hydroxybutyrate)-porphyrin, are of particular interest because of the combination of their unique properties. The objective of this work are electrospun fibrous material based on poly(-3-hydroxybutyrate) (PHB), hemin (Hmi), and tetraphenylporphyrin with iron (Fe(TPP)Cl). The structure of these new materials was investigated by methods such as optical and scanning electron microscopy, X-ray diffraction analysis, Electron paramagnetic resonance method, and Differential scanning calorimetry. The properties of the electrospun materials were analyzed by mechanical and biological tests, and the wetting contact angle was measured. In this work, it was found that even small concentrations of porphyrin can increase the antimicrobial properties by 12 times, improve the physical and mechanical properties by at least 3.5 times, and vary hydrophobicity by at least 5%. At the same time, additives similar in the structure had an oppositely directed effect on the supramolecular structure, the composition of the crystalline, and the amorphous phases. The article considers assumptions about the nature of such differences due to the influence of Hmi and Fe(TPP)Cl on the macromolecular and fibrous structure of PHB.

Keywords: poly(3-hydroxybutyrate); porphyrin complex; hemin; tetraphenylporphyrin with iron; electrospun fibrous materials; molecular mobility; supramolecular structure; antibacterial effect

Citation: Tyubaeva, P.; Varyan, I.; Krivandin, A.; Shatalova, O.; Karpova, S.; Lobanov, A.; Olkhov, A.; Popov, A. The Comparison of Advanced Electrospun Materials Based on Poly(-3-hydroxybutyrate) with Natural and Synthetic Additives. *J. Funct. Biomater.* 2022, *13*, 23. https://doi.org/10.3390/jfb13010023

Academic Editors: Sandra Varnaitė-Žuravliova, Jolanta Sereikaitė and Julija Baltušnikaitė-Guzaitienė

Received: 3 February 2022
Accepted: 27 February 2022
Published: 28 February 2022

Publisher's Note: MDPI stays neutral with regard to jurisdictional claims in published maps and institutional affiliations.

Copyright: © 2022 by the authors. Licensee MDPI, Basel, Switzerland. This article is an open access article distributed under the terms and conditions of the Creative Commons Attribution (CC BY) license (https:// creativecommons.org/licenses/by/ 4.0/).

1. Introduction

One of the most effective ways to create binary compositions based on biocompatible polymers for biomedical purposes is electrospinning (ES) [1]. Electrospun materials have an extremely high specific surface area, which is a big advantage for biomedicine [2]. ES is a unique technique used for the effective introduction of various modifying additives of natural and synthetic origin [3]. The use of a large number of different additives makes it possible to solve many problems associated with obtaining materials with controlled properties and structure [4]. Of particular interest is the selection of such additives, which can significantly increase the characteristics of known biocompatible polymers, even at low concentrations.

Among different additives, close attention is paid to porphyrins, which a widely used in tumor and gene therapy [5–7], biomedicine [8,9], chemotherapy [10], and drug delivery [11]. Due to their chemical structure, porphyrins possess excellent chemical and thermal stabilities, photophysical and electrochemical performances, and biological compatibility [12]. Moreover, close attention should also be paid to the antimicrobial and antiviral activities of porphyrins [13]. In terms of the chemical origin and structural differences, two classes of porphyrins—natural and synthetic—can be distinguished [14].

Both of these classes are widely used in composition with biocompatible, biodegradable polymeric materials because of the ease of producing, simplicity, and lower investment costs compared to other nanoparticulate systems [15]. There is a wide variety of approaches to creating a porphyrin-polymer system: weak interactions, such as hydrophobic, electrostatic forces, coordination interaction, and hydrogen bonding [16].

Among a wide sample of polymer matrices, poly(3-hydroxybutyrate) (PHB) is of particular interest as a promising polymer for therapeutic applications. PHB is characterized by a high melting point, a high degree of crystallinity, and low permeability to oxygen, water, and carbon dioxide [17]. This bio-based polymer is biocompatible [18], obtained from renewable sources [19], and degrades in the biologically active environment [20].

There is a wide number of PHB-based composites biomaterials with poly(ethylene glycol) [21], polylactide [22], polycaprolactone [23], chitosan [24], elastomers [25], nanoparticles [26], carbon nanotubes [27], catalysts and enzymes [28], and bioactive molecules [29]. There are several works in which electrospun PHB-porphyrin composite materials were obtained: Polystyrene/Polyhydroxybutyrate/Graphene/Tetraphenylporphyrin [30], Polyhydroxybutyrate/Hemin [31], Polyhydroxybutyrate/Tetraphenylporphyrin with iron [32], and Polyhydroxybutyrate/5,10,15,20-*tetrakis*(4-hydroxy-phenyl)-21H,23H-porphine [33].

In a large number of works it was shown that PHB-based materials are immunologically inert, which allows using these materials as biocompatible for different biomedical applications [34–37]. Composites based on PHB and various additives did not cause any inflammatory reaction accompanied by leukocyte migration [38] and they had no hemolytic effect on the red cell suspension, so were they suitable for the blood-contacting applications [39].

The purpose of this work was to compare the effect of two porphyrins of natural and synthetic origin, containing the same metal atom in order to identify the possibilities of creating highly effective materials for biomedicine. As objects for polymer-porphyrin systems, the following hemin (Hmi) and tetraphenylporphyrin with iron (Fe(TPP)Cl) were selected.

Hmi is thermally stable [40], antimicrobial active against *Staphylococcus aureus* [41], biocompatible, and can be used for medical purposes [42,43]. Moreover, there are several successful works of creating electrospun materials with Hmi [44,45]. Materials containing Hmi can be successfully used for different biomedical applications [46], including the containers for drug delivery systems [47] because of their biocompatibility. The most important properties of hemin include the catalytic functions of heme and its oxidized form [48] and thermal stability [49]. Owing to the functionality of the porphyrin ring composed of a tetrapyrrole scaffold, Hmi can easily coordinate with many transition metal ions [48] and its redox properties could be easily controlled [50].

On the other hand, Fe(TPP)Cl is known for its catalytic effect [51,52]. This porphyrin complex is used in various fields of chemistry and biomedicine [53]. The magnetic properties of Fe(TPP)Cl are good enough for biomedical and therapeutic application [54]. It is antimicrobial active against Gram-positive Staphylococcus aureus and Gram-negative *Escherichia coli* [55]. Fe(TPP)Cl has low toxicity to eukaryotic cells [13]. Investigating the geometry of this complex allows us to significantly influence its electronic and electrical properties [51].

2. Materials and Methods

2.1. Materials

Poly(3-hydroxybutyrate) (PHB) 16F series (BIOMER, Frankfurt am Main, Germany) with molecular weight of 206 kDa, density of 1.248 g/cm^3, and crystallinity of 59% was used (Figure 1a) as a polymeric matrix. Hemin (Hmi) isolated from bovine blood (Moscow, Russia) was used as a modifying additive of natural origin (Figure 1b) [56]. Tetraphenylporphyrin with iron (Fe(TPP)Cl) (Moscow, Russia) was used as a modifying additive of synthetic origin (Figure 1c) [56]. Both tetrapyrroles are coordination complexes of iron (oxidation state: III) [57].

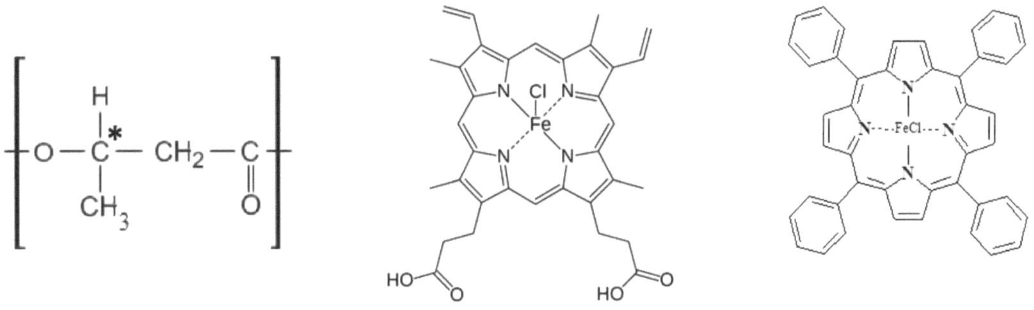

(a)   (b)   (c)

Figure 1. Structural formulas of PHB (**a**) [31], where * is the designation of the chiral carbon atom, Hmi (**b**) [31], Fe(TPP)Cl (**c**).

2.2. Preparation of the Electrospun Materials

Polymer nanofibrous materials based on PHB-porphyrin composition were obtained by electrospinning method (ES) on a single-capillary laboratory unit EFV-1 (Moscow, Russia). A photo and schematic view of the laboratory unit are shown in Supplementary Materials Figure S6. The diameter of the capillary was 0.1 mm. The collector was stable, 300 × 300 mm. The consumption of the forming solution was 5–7×10^{-5} g/s.

The PHB forming solution was prepared by dissolving the well-dispersed powder in chloroform at 60 °C. The content of PHB was 7% wt. The electrical conductivity of the 7% PHB in chloroform was 8 µS/cm and the viscosity was 1.0 Pa s.

The PHB-Fe(TPP)Cl materials were obtained from the forming solution with a content of Fe(TPP)Cl—1, 3, and 5% wt. of the PHB. Fe(TPP)Cl was dissolved in chloroform in the PHB solution at 25 °C. The electrical conductivity of the PHB-Fe(TPP)Cl forming solution was 12–15 µS/cm and viscosity was 1.0–1.2 Pa s [32]. The voltage of the ES for the PHB-Fe(TPP)Cl solution was 17–19 kV, the distance between the electrodes was 200–210 mm, and the gas pressure on the solution was 5–10 kg(f)/cm^2.

The PHB-Hmi materials were obtained by the method of double-solution electrospinning [58,59]. Hmi was dissolved in N,N-dimethylformamide at a temperature of 25 °C and homogenized with the PHB solution. The electrical conductivity of the forming solution was 10–14 µS/cm and the viscosity was 1.4–1.9 Pa s [31]. The voltage of the ES for the PHB-Hmi solution was 19–20 kV, the distance between the electrodes was 210–220 mm, and the gas pressure on the solution was 5–7 kg(f)/cm^2.

2.3. Methods

2.3.1. Microscopy

Primary data of morphology and topology of the fibrous materials with different content of Fe(TPP)Cl and Hmi was obtained by optical microscope Olympus BX43 (Tokyo, Japan) and Scanning electron microscopy (SEM) by Tescan VEGA3 microscope (Wurttemberg, Germany) in order to characterize the morphology. Optical microphotographs were obtained in the reflected light at a magnification of 200 times. The SEM microphotographs were obtained at an accelerating voltage of 20 kV at a magnification of 500 times. PHB-Fe(TPP)Cl and PHB-Hmi samples 10 × 10 mm were covered with a platinum layer for the SEM method.

2.3.2. Morphology and Density Analysis

The structure of fibrous materials was evaluated by the counting method using the software Olympus Stream Basic (Tokyo, Japan). PHB-Fe(TPP)Cl and PHB-Hmi samples 100 × 100 mm were used for the counting method.

The average diameter of fibers was determined manually for each fiber on the z-stack at 10 different points of the sample excluding defective areas.

Average density characterizes the mass per unit volume of the material. The data were averaged over 10 samples. The density, δ, was defined as:

$$\delta = \frac{m}{l \times B \times b}, \quad (1)$$

where m is the mass; l is the length; B is the width; and b is the thickness.

Theoretical porosity is the percentage of the mass of the material and the fiber-free volume. The data were averaged over five samples.

2.3.3. Differential Scanning Calorimetry

Thermal properties of the PHB-Fe(TPP)Cl and PHB-Hmi samples were studied using differential scanning calorimeter (DSC) by Netzsch 214 Polyma (Selb, Germany), in an air atmosphere, with a heating rate of 10 °K/min and with a cooling rate of 10 °K/min. The results of scanning in the air atmosphere corresponded to the values of scanning in the argon atmosphere due to the low oxidation of the samples. The samples weight was 6–7 mg. The samples were heated from 20 °C to 220 °C and then cooled to 20 °C twice. The average statistical error in measuring thermal effects was ±2.5%.

Enthalpy of melting, ΔH, was calculated by NETZSCH Proteus software according to the standard technique [60].

Crystallinity degree, χ, was defined from the melting peak as:

$$\chi = \frac{\Delta H}{H_{PHB}} \times 100\%, \quad (2)$$

where ΔH is the melting enthalpy; H_{PHB} is the melting enthalpy of the ideal crystal of the PHB, 146 J/g [61]; C is the content of the PHB in the composition.

2.3.4. Electron Paramagnetic Resonance

The state of the amorphous phase of PHB in the polymer matrix was studied using electron paramagnetic resonance (EPR) by EPR-V automatic spectrometer (Moscow, Russia). The modulation amplitude was < 0.5 G. The spin probe was 2,2,6,6-tetramethylpiperidine-1-oxyl (TEMPO). TEMPO was introduced into the samples from the gas phase at 50 °C.

Radical concentration in the polymer was determined by the Bruker WinEPR software (the reference was CCl4 with the radical concentration not exceeding 10^{-3} mol/L). The average statistical error in measuring thermal effects was ±5%.

The experimental spectra of the spin probe in the region of slow motions ($\tau > 10^{-10}$ s) were analyzed within the model of isotropic Brownian rotation using the program described in [62]. Probe rotation correlation time, τ, in the region of fast rotations ($5 \times 10^{-11} < \tau < 10^{-9}$ s) was found based on the ESR spectra from the formula [63]:

$$\tau = \Delta H_+ \times \left(\sqrt{\frac{I_+}{I_-}} - 1 \right) \times 6.65 \times 10^{-10}, \quad (3)$$

where ΔH_+ is the width of the spectrum component located in a weak field and $\frac{I_+}{I_-}$ is the ratio of the component intensities in the weak and strong fields.

2.3.5. X-ray Diffraction Analysis

The state of the crystalline phase of PHB in the polymer matrix was studied by X-ray diffraction analysis. The intensity of wide- and small-angle X-ray scattering was measured in transmission geometry on a diffractometer with optical focusing of the X-ray beam and a linear coordinate (position-sensitive) detector [64,65] (X-ray tube with a copper anode, Ni filter) and was corrected for background scattering. The intensity of wide-angle X-ray scattering was also measured in Bragg–Brentano reflection geometry on an HZG4 diffractometer (Freiberger Präzisionsmechanik, Germany) with a diffracted beam graphite monochromator (CuKα radiation).

Degree of crystallinity of PHB, ξ, was calculated with the diffraction patterns obtained in transmission geometry, as [66]:

$$\xi = (I_{exp} - I_{am})/(I_{exp} - I_b) \times 100\% \quad (4)$$

where I_{exp} is the integral experimental intensity of the diffractogram of the sample; I_{am} is the integral intensity of the hypothetical diffractogram of the amorphous phase passing through the points of minima between the diffraction maxima, I_b is the integral intensity of a baseline underlying experimental diffractogram.

Average sizes of PHB crystallites were calculated from diffractograms obtained with the Bragg-Brentano method using the Selyakov-Scherrer formula as:

$$L_{hkl} = \lambda/(\beta_{hkl} \cdot \cos\theta_{hkl}), \quad (5)$$

where L_{hkl} is the average size of the crystallites calculated with the diffraction line (hkl), λ is the wavelength of X-ray radiation, β_{hkl} and θ_{hkl} are the integral width (in radians on the 2θ scale) corrected for instrumental broadening and half of the scattering angle for the diffraction line (hkl), respectively.

2.3.6. Mechanical Analysis

Mechanical properties were examined by a tensile compression testing machine Devotrans DVT GP UG 5 (Istambul, Turkey). The stretching speed was 25 mm/min. The preload pressure was absent. PHB-Fe(TPP)Cl and PHB-Hmi samples size were 10 × 40 mm. The data were averaged over five samples.

Tensile strength was registered automatically by Devotrans software. The average statistical error in measuring thermal effects was ±0.02 MPa.

Elongation at break, ε, was calculated as:

$$\varepsilon = \frac{\Delta l}{l_0} \times 100\%, \quad (6)$$

where Δl is the difference between the final and initial length of the sample; l_0 is the initial length of the sample. The average statistical error in measuring thermal effects was ±0.2%.

2.3.7. Wetting Contact Angle Measurement

The wettability and degree of hydrophilicity of the surface of the PHB-Fe(TPP)Cl and PHB-Hmi samples were evaluated by measuring the contact angle of wetting formed between a drop of water and the surface of the sample. Water drops (2 µL) were applied to three different areas of the film surface by an automatic dispenser.

The marginal wetting angle of the surface of the samples was measured using an optical microscope M9 No. 63649, lens FMA050 (Moscow, Russia). Image processing was done using Altami studio 3.4 software. The result is the average of three measurements from different parts of the sample. The relative measurement error was ±0.5%

2.3.8. Biological Analysis

The antimicrobial activity of PHB-Fe(TPP)Cl and PHB-Hmi samples was studied by biomedical tests on cellular material. *Staphylococcu. aureus p 209, Salmonella. typhimurium*, and *Escherichia coli 1257* were used as test cultures. Samples of initial PHB were served as a control. Cultures of test microorganisms were transplanted onto meat-peptone agar and incubated for 24 h at 37 °C. Then, a suspension of each microorganism was prepared in saline solution and the concentration of microbial cells was determined according to the turbidity standard of 104 mk/mL. PHB-Fe(TPP)Cl and PHB-Hmi samples size were 20 × 20 mm. Samples were placed in sterile Petri dishes, to which 1 mL of a test culture suspension was added and kept at room temperature for 30 min. After that, 10 mL of sterile saline solution was poured into the cup and kept for 15 min to elute the test culture from the samples of the test material. After the exposure, the suspension from the cups in the amount of 100 mL was sown on the surface of meat-peptone agar, previously poured into Petri dishes. The crops were incubated for 48 h at 37 °C. In parallel, the test culture suspensions used in the experiment were seeded to control the concentration of viable microorganisms. Then, the colonies of viable microorganisms grown on the surface of the agar were counted.

3. Results and Discussion

Poly(-3-hydroxybutyrate) has many advantages, which are enhanced by the development of the electrospinning method: the material degrades rapidly in the soil, remaining stable in the air [67]; the fibrous structure compensates for the fragility of the semicrystalline polymer [68]; it is possible to introduce modifying additives evenly into the fiber's structure [31]. Therefore, three levels of structural organization for electrospun materials could be distinguished: macroscopic (whole system), mesoscopic (fiber contact area), and microscopic (structure of the fiber) scale [69].

3.1. Electrospun Material Structure

To describe the morphology and mutual orientation of fibrous materials, it is convenient to use parameters that reliably characterize it: density; average diameter of the fibers; and porosity [31]. The formation of a unique highly developed structure of electrospun materials is influenced by a complex of parameters and depends on the type of polymer solution, processing parameters, and environmental conditions [70]. The introduction of additives into the polymer solution made it possible to significantly affect electrical conductivity, viscosity (which also affected the voltage), flow rate, Taylor's cone shape, and the evaporation rate of the solvent. All these aspects had a great impact on the appearance of the produced fibers: color, surface character, morphology, the presence of inclusions, and defects.

3.1.1. Optical Microscopy

The microphotographs of the material based on the PHB with a different content of hemin and tetraphenylporphyrin with iron are shown in Figure 2.

The introduction of small concentrations (1, 3, and 5% wt. of PHB) of porphyrins of natural and synthetic origin containing a trivalent iron atom had a significant effect on the formation of the fibrous layer. Characteristics of the fibrous layer are presented in Table 1. It is important to note the formation of black inclusions 4–32 µm for 1% wt. of Hmi, 0.7–17 µm for 3% wt. of Hmi and 1% wt. of Fe(TPP)Cl. In the case of 5% wt. of Hmi (Figure 2c), inclusions are practically absent, their size was 1–4 µm. And in the case of 3 and 5% wt. of Fe(TPP)Cl (Figure 2e,f) inclusions are completely absent.

Table 1 shows that the presence of additives leads to a decrease in the density of the material: porosity increases markedly by 9–15% depending on the additive, and the density decreases by an average of 30–45% depending on the additive. At the same time, the trend of density changes coincides for both PHB-Hmi and PHB-Fe(TPP)Cl, but the trend of the change in the average size of the fibers differs significantly. The addition of Hmi leads to a decrease in the average diameter of the fibers by 42–50%. The addition of 1% wt. of Fe(TPP)Cl leads to a decrease in the average diameter of the fibers by 40% and the addition of 3 and 5% wt. of Fe(TPP)Cl leads to a small increase in the average diameter of the fibers by 1–2%, which is an extremely small impact.

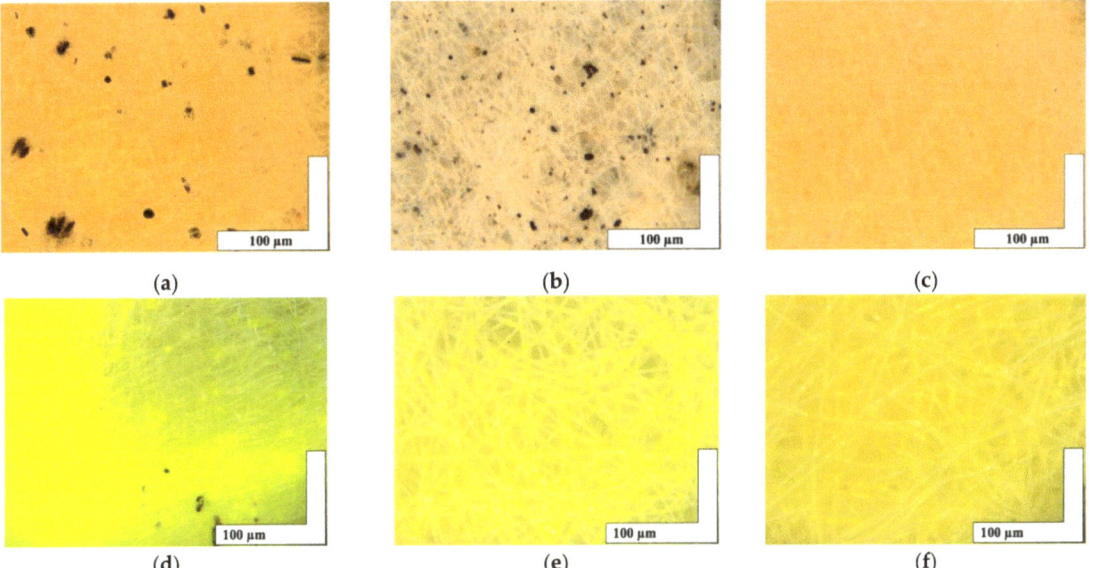

Figure 2. The microphotographs of PHB with different content of Hmi [31]: 1% wt. (**a**), 3% wt. (**b**) and 5% wt. (**c**) and Fe(TPP)Cl: 1% wt. (**d**), 3% wt. (**e**) and 5% wt. (**f**).

Table 1. Average values of the characteristics of the fibrous layer of PHB-Hmi and PHB-Fe(TPP)Cl composites.

Sample	Concentration of Additive, %	Density, g/cm^3 (±S.D., n = 10)	Average Diameter, μm (±S.D., n = 100)	Porosity, % (±S.D., n = 50)
PHB	0	0.30 ± 0.01	3.50 ± 0.08	80 ± 2.0
PHB-Hmi	1	0.20 ± 0.02	2.06 ± 0.07	92 ± 1.5
PHB-Hmi	3	0.20 ± 0.01	1.77 ± 0.04	92 ± 1.5
PHB-Hmi	5	0.17 ± 0.01	1.77 ± 0.04	94 ± 1.2
PHB-Fe(TPP)Cl	1	0.21 ± 0.02	2.07 ± 0.07	93 ± 1.4
PHB-Fe(TPP)Cl	3	0.20 ± 0.02	3.55 ± 0.04	95 ± 1.2
PHB-Fe(TPP)Cl	4	0.16 ± 0.01	3.54 ± 0.04	89 ± 1.2

3.1.2. Scanning Electron Microscopy

A detailed study of the fiber's surface was carried out by the SEM method. The SEM microphotograph of the material based on the initial PHB is shown in Figure 3.

Figure 3 shows that the initial PHB electrospun material is characterized by a large number of defects: thickenings, fiber irregularities, and gluings. There are areas where the fibers are unevenly distributed or glued together. The size of the pear-shaped thickenings is 14–25 μm in diameter and their length is 20–70 μm on average. Such defects are mainly due to the insufficient balance of viscosity and electrical conductivity of the polymer forming solution [32]. Imbalanced polymer solution unevenly passes through the capillary, forming local thickenings on the surface of the fibers. Such fibers do not have time to fully cure at the stage of movement from the capillary to the collector, as a result of which glues and individual thickenings are formed. It is important to note that such areas can negatively affect the mechanical and diffusion properties. Moreover, such areas make the properties of the whole material inhomogeneous along the surface of the web, preventing reliable prediction of operational characteristics. Based on this, the effect of the porphyrins considered on the morphology of PHB electrospun materials can be assessed as positive.

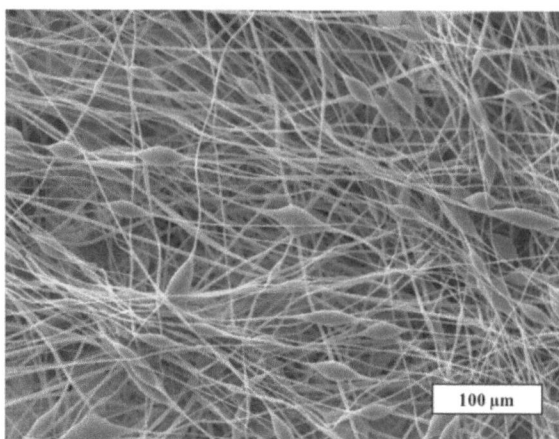

Figure 3. The SEM microphotograph of PHB electrospun material.

The SEM microphotographs of the material based on the PHB with different content of hemin and tetraphenylporphyrin with iron are shown in Figure 4.

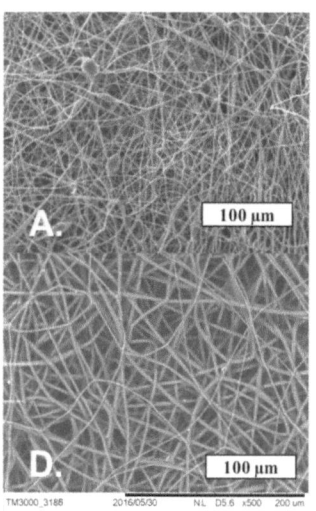 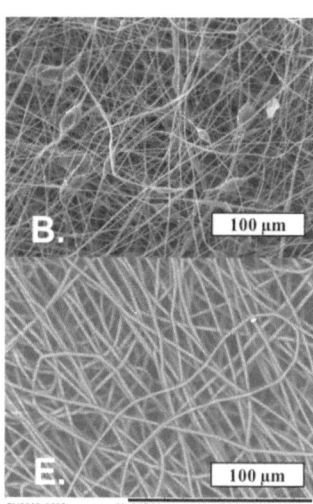 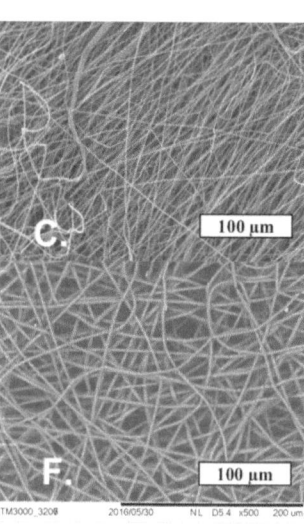

Figure 4. The microphotographs of PHB with a different content of Hmi: 1% wt. (**A**), 3% wt. (**B**) and 5% wt. (**C**) and Fe(TPP)Cl: 1% wt. (**D**), 3% wt. (**E**) and 5% wt. (**F**).

Figure 4 shows that Hmi and Fe(TPP)Cl contribute to the formation of smooth uniform fibers, reducing the number of defects. With the introduction of 1 and 3% wt. of Hmi, the number of defects and glues is reduced by 50–60%, and the introduction of 5% wt. of Hmi allowed to obtain completely faultless fibers. The thickness and tortuosity of the fibers decreased with an increase in the concentration of Hmi, which has a positive effect on the uniformity of operational properties. Fe(TPP)Cl in all concentrations had a strong effect on the fibrous structure, making the fibers uniform and smooth without snagging and thickening. A total of 1% wt. of Fe(TPP)Cl allowed obtaining thinner fibers than 3 and 5%

wt. of Fe(TPP)Cl, which differed little from each other. However, those fibers (Figure 4D) were characterized by a large thickness difference with local thickness distinctness of 70–75%.

3.2. Supramolecular Structure

Semi-crystalline polymers have a metastable structure, where various nanophases can be crystalline, liquid, glass, or mesophase. This multi-level structure is installed during material processing [71]. In biomedical applications, the supramolecular structure plays a significant role in the degradation, stability, and properties of the final product [70]. The predicted control of different parameters of the supramolecular structure is very important for constructing biomedical material with regulated properties.

The supramolecular structure of PHB is well known. PHB crystallizes into α-form crystal modification from the melt, which has an orthorhombic unit cell with $a = 5.76$ A°, $b = 13.20$ A°, c (fiber axis) = 5.96 Å [72]. Crystallites of PHB tend to be laid in lamellae [73] and PHB spherolites are possible in case of sufficient time and optimal conditions for repeated cold crystallization [74]. The ES process promotes rapid curing of the dissolved polymer. The polymeric fibers are fixed in the material, having a predominant orientation if an optimal balance of electrical conductivity, viscosity, and molding conditions is found for the polymer [75].

3.2.1. X-ray Diffraction Analysis

By X-ray diffraction analysis, it was found that the introduction of additives did not affect the parameters of the orthorhombic crystal lattice of PHB ($a = 0.576$ nm, $b = 1.320$ nm, $c = 0.596$ nm, space group symmetry of $P2_12_12_1$). The values of the long period were close to each other, which were between 5.2 and 5.4 nm for samples containing Hmi and between 5.7 and 5.9 nm for samples containing Fe(TPP)Cl. As a result of the analysis of X-ray diffractograms, the values characterizing the crystalline phase of PHB were obtained (Figure 5). X-ray diffractograms are shown in Supplementary Materials Figures S2 and S3 for the PHB-Hmi system and X-ray diffractograms for the PHB-Fe(TPP)Cl system were discussed in previous work [32].

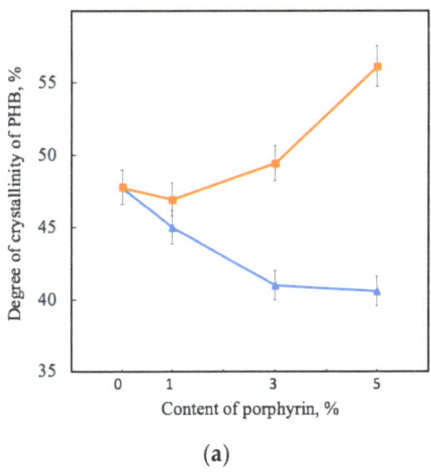
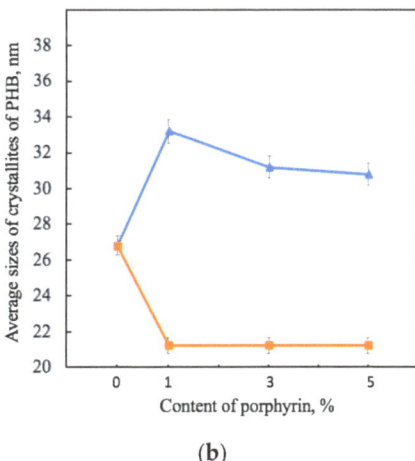

Figure 5. Dependence of the degree of crystallinity (**a**) and average sizes of PHB crystallites L_{020} (**b**) on the amount of porphyrin according to X-ray diffraction analysis for PHB-Hmi (blue) and PHB-Fe(TPP)Cl (orange).

Figure 5 shows that the effect of Hmi and Fe(TPP)Cl on PHB was significantly different. The gradual decrease in the proportion of crystallites of PHB-Hmi samples was between 6 and 17%, and the gradual decrease in the proportion of crystallites of PHB-Fe(TPP)Cl samples was between 3 and 17%. Addition of 1% wt. of Fe(TPP)Cl has a low effect on the degree of crystallinity of PHB. The introduction of Hmi contributes to a slight increase in the size of PHB crystallites by 14–20%, while the introduction of Fe(TPP)Cl leads to a decrease in their size by 40% (Figure 5b). It should be mentioned that the longitudinal size of PHB crystallites L020 in PHB-Fe(TPP)Cl did not change with different content of additive (Figure 5b). However, at the same time the transverse size of the PHB crystallites L002 in PHB-Fe(TPP)Cl increased from 9.8 (0 and 1% wt.) up to 12.7 nm (3% wt.) and 12.5 nm (5% wt.).

3.2.2. Electron Paramagnetic Resonance Analysis

The EPR method was used for characterizing the amorphous phase. EPR spectra of the spin probe TEMPO in structure of samples PHB-Hmi and PHB-Fe(TPP)Cl are shown in Supplementary Materials Figure S4.

Figure 6 shows that the effect of Hmi and Fe(TPP)Cl on PHB's amorphous region was consistent with the effects shown previously by the X-ray diffraction method. The correlation time of the probe in PHB-Hmi samples decreases. TEMPO mobility becomes less by 18–80%. At the same time in the same conditions the correlation time of the probe in PHB-Fe(TPP)Cl samples increases by 19–140% (Figure 6a). These results are related to the concentration of the radical entering the samples of the material (Figure 6b). Fe(TPP)Cl most likely occupies space in the amorphous phase and prevents the penetration of the radical into the material, reducing the concentration of the radical by 40–70%. At the same time, Hmi does not prevent the penetration of TEMPO into the amorphous region. The more of the radical enters, the less mobility it has in the PHB-Hmi composition. The main reason for such an effect is the localization of Hmi in the amorphous phase, as there are no obstacles filling it with a radical.

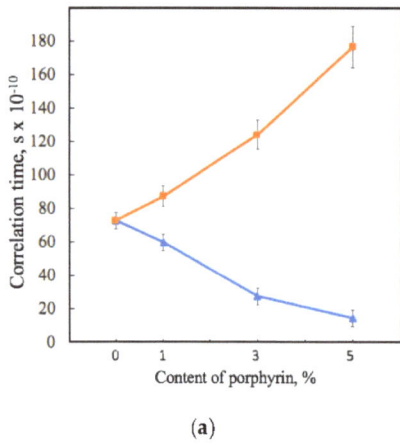
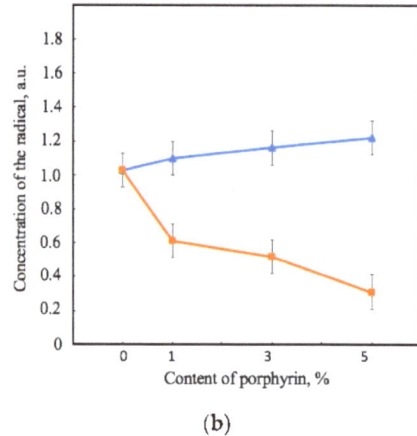

(a) (b)

Figure 6. Dependence of the correlation time of the spin robe TEMPO in the structure of the samples (**a**) and the concentration of the spin probe in relation to the mass of the material's sample (**b**) on the amount of porphyrin according to EPR analysis for PHB-Hmi (blue) and PHB0-Fe(TPP)Cl (orange).

3.2.3. Differential Scanning Calorimetry Analysis

The structure of many semi-crystalline polymers, including PHB, cannot simply be described by a conventional two-phase model consisting of crystalline and amorphous phases [76]. Decoupling between the crystalline and amorphous phases is generally incomplete due to the length of the polymer molecules, which far exceeds the dimensions, at least in one direction, of the crystalline phase, and due to possible geometric limitations [77]. The intermediate phase is non-crystalline and includes amorphous sections of macromolecules, whose mobility is hindered by near-crystalline structures [77].

The DSC method is an effective instrument for studying whole crystalline structure including even near-crystalline structures, which melt at temperatures, close to the melting temperature of PHB crystallites. DSC thermograms of PHB-Hmi and PHB-Fe(TPP)Cl are shown in Supplementary Materials Figure S1.

Table 2 shows that the melting temperature changes very slightly, with 3–6 °C, which is consistent with the assumption that PHB crystallites do not change much in size, since it significantly depends on this parameter. Of great interest are the values of crystallinity, with trends fully consistent with the results of the X-ray diffraction method (Figure 7).

Table 2. Results of the DCS analysis, where χ—crystallinity degree Δ ± 2.5%, ΔH – melting enthalpy Δ ± 2.5%, T_m—melting temperature Δ ± 2%.

Sample	Concentration of Additive, %	First Heating Run		χ PHB, %	Second Heating Run		χ PHB, %
		T_m, °C	ΔH, J/g		T_m, °C	ΔH, J/g	
PHB	0	175	93.1	65.2	170	90.8	63.9
PHB-Hmi	1	172	81.8	57.0	168	78.7	54.9
PHB-Hmi	3	173	77.8	53.1	170	75.4	51.5
PHB-Hmi	5	174	75.3	50.4	170	72.7	48.6
PHB-Fe(TPP)Cl	1	170	92.9	65.5	148	67.4	41.5
PHB-Fe(TPP)Cl	3	169	96.8	68.3	157	73.2	51.6
PHB-Fe(TPP)Cl	5	169	119.0	84.5	156	76.3	53.7

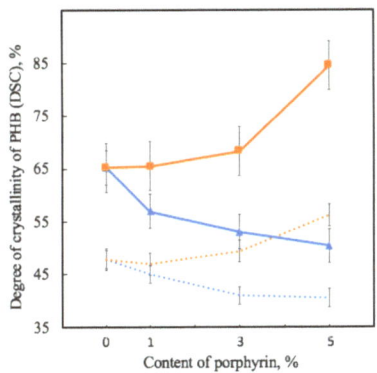

Figure 7. Dependence of the degree of crystallinity of samples on the amount of porphyrin according to DSC analysis for PHB-Hmi (blue line) and PHB-Fe(TPP)Cl (orange line) and X-ray diffraction analysis for PHB-Hmi (blue dots) and PHB-Fe(TPP)Cl (orange dots).

As the degree of crystallinity in DSC is understood as the total fraction of the crystalline phase in a semi-crystalline polymer, which includes both well-crystallized crystallites and uncrystallized, defective and paracrystalline formations, it is seen that crystallinity in DSC decreases by 13–25% for PHB-Hmi and increases by 2–28% for PHB-Fe(TPP)Cl at the first heating.

3.3. Properties of Electrospun Materials

3.3.1. Mechanical Analysis

The physical and mechanical properties of composite materials are an important class of operational properties, but they are also to a large extent an indicator of the state of the polymer-additive molecular system. The results of the mechanical tests are shown in Table 3. Stress-strain curves are shown in Supplementary Materials Figure S5.

Table 3. Results of the mechanical analysis.

Sample	Concentration of Additive, %	Tensile Strength, MPa ±0.02 MPa	Elongation at Break, % ±0.2 %
PHB	0	1.7	3.6
PHB-Hmi	1	0.7	4.7
PHB-Hmi	3	1.9	4.7
PHB-Hmi	5	5.5	6.1
PHB-Fe(TPP)Cl	1	2.1	3.5
PHB-Fe(TPP)Cl	3	1.6	3.5
PHB-Fe(TPP)Cl	5	1.4	3.6

Table 3 shows that high results in improving physical and mechanical properties were provided by 5% wt. of Hmi and 1% wt. of Fe(TPP)Cl. All other combinations of additives led to a reduction of the mechanical parameters of the material.

Mechanical properties are complex characteristics that depend on all levels of organization of nonwoven fibrous material. The contribution is made by defects of the fibrous layer, fiber bondings, and features of the supramolecular structure. There are two components that cause the growth of physical and mechanical characteristics. Firstly is the contribution of porphyrin with an atom of the metal to the formation of well-cured fibers without defects. These fibers form the layer with a higher possibility of withstanding loads due to the mobility of fibers in the whole system. Secondly, the addition of the porphyrin complexes affect the crystallization process, which can lead to a greater flexibility of the amorphous phase in the fiber and is capable of compensating for the high fragility of the initial PHB.

3.3.2. Wetting Contact Angle Analysis

Wetting contact angles of the fibrous materials were determined to evaluate the hydrophobicity of the surface area. The results are shown in Figure 8.

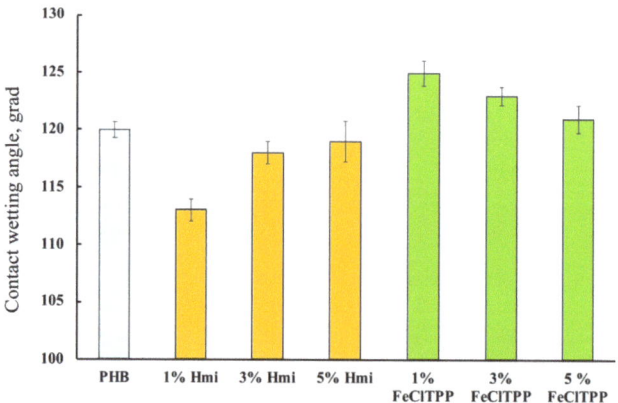

Figure 8. Contact wetting angles of the fibrous materials with a different amount of additives.

PHB is hydrophobic [78]. This property persists after the ES process. The introduction of porphyrin complexes allowed to influence this property of the material to some extent. Interestingly, in both cases, the greatest effect was found for 1% of the additive, regardless of its nature.

3.3.3. Biological Analysis

Biological tests made it possible to test the effectiveness of Hmi and Fe(TPP)Cl against Gram-positive and Gram-negative cultures. The results of the biological analysis tests are shown in Table 4.

Table 4. Results of the biological analysis.

Test Culture	Initial Test Culture, CFU/mL	Sample, CFU/mL	Control, CFU/mL
		PHB with 3% wt. Hmi	
S. aureus p 209	2.1×10^4	1.8×10^3	8.6×10^3
E. coli 1257	2.0×10^4	$<1 \times 10^2$	9.8×10^3
S. typhimurium	2.0×10^4	2.1×10^3	8.1×10^3
		PHB with 3% wt. Fe(TPP)Cl	
S. aureus p 209	2.0×10^4	1.8×10^3	4.0×10^3
E. coli 1257	2.0×10^4	$<1 \times 10^2$	9.0×10^3
S. typhimurium	2.2×10^4	1.0×10^3	6.0×10^3

The antibacterial properties of the Hmi and Fe(TPP)Cl are similarly high. In the PHB-Hmi and PHB-Fe(TPP)Cl, porphyrins are primarily associated with their effect on the cell walls of microorganisms by changing the charge of the bacterial cell. As a consequence, porphyrin molecules can suppress the function of adhesion and colonization of pathogens. Apparently, metal complexes are capable of disrupting the ionic balance of a living cell. In particular, this effect is enhanced in nanoscale fibrous materials. In addition, complexes containing metals of variable valences, such as iron, stimulate the formation of reactive oxygen species in aqueous media, which in turn also negatively affects the viability of pathogenic microorganisms. In general, the data obtained suggest that in the case of creating an antimicrobial material for biomedical purposes, the inclusion of Hmi or Fe(TPP)Cl in the composition positively affects the ability of the material to suppress the viability of bacteria and fungi.

4. Discussion

At the stage of the ES process, the introduction of both modifying additive at low concentrations (1, 3, and 5% wt.) can have a significant effect on the formation of the material's structure. The addition of Hmi to the forming solution of PHB increases viscosity by 40–90% and electrical conductivity by 25–75%. These parameters let the jet of forming solution move fast enough to form regular uniform fibers during the ES process. As a result, the optimum of balance viscosity-electrical conductivity is obtained at a concentration of 5% wt. of Hmi (Figure 4C). The addition of Fe(TPP)Cl to the forming solution of PHB increases viscosity by 20% and electrical conductivity by 50–87%. As a result, the optimum of balance viscosity-electrical conductivity is obtained at a concentration of 1, 3, and 5% wt. of Fe(TPP)Cl (Figure 4D–F). Both additives provide such an effect, mostly because of the atom of the metal in its structure (Figure 1b,c), and serves as a good current conductor to ensure higher efficiency of the ES process.

Hmi and Fe(TPP)Cl contribute to the reduction of the density of the fibrous layer by 34–44% and by 30–47%, respectively, increasing the porosity no less than by 9%, which is a great advantage for producing material with a higher developed surface area. The change in average diameters occurs differently, but the main aspect is reducing the number of caverns, pear-shaped defects, smudges, and other negative consequences of insufficient forming properties of the forming solution.

Of interest are the black inclusions found on the surface of the 1 and 3% wt. of Hmi on the surface PHB-Hmi and 1% wt. of Fe(TPP)Cl on the surface PHB-Fe(TPP)Cl. In previous work [31] EDX elemental analysis showed that these inclusions were particles of Hemin. A lower concentration of Hmi and Fe(TPP)Cl leads to the formation of agglomerates during the curing of the solution. They can diffuse onto the surface of the fibers with a high probability, due to their small size, sticking together into larger formations. With an increase in their concentration, agglomerates of particles remain in the structure of the polymer fiber, discovered using X-ray diffraction, which showed the presence of a large-crystalline phase of hemin (crystallite sizes more than 50–100 nm) in 5% wt. of Hmi.

These changes are closely related to the supramolecular structure of PHB [79]. PHB preserves the order of the crystal lattice. At the same time, we see noticeable changes in the structure of the crystalline phase.

PHB macromolecules could be considered as the alternation of sections of the crystalline and amorphous phases that affect a number of physical and mechanical properties of PHB fibers [80]. The addition of the additives leads to an introduction of the crystallization centers into the polymer system. These centers allow macromolecules of PHB to take an advantageous position, which lets the crystalline phase form better organized structures. For both additives, the realization of this effect is observed. In the case of Hmi, the crystallites have a larger size with a smaller number of them (Figure 5). In the case of Fe(TPP)Cl, the crystallites have a larger number and a smaller size (Figure 5). This may indicate the different nature of the interaction of porphyrin particles with each other during the formation of a PHB fiber from a solution.

A significant increase in the degree of crystallinity of PHB (detected by X-ray method) in the system PHB-Fe(TPP)Cl was accompanied by a significant increase in the number of irregular crystal formations that contributed to the DSC signal. It must be the structures that most likely affect the decrease in physical and mechanical properties of the material. Such a large proportion of the crystalline phase leads to embrittlement of the fibers, despite the fact that they have fewer defects compared to the initial ones and a more refined morphology.

These assumptions are in good agreement with the EPR results. An increase in the number of crystallites leads to a decrease in the proportion of a loose and mobile amorphous phase, into which a radical can enter and rotate. In the case of Fe(TPP)Cl, the diffusion of the radical into the amorphous phase was hindered, most likely by particles of tetrapyrrole rings, which could most likely occupy the free volume of the amorphous phase (Figure 6). At the same time, Hmi contributes to the free rotation of the radical and its greater concentration in the fiber structure (Figure 6). This effect increases with the increase of the Hmi concentration. Considering the detected Hmi crystal formations, they are most likely localized at the boundary of the crystalline phase of the PHB or in the zones of rigid amorphous areas.

Hmi and Fe(TPP)Cl are thermally stable, so they have no contribution to the melting behavior of PHB, except for their important role in the formation of the supramolecular structure. The first heating shows the state of the PHB structure as a consequence of the ES process, while the second heating shows the initial polymer structure. This explains the decrease in temperature and enthalpy of melting by 3–5% on average for both. Each subsequent heating leads to a decrease in the fraction of regular crystallites that have managed to crystallize well even during fast cooling (10 °K/min during 20 min), converting poorly crystallized areas into a low-molecular fraction. Changes in the supramolecular structure and in the morphology and surface of fibers significantly affect the properties of materials.

The wetting angle changes slightly with the introduction of additives, however, hydrophobic PHB, most likely due to a change in the state of the surface, is slightly hydrophilized by 3–5% with the introduction of Hmi. The effect of Hmi should be due to the polar groups –COOH (Figure 1b) located in the structure of the tetrapyrrole ring. This can be a notable advantage for planning certain types of biomedical materials. At the same time, Fe(TPP)Cl slightly decreases the contact wetting angle by 2–4%, which leads to an increase in the hydrophobicity of the material. The absence of polar groups and the localization of Fe(TPP)Cl lead to an improvement in the morphology of the surface, providing a more complete and organized structure with slightly higher hydrophobicity.

The antimicrobial tests show antimicrobial activity against drug-resistant and Gram-negative *E. coli* and Gram-positive *S. aureus* and *S. typhimurium*. These results are a second very important advantage of these new materials based on biocompatible PHB.

The third advantage is the growth of physical and mechanical characteristics of materials with the 3 and 5% wt. of Hmi and 1% of Fe(TPP)Cl. The effects of Hmi are certainly higher than those of Fe(TPP)Cl, but for biomedical purposes where an increase in the strength of a non-woven material is not required, Fe(TPP)Cl could be recommended as well.

5. Conclusions

The effect of natural and synthetic molecular complexes on the structure and properties of the electrospun composite materials based on PHB was investigated. The possibility of obtaining fibrous materials with high mechanical properties, high antibacterial activity, and controlled wettability was shown in the work. The introduction of 1–5% wt. of hemin and tetraphenylporphyrin with iron has an effect on the supramolecular structure, morphology, and properties of PHB-based fibers due to crystallization processes occurring at the stage of forming and curing of the fiber. The addition of metal atom (trivalent iron) contained in the tetrapyrrole ring of chosen complexes makes it possible to obtain an optimal balance of electrical conductivity and viscosity for forming defect-free uniform fibers. However, the influence of porphyrin complexes on the supramolecular structure had the opposite effect, with similar trends. This observation serves as a basis for the modification and directional design of the supramolecular structure of semi-crystalline polymers and properties of the fibrous material.

Supplementary Materials: The following supporting information can be downloaded at: https://www.mdpi.com/article/10.3390/jfb13010023/s1, Figure S1: DSC thermograms of PHB-Hmi composites: first heating run (a), second heating run (b) and PHB-Fe(TPP)Cl composites: first heating run (c), second heating run (d).; Figure S2: X-ray diffractograms of PHB-Hmi composites: 0% wt.y (b), 3% wt. (c), 5% wt. (d); Figure S3: Small Angle X-ray Scattering of PHB-Hmi composites: 0% wt. (1), 1% wt. (2), 3% wt. (3), 5% wt. (4); Figure S4: EPR spectra of the spin probe TEMPO in structure of samples PHB-Hmi (a) and PHB-Fe(TPP)Cl (b), where 1—0% wt., 2—1% wt., 3—3% wt., 4—5% wt. of the additive; Figure S5: Mechanical tests curves of samples PHB-Hmi (a) and PHB-Fe(TPP)Cl (b), where blue—0% wt., yellow—1% wt., grey—3% wt., red—5% wt. of the additive. Figure S6: Photo (a) and schematic view of the single-capillary laboratory unit for the electrospinning process (b), where: 1—protective installation box; 2—bin with a polymer solution and a capillary; 3—high voltage source; 4—stable precipitating electrode, 5—air pressure regulator.

Author Contributions: Conceptualization, P.T.; methodology, P.T., A.K., and A.L.; software, I.V.; validation, A.P.; formal analysis, S.K.; investigation, P.T., A.K., O.S., S.K., and A.L.; resources, A.P.; data curation, I.V., A.O., A.P., and O.S.; writing—original draft preparation, P.T., I.V., and A.K.; writing—review and editing, P.T.; visualization, I.V.; supervision, A.O.; project administration, I.V. All authors have read and agreed to the published version of the manuscript.

Funding: This research received no external funding.

Institutional Review Board Statement: Not applicable.

Informed Consent Statement: Not applicable.

Acknowledgments: The study was carried out using scientific equipment of the Center of Shared Usage «New Materials and Technologies» of Emanuel Institute of Biochemical Physics and the Common Use Centre of Plekhanov Russian University of Economics.

Conflicts of Interest: The authors declare no conflict of interest.

References

1. Greiner, A.; Wendorff, J.H. Electrospinning: A Fascinating Method for the Preparation of Ultrathin Fibers. *Angew. Chem. Int. Ed.* **2007**, *46*, 5670–5703. [CrossRef] [PubMed]
2. Ding, J.; Zhang, J.; Li, J.; Li, D.; Xiao, C.; Xiao, H.; Yang, H.; Zhuang, X.; Chen, X. Electrospun polymer biomaterials. *Prog. Polym. Sci.* **2019**, *90*, 1–34. [CrossRef]
3. Thenmozhi, S.; Dharmaraj, N.; Kadirvelu, K.; Kim, H.Y. Electrospun nanofibers: New generation materials for advanced applications. *Mater. Sci. Eng. B* **2017**, *217*, 36–48. [CrossRef]
4. Munj, H.R.; Nelson, M.T.; Karandikar, P.S.; Lannutti, J.J.; Tomasko, D.L. Biocompatible electrospun polymer blends for biomedical applications. *J. Biomed. Mater. Res. Part B Appl. Biomater.* **2014**, *102*, 1517–1527. [CrossRef]
5. Waghorn, P.A. Radiolabelled porphyrins in nuclear medicine. *J. Label. Comp. Radiopharm.* **2013**, *57*, 304–309. [CrossRef] [PubMed]
6. Chen, Z.; Mai, B.; Tan, H.; Chen, X. Nucleic acid based nanocomposites and their applications in biomedicine. *Compos. Commun.* **2018**, *10*, 194–204. [CrossRef]
7. Suo, Z.; Chen, J.; Hou, X.; Hu, Z.; Xing, F.; Feng, L. Growing prospects of DNA nanomaterials in novel biomedical applications. *RSC Adv.* **2019**, *9*, 16479–16491. [CrossRef]
8. Wu, J.; Li, S.; Wei, H. Integrated nanozymes: Facile preparation and biomedical applications. *Chem. Commun.* **2018**, *54*, 6520–6530. [CrossRef]
9. Ruthard, C.; Schmidt, M.; Gröhn, F. Porphyrin-polymer networks, worms, and nanorods: pH-triggerable hierarchical self-assembly. *Macromol. Rapid Commun.* **2011**, *32*, 706–711. [CrossRef]
10. Yu, W.; Zhen, W.; Zhang, Q.; Li, Y.; Luo, H.; He, J.; Liu, Y.M. Porphyrin-Based Metal-Organic Frameworks compounds as a promising nanomedicine in photodynamic therapy. *ChemMedChem* **2020**, *15*, 1766–1775. [CrossRef]
11. Imran, M.; Ramzan, M.; Qureshi, A.; Khan, M.; Tariq, M. Emerging Applications of Porphyrins and Metalloporphyrins in Biomedicine and Diagnostic Magnetic Resonance Imaging. *Biosensors* **2018**, *8*, 1766–1775. [CrossRef] [PubMed]
12. Zhu, Y.; Chen, J.; Kaskel, S. Porphyrin-Based Metal-Organic Frameworks for Biomedical Applications. *Angew. Chem. Int. Engl.* **2021**, *60*, 5010–5035. [CrossRef]
13. Stojiljkovic, I.; Evavold, B.D.; Kumar, V. Antimicrobial properties of porphyrins. *Expert Opin. Investig. Drugs* **2001**, *10*, 309–320. [CrossRef] [PubMed]
14. Falk, J.E. *Porphyrins and Metalloporphyrins*; Elsevier Pub. Co.: Amsterdam, The Netherlands; New York, NY, USA, 1964.
15. Massiot, J.; Rosilio, V.; Makky, A. Photo-triggerable liposomal drug delivery systems: From simple porphyrin insertion in the lipid bilayer towards supramolecular assemblies of lipid–porphyrin conjugates. *J. Mater. Chem. B* **2019**, *7*, 1805–1823. [CrossRef] [PubMed]
16. Zhao, L.; Qu, R.; Li, A.; Ma, R.; Shi, L. Cooperative self-assembly of porphyrins with polymers possessing bioactive functions. *Chem. Commun.* **2016**, *52*, 13543–13555. [CrossRef]
17. Rajan, K.P.; Thomas, S.P.; Gopanna, A.; Chavali, M. Polyhydroxybutyrate (PHB): A Standout Biopolymer for Environmental Sustainability. In *Handbook of Ecomaterials*; Martínez, L.M.T., Kharissova, O.V., Kharisov, B.I., Eds.; Springer International Publishing AG: Cham, Switzerland, 2017; pp. 1–23.
18. Pati, S.; Maity, S.; Dash, A.; Jema, S.; Mohapatra, S.; Das, S.; Samantaray, D.P. Biocompatible PHB production from Bacillus species under submerged and solid-state fermentation and extraction through different downstream processing. *Curr. Mi-crobiol.* **2020**, *77*, 1203–1209. [CrossRef]
19. Amadu, A.A.; Qiu, S.; Ge, S.; Addico, G.N.D.; Ameka, G.K.; Yu, Z.; Xia, W.; Abbew, A.W.; Shao, D.; Champagne, P.; et al. A review of biopolymer (Poly-β-hydroxybutyrate) synthesis in microbes cultivated on wastewater. *Sci. Total Environ.* **2021**, *756*, 143729. [CrossRef]
20. Woolnough, C.A.; Yee, L.H.; Charlton, T.S.; Foster, L.J.R. Environmental degradation and biofouling of green plastics including short and medium chain length polyhydroxyalkanoates. *Polym. Int.* **2010**, *59*, 658–667. [CrossRef]
21. Sreedevi, S.; Unni, K.N.; Sajith, S.; Priji, P.; Josh, M.S.; Benjamin, S. Bioplastics: Advances in polyhydroxybutyrate research. In *Advances in Polymer Science*; Springer: Berlin, Germany, 2014; pp. 1–30.
22. Arrieta, M.P.; López, J.; Hernández, A.; Rayón, E. Ternary PLA–PHB–limonene blends intended for biodegradable food packaging applications. *Eur. Polym. J.* **2013**, *50*, 255–270. [CrossRef]
23. Kumara Babu, P.; Maruthi, Y.; Veera Pratap, S.; Sudhakar, K.; Sadiku, R.; Prabhakar, M.N.; Song, J.I.; Subha, M.C.S.; Chowdoji Rao, K. Development and characterization of polycaprolactone (PCL)/poly ((R)-3-hydroxybutyric acid) (PHB) blend microspheres for tamoxifen drug relese studies. *Int. J. Pharm. Pharmac. Sci.* **2015**, *7*, 95–100.
24. Karimi, A.; Karbasi, S.; Razavi, S.; Zargar, E. Poly(hydroxybutyrate)/chitosan aligned electrospun scaffold as a novel sub-strate for nerve tissue engineering. *Adv. Biomed. Eng.* **2018**, *7*, 44. [CrossRef]
25. Saad, B.; Neuenschwander, P.; Uhlschmid, G.; Suter, U. New versatile, elastomeric, degradable polymeric materials for medicine. *Int. J. Biol. Macromol.* **1999**, *25*, 293–301. [CrossRef]
26. Kim, G.M.; Wutzler, A.; Radusch, H.J.; Michler, G.H.; Simon, P.; Sperling, R.A.; Parak, W.J. One-dimension arrangement of gold nano-particles by electrospinning. *Chem. Mater.* **2005**, *17*, 4949–4957. [CrossRef]
27. Dror, Y.; Salalha, W.; Khalfin, R.L.; Cohen, Y.; Yarin, A.L.; Zussman, E. Carbon nanotubes embeded in oriented polymer nano-fibers by electrospinning. *Langmuir* **2003**, *19*, 7012–7020. [CrossRef]

28. Jun, Z.; Aigner, A.; Czubayko, F.; Kissel, T.; Wendorff, J.H.; Greiner, A. Poly (vinyl alcohol) nanofibers by electrospinning as a protein delivery system and retardation of enzyme release by additional polymer coatings. *Biomacromolecules* **2005**, *6*, 1484–1488. [CrossRef]
29. Joung, K.; Bae, J.W.; Park, K.D. Controlled release of heparin-binding growth factors using heparin-containing particulate systems for tissue regeneration. *Expert Opin. Drug Deliv.* **2008**, *5*, 1173–1184. [CrossRef]
30. Avossa, J.; Paolesse, R.; Di Natale, C.; Zampetti, E.; Bertoni, G.; De Cesare, F.; Macagnano, A. Electrospinning of Polystyrene/Polyhydroxybutyrate Nanofibers Doped with Porphyrin and Graphene for Chemiresistor Gas Sensors. *Nanomaterials* **2019**, *9*, 280. [CrossRef]
31. Tyubaeva, P.; Varyan, I.; Lobanov, A.; Olkhov, A.; Popov, A. Effect of the Hemin Molecular Complexes on the Structure and Properties of the Composite Electrospun Materials Based on Poly(3-hydroxybutyrate). *Polymers* **2021**, *13*, 4024. [CrossRef]
32. Olkhov, A.A.; Tyubaeva, P.M.; Zernova, Y.N.; Kurnosov, A.S.; Karpova, S.G.; Iordanskii, A.L. Structure and Properties of Biopolymeric Fibrous Materials Based on Polyhydroxybutyrate–Metalloporphyrin Complexes. *Russ. J. Gen. Chem.* **2021**, *91*, 546–553. [CrossRef]
33. Pramual, S.; Assavanig, A.; Bergkvist, M.; Batt, C.A.; Sunintaboon, P.; Lirdprapamongkol, K.; Niamsiri, N. Development and characterization of bio-derived polyhydroxyalkanoate nanoparticles as a delivery system for hydrophobic photodynamic therapy agents. *J. Mater. Sci. Mater. Med.* **2015**, *27*, 40. [CrossRef]
34. Bonartzev, A.P.; Bonartzeva, G.A.; Shaitari, K.V.; Kirpichnikov, M.P. Poly(3-Hydroxybutyrate) and Biopolymer Systems on the Basis of This Polyester. *Biomed. Khimiya* **2011**, *57*, 374–391. [CrossRef]
35. Williams, S.F.; Martin, D.P.; Horowitz, D.M.; Peoples, O.P. PHA Applications: Addressing the Price Performance Issue I. Tissue Engineering. *Int. J. Biol. Macromol.* **1999**, *25*, 111–121. [CrossRef]
36. Chen, G.; Wang, Y. Medical applications of biopolyesters polyhydroxyalkanoates. *Chin. J. Polym. Sci.* **2013**, *31*, 719–736. [CrossRef]
37. Volova, T.; Shishatskaya, E.; Sevastianov, V.; Efremov, S.; Mogilnaya, O. Results of Biomedical Investigations of PHB and PHB/PHV Fibers. *Biochem. Eng. J.* **2003**, *16*, 125–133. [CrossRef]
38. Vieyra, H.; Juárez, E.; López, U.F.; Morales, A.G.; Torres, M. Cytotoxicity and Biocompatibility of Biomaterials Based in Polyhydroxybutyrate Reinforced with Cellulose Nanowhiskers Determined in Human Peripheral Leukocytes. *Biomed. Mater.* **2018**, *13*, 045011. [CrossRef] [PubMed]
39. Chen, C.; Cheng, Y.C.; Yu, C.H.; Chan, S.W.; Cheung, M.K.; Yu, P.H.F. In vitrocytotoxicity, hemolysis assay, and biodegradation behavior of biodegradable poly(3-hydroxybutyrate)-poly(ethylene glycol)-poly(3-hydroxybutyrate) nanoparticles as potential drug carriers. *J. Biomed. Mater. Res. A* **2008**, *87A*, 290–298. [CrossRef]
40. Zhao, Y.; Zhang, L.; Wei, W.; Li, Y.; Liu, A.; Zhang, Y.; Liu, S. Effect of annealing temperature and element composition of tita-nium dioxide/graphene/hemin catalysts for oxygen reduction reaction. *RSC Adv.* **2015**, *5*, 82879–82886. [CrossRef]
41. Nitzan, Y.; Ladan, H.; Gozansky, S.; Malik, Z. Characterization of hemin antibacterial action on Staphylococcus aureus. *FEMS Microbiol. Lett.* **1987**, *48*, 401–406. [CrossRef]
42. Dell'Acqua, S.; Massardi, E.; Monzani, E.; Di Natale, G.; Rizzarelli, E.; Casella, L. Interaction between hemin and prion pep-tides: Binding, oxidative reactivity and aggregation. *Int. J. Mol. Sci.* **2020**, *21*, 7553. [CrossRef]
43. Zozulia, O.; Korendovych, I.V. Semi-rationally designed short peptides self-assemble and bind hemin to promote cyclopro-panation. *Angew. Chem. Int. Ed.* **2020**, *59*, 8108–8112. [CrossRef]
44. Dong, L.; Zang, J.; Wang, W.; Liu, X.; Zhang, Y.; Su, J.; Li, J. Electrospun single iron atoms dispersed carbon nanofibers as high performance electrocatalysts toward oxygen reduction reaction in acid and alkaline media. *J. Colloid Interface Sci.* **2019**, *564*, 134–142. [CrossRef] [PubMed]
45. Hsu, C.C.; Serio, A.; Amdursky, N.; Besnard, C.; Stevens, M.M. Fabrication of Hemin-Doped Serum Albumin-Based Fibrous Scaffolds for Neural Tissue Engineering Applications. *ACS Appl. Mater. Interfaces* **2018**, *10*, 5305–5317. [CrossRef] [PubMed]
46. Lu, Y.; Berry, S.M.; Pfister, T.D. Engineering Novel Metalloproteins: Design of Metal-Binding Sites into Native Protein Scaffolds. *Chem. Rev.* **2001**, *101*, 3047–3080. [CrossRef] [PubMed]
47. Zhang, Y.; Xu, C.; Li, B. Self-Assembly of Hemin on Carbon Nanotube as Highly Active Peroxidase Mimetic and Its Application for Biosensing. *RSC Adv.* **2013**, *3*, 6044. [CrossRef]
48. Alsharabasy, A.M.; Pandit, A.; Farràs, P. Recent Advances in the Design and Sensing Applications of Hemin/Coordination Polymer-Based Nanocomposites. *Adv. Mater.* **2021**, *33*, 2003883. [CrossRef]
49. Yang, J.; Xiong, L.; Li, M.; Xiao, J.; Geng, X.; Wang, B.; Sun, Q. Preparation and Characterization of Tadpole- and Sphere-Shaped Hemin Nanoparticles for Enhanced Solubility. *Nanoscale Res. Lett.* **2019**, *14*, 47. [CrossRef]
50. Tomat, E. Coordination Chemistry of Linear Tripyrroles: Promises and Perils. *Comments Mod. Chem. A Comments Inorg. Chem.* **2016**, *36*, 327–342. [CrossRef]
51. Nishi, M.; Ishii, R.; Ikeda, M.; Hanasaki, N.; Hoshino, N.; Akutagawa, T.; Sumimoto, M.; Matsuda, M. An Electrically Con-ducting Crystal Composed of an Octahedrally Ligated Porphyrin Complex with High-Spin Iron(III). *Dalton Trans.* **2018**, *47*, 4070–4075. [CrossRef]
52. Pegis, M.L.; Martin, D.J.; Wise, C.F.; Brezny, A.C.; Johnson, S.I.; Johnson, L.E.; Kumar, N.; Raugei, S.; Mayer, J.M. The Mechanism of Catalytic O2 Reduction by Iron Tetraphenylporphyrin. *J. Am. Chem. Soc.* **2019**, *141*, 8315–8326. [CrossRef]
53. Sun, Z.C.; She, Y.B.; Zhou, Y.; Song, X.F.; Li, K. Synthesis, Characterization and Spectral Properties of Substituted Tetraphenylpor-phyrin Iron Chloride Complexes. *Molecules* **2011**, *16*, 2960–2970. [CrossRef]

54. McCann, S.W.; Wells, F.V.; Wickman, H.H.; Sorrell, T.N.; Collman, J.P. Magnetic properties of a (tetraphenylporphyrin)iron(III) thiolate: Fe(TPP)(SC6H5)(HSC6H5). *Inorg. Chem.* **1980**, *19*, 621–628. [CrossRef]
55. Tovmasyan, A.; Batinic-Haberle, I.; Benov, L. Antibacterial Activity of Synthetic Cationic Iron Porphyrins. *Antioxidants* **2020**, *9*, 972. [CrossRef] [PubMed]
56. Adler, A.D.; Longo, F.R.; Kampas, F.; Kim, J. On the preparation of metalloporphyrins. *J. Radioanal. Nucl. Chem.* **1970**, *32*, 2443–2445. [CrossRef]
57. Shalit, H.; Libman, A.; Pappo, D. Meso-Tetraphenylporphyrin Iron Chloride Catalyzed Selective Oxidative Cross-Coupling of Phenols. *J. Am. Chem. Soc.* **2017**, *139*, 13404–13413. [CrossRef]
58. Lubasova, D.; Martinova, L. Controlled Morphology of Porous Polyvinyl Butyral Nanofibers. *J. Nanomater* **2011**, *2011*, 292516. [CrossRef]
59. You, Y.; Youk, J.H.; Lee, S.W.; Min, B.M.; Lee, S.J.; Park, W.H. Preparation of porous ultrafine PGA fibers via selective dissolu-tion of electrospun PGA/PLA blend fibers. *Mater. Lett.* **2006**, *60*, 757–760. [CrossRef]
60. Vyazovkin, S.; Koga, N.; Schick, C.V. *Handbook of Thermal Analysis and Calorimetry, Applications to Polymers and Plastics*; Elsevier Pub. Co.: Amsterdam, The Netherlands, 2002.
61. Scandola, M.; Focarete, M.L.; Adamus, G.; Sikorska, W.; Baranowska, I.; Świerczek, S.; Jedliński, Z. Polymer blends of natural poly(3-hydroxybutyrate-co-3-hydroxyvalerate) and a synthetic atactic poly(3-hydroxybutyrate). Characterization and bio-degradation studies. *Macromolecules* **1997**, *30*, 2568–2574. [CrossRef]
62. Liang, Z.; Freed, J.H. An Assessment of the Applicability of Multifrequency ESR to Study the Complex Dynamics of Biomolecules. *J. Phys. Chem.* **1999**, *103*, 6384–6396. [CrossRef]
63. Sezer, D.; Freed, J.H.; Roux, B. Simulating electron spin resonance spectra of nitroxide spin labels from molecular dynamics and stochastic trajectories. *J. Chem. Phys.* **2008**, *128*, 165106. [CrossRef]
64. Krivandin, A.V.; Solov'eva, A.B.; Glagolev, N.N.; Shatalova, O.V.; Kotova, S.L. Structure alterations of perfluorinated sulfocationic membranes under the action of ethylene glycol (SAXS and WAXS studies). *Polymer* **2003**, *44*, 5789–5796. [CrossRef]
65. Krivandin, A.V.; Fatkullina, L.D.; Shatalova, O.V.; Goloshchapov, A.N.; Burlakova, E.B. Small-angle X-ray scattering study of the incorporation of ICHPHAN antioxidant in liposomes. *Russ. J. Phys. Chem. B* **2013**, *7*, 338–342. [CrossRef]
66. Shibryaeva, L.S.; Shatalova, O.V.; Krivandin, A.V.; Tertyshnaya, Y.V.; Solovova, Y.V. Specific structural features of crystalline re-gions in biodegradable composites of poly-3-hydroxybutyrate with chitosan. *Russ. J. Appl. Chem.* **2017**, *90*, 1443–1453. [CrossRef]
67. Altaee, N.; El-Hiti, G.A.; Fahdil, A.; Sudesh, K.; Yousif, E. Biodegradation of different formulations of polyhydroxybutyrate films in soil. *Springerplus* **2016**, *5*, 762. [CrossRef]
68. Rashid, T.U.; Gorga, R.E.; Krause, W.E. Mechanical Properties of Electrospun Fibers—A Critical Review. *Adv. Eng. Mater.* **2021**, *23*, 2100153. [CrossRef]
69. Syerko, E.; Comas-Cardona, S.; Binetruy, C. Models of mechanical properties/behavior of dry fibrous materials at various scales in bending and tension: A review. *Compos. Part A Appl. Sci. Manuf.* **2012**, *43*, 1365–1388. [CrossRef]
70. Szewczyk, P.K.; Stachewicz, U. The impact of relative humidity on electrospun polymer fibers: From structural changes to fiber morphology. *Adv. Colloid Interface Sci.* **2020**, *286*, 102315. [CrossRef] [PubMed]
71. Di Lorenzo, M.L.; Gazzano, M.; Righetti, M.C. The Role of the Rigid Amorphous Fraction on Cold Crystallization of Poly(3-hydroxybutyrate). *Macromolecules* **2012**, *45*, 5684–5691. [CrossRef]
72. Cornibert, J.; Marchessault, R.H. Conformational analysis and crystalline structure. *J. Mol. Biol.* **1972**, *71*, 735–756. [CrossRef]
73. Mota, C.; Puppi, D.; Dinucci, D.; Gazzarri, M.; Chiellini, F. Additive manufacturing of star poly(ε-caprolactone) wet-spun scaffolds for bone tissue engineering applications. *J. Bioact. Compat. Polym.* **2013**, *28*, 320–337. [CrossRef]
74. Hoffman, J.D.; Davis, G.T.; Lauritzen, J.I. *Treatise on Solid State Chemistry, Crystalline and Noncrystalline Solids*, 3rd ed.; Plenum Press: New York, NY, USA, 1976; pp. 497–498.
75. Reneker, D.H.; Yarin, A.L.; Fong, H.; Koombhongse, S. Bending instability of electrically charged liquid jets of polymer solu-tions in electrospinning. *J. Appl. Phys.* **2000**, *87*, 4531–4547. [CrossRef]
76. Wunderlich, B. Reversible crystallization and the rigid–amorphous phase in semicrystalline macromolecules. *Prog. Polym. Sci.* **2013**, *28*, 383–450. [CrossRef]
77. Righetti, M.C.; Tombari, E. Crystalline, mobile amorphous and rigid amorphous fractions in poly(L-lactic acid) by TMDSC. *Thermochim. Acta* **2011**, *522*, 118–127. [CrossRef]
78. Chan, S.Y.; Chan, B.Q.Y.; Liu, Z.; Parikh, B.H.; Zhang, K.; Lin, Q.; Su, X.; Kai, D.; Choo, W.S.; Young, D.J.; et al. Electrospun Pectin-Polyhydroxybutyrate Nanofibers for Retinal Tissue Engineering. *ACS Omega* **2017**, *2*, 8959–8968. [CrossRef] [PubMed]
79. Reneker, D.H.; Yarian, A.L.; Zussman, E.; Xu, H. Electrospinning of nanofibers from polymer solutions and melts. *Adv. Appl. Mech.* **2007**, *41*, 43–195, 345–346. [CrossRef]
80. Greenfeld, I.; Arinstein, A.; Fezzaa, K.; Rafailovich, M.H.; Zussman, E. Polymer dynamics in semidilute solution during electrospinning: A simple model and experimental observations. *Phys. Rev.* **2011**, *84*, 041806. [CrossRef] [PubMed]

Article

Design of a Naturally Dyed and Waterproof Biotechnological Leather from Reconstituted Cellulose

Claudio José Galdino da Silva Junior [1,2,3], Julia Didier Pedrosa de Amorim [1,2,3], Alexandre D'Lamare Maia de Medeiros [1,2,3], Anantcha Karla Lafaiete de Holanda Cavalcanti [4], Helenise Almeida do Nascimento [5], Mariana Alves Henrique [5], Leonardo José Costa do Nascimento Maranhão [6], Glória Maria Vinhas [5], Késia Karina de Oliveira Souto Silva [6], Andréa Fernanda de Santana Costa [2,7] and Leonie Asfora Sarubbo [1,2,3,*]

[1] Rede Nordeste de Biotecnologia (RENORBIO), Universidade Federal Rural de Pernambuco (UFRPE), Rua Dom Manuel de Medeiros, Dois Irmãos, Recife 52171-900, PE, Brazil; claudiocjg@gmail.com (C.J.G.d.S.J.); julia_amorim@hotmail.com (J.D.P.d.A.); alexandre_dlamare@outlook.com (A.D.M.d.M.)

[2] Instituto Avançado de Tecnologia e Inovação (IATI), Rua Potyra, n. 31, Prado, Recife 52171-900, PE, Brazil; andrea.santana@ufpe.br

[3] Escola Icam Tech, Universidade Católica de Pernambuco (UNICAP), Rua do Príncipe, n. 526, Boa Vista, Recife 52171-900, PE, Brazil

[4] Centro de Tecnologia em Design de Moda, Faculdade Senac Pernambuco, Rua do Pombal, n. 57, Santo Amaro, Recife 52171-900, PE, Brazil; anantchalafaiete@gmail.com

[5] Centro de Tecnologia e Geociências, Departamento de Engenharia Química, Universidade Federal de Pernambuco (UFPE), Cidade Universitária, s/n, Recife 52171-900, PE, Brazil; helenise_almeida@hotmail.com (H.A.d.N.); mariana.ahenrique@ufpe.br (M.A.H.); gloria.vinhas@ufpe.br (G.M.V.)

[6] Centro de Tecnologia, Departamento de Engenharia Têxtil, Universidade Federal do Rio Grande do Norte (UFRN), Avenida Senador Salgado Filho, n. 3000, Lagoa Nova, Natal 59078-970, RN, Brazil; leonardo.nascimento.063@ufrn.edu.br (L.J.C.d.N.M.); kesia.souto@ufrn.br (K.K.d.O.S.S.)

[7] Centro de Comunicação e Design, Centro Acadêmico da Região Agreste, Universidade Federal de Pernambuco (UFPE), BR 104, Km 59, s/n, Nova Caruaru, Caruaru 50670-901, PE, Brazil

* Correspondence: leonie.sarubbo@unicap.br

Abstract: Consumerism in fashion involves the excessive consumption of garments in modern capitalist societies due to the expansion of globalisation, especially at the beginning of the 21st Century. The involvement of new designers in the garment industry has assisted in creating a desire for new trends. However, the fast pace of transitions between collections has made fashion increasingly frivolous and capable of generating considerable interest in new products, accompanied by an increase in the discarding of fabrics. Thus, studies have been conducted on developing sustainable textile materials for use in the fashion industry. The aim of the present study was to evaluate the potential of a vegan leather produced with a dyed, waterproof biopolymer made of reconstituted bacterial cellulose (BC). The dying process involved using plant-based natural dyes extracted from *Allium cepa* L., *Punica granatum*, and *Eucalyptus globulus* L. The BC films were then shredded and reconstituted to produce uniform surfaces with a constant thickness of 0.10 cm throughout the entire area. The films were waterproofed using the essential oil from *Melaleuca alternifolia* and wax from *Copernicia prunifera*. The characteristics of the biotechnological vegan leather were analysed using scanning electron microscopy (SEM), thermogravimetric analysis (TGA), flexibility and mechanical tests, as well as the determination of the water contact angle (°) and sorption index (s). The results confirmed that the biomaterial has high tensile strength (maximum: 247.21 ± 16.52 N) and high flexibility; it can be folded more than 100 times at the same point without breaking or cracking. The water contact angle was 83.96°, indicating a small water interaction on the biotextile. The results of the present study demonstrate the potential of BC for the development of novel, durable, vegan, waterproof fashion products.

Keywords: fashion; design; sustainable clothing; bacterial cellulose

1. Introduction

The exponential increase in the world population has led to an increasing demand for products, which has pressured industries and supply chains to develop cheaper products at a fast pace and on a large scale. Such products generally have low durability, resulting in considerable socioenvironmental harm. The increase in industrial production also contributes to the intensification of pollution. Thus, numerous consumers, researchers, and specialists have become concerned with issues of production and consumption and the sustainable management of the supply chain [1]. Moreover, the awareness of socio-environmental problems, such as climate change, resource scarcity, the exploitation of labour, and the pollution of water resources, has increased expectations on the part of consumers regarding how companies develop their brands and products [2].

The textile industry is the second-largest polluter globally, behind only the oil industry. Thus, textile companies have been looking for safer materials to replace current highly polluting manufacturing methods, and studies on sustainable alternative raw materials are currently being conducted worldwide [3–5]. Bacterial cellulose is one of these biomaterials with considerable potential for use in the fashion industry [6].

Although plants are the major source of cellulose, several genera of microorganisms are also capable of producing this substance, which is known as bacterial cellulose or biocellulose (BC). As BC is considered a low-cost, extremely versatile, ecologically correct biopolymer, studies involving different applications of this biomaterial have increased over the years [6–10].

The molecular formula of BC is the same as that of plant cellulose. However, the fibrillar structure of BS is smaller and has a larger surface area. The fibres of plant cellulose have a diameter of approximately 13 to 22 μm, with a crystallinity of 44 to 65% [11,12]. In contrast, BC fibrils are naturally nanometric, with a diameter of 10 to 100 nm and crystallinity close to 90%. Moreover, BC has a high tensile strength (~70 N) and has a hydrophilic nature due to the high number of hydroxyl groups on its surface [12,13].

Due to these characteristics, microbial cellulose can serve as the basis for the sustainable textiles that the fashion industry seeks. The biomaterial can be produced in different shapes and thicknesses, taking on the shape of the recipient in which the microbial fermentation is performed, thereby avoiding waste in the modelling of parts. It is also possible to use small patches of BC to make larger pieces through homogenisation and remodelling of the material. The production of BC can also use agro-industrial waste products, thereby lowering costs and making the product more accessible. Moreover, BC is easily degradable when discarded in nature [3,14,15].

Another factor that can aggregate value to biocellulose in fashion is dyeing using natural pigments, which provide colour, tonality, aesthetics, and sustainability to the products. Such dyes can be extracted from different parts of plants (leaves, flowers, fruits, stems, and roots). Besides being sustainable, natural dyes can be successfully used with BC when the colourisation conditions are controlled. Thus, non-polluting, biodegradable products, such as plant-based dyes and biotextiles, have considerable potential for use as novel biotechnological products that meet the needs of the world market [6,16].

However, as a production process that occurs by natural fermentation, some challenges need to be overcome to apply biocellulose as a textile material. It is necessary to ensure a uniform structure with a constant thickness, attractive textures, adequate strength, fit, comfort, water resistance, and durability, along with the maintenance of attractive aesthetics to create novel products [17].

Considering the concepts, trends, and possibilities of biocellulose and natural dyes for use in the fashion industry, the aim of the present study was to produce a biocellulose-based vegan leather using an innovative shredding and reconstitution process to create a more uniform structure with a constant thickness. The biotextile was then dyed naturally with pigments extracted from plants and waterproofed to ensure its applicability in developing novel textile products for increasingly conscientious consumers.

2. Materials and Methods

2.1. Microorganisms and Means of Maintenance

Microorganisms in a symbiotic culture of bacteria and yeast (SCOBY), obtained from the culture collection of Nucleus of Resource in Environmental Sciences, Catholic University of Pernambuco, Brazil, were used to produce BC. According to Villarreal-Soto et al. (2018) [18], the microorganisms in the microbiological composition of the consortium include acetic acid bacteria (*Komagataeibacter* sp. and *Acetobacter* sp.), lactic acid bacteria (*Lactococcus* sp. and *Lactobacillus* sp.) and yeasts (*Zygosaccharomyces bailii*, *Saccharomyces cerevisiae* and *Schizosaccharomyces pombe*). The maintenance medium, called green tea medium, is constituted of 50.00 g/L of sucrose and 1.15 g/L of citric acid, acquired from MERTEC (Brazil), and 10.00 g/L of green tea leaves (*Camellia sinensis*) from *Chá Leão* (Brazil), adjusted to pH 6.

2.2. BC Culture Conditions, Purification, and Yield

BC production was performed by transferring 10% (v/v) of a pre-inoculum containing the microorganisms in the consortium to 2500-mL Schott flasks containing 2000 mL of the green tea medium. Static cultivation was performed at 30 °C for 14 days. The BC was rinsed in running water, and purification was achieved by immersion in a 0.1 M NaOH solution at 70 °C for 1 h. The BC films were then neutralised and weighed, followed by calculating the yield.

2.3. Water Retention Capacity (WRC)

WRC is linked to moisture. The analysis of this measure determines the capacity of the biomaterial to adsorb and fix dyes. The BC films were weighed (25 °C, 1 atm) and dried in an oven at 60 °C until reaching a constant weight, indicating the complete removal of water. The WRC was then obtained using Equation (1):

$$\text{WRC } (\%) = \frac{\text{Mean of wet weights} - \text{Mean of dry weights}}{\text{Mean of wet weights}} \quad (1)$$

2.4. Natural Dye Extraction

An infusion was made at room temperature in a solution composed of 1000 mL of deionised water and 250 mL of 70% ethanol to extract dye from *Eucalyptus globulus* L. (50 g of dry leaves), *Allium cepa* L. (50 g of bulb bark) and *Punica granatum* (50 g of dried fruit peel) obtained at public markets in the city of Recife, Pernambuco, Brazil. After 24 h, the infusions were boiled for 30 min and filtered to remove the vegetable matter. The liquids were then used for dyeing.

2.5. Preparation of BC Films for Dyeing

A water solution was prepared with the fixative (potassium alum 99.5%) at 20 g/L. The BC films were submerged in the solution and heated at 90 °C for 30 min under agitation for the penetration of the potassium alum and fixation in the fibres of the films.

2.6. Dyeing and Natural Dye Fixation Procedure

The volumes obtained from the dye extracts were kept at 90 °C and used for dyeing 1000 g of BC fibres (wet mass). The films were submerged in the heated extracts for 1 h with light agitation. The films dyed with *Allium cepa* L., *Punica granatum*, and *Eucalyptus globulus* L. were then rinsed in running water and placed in a fixative bath containing 20 g of NaCl/L in water for 30 min.

2.7. Shredding, Reconstitution, and Drying

Each dyed BC film was shredded in wet condition with the aid of an industrial blender at 18,000 rpm for 2 min to form a homogeneous mass. This mass was uniformly distributed on a 20 cm × 20 cm silkscreen for reconstitution, and the fibres were dried at room temperature. This process lasted three to six days and only ended when the total reconstitution of the biotextile was observed. At the end of the process, the biotextile had the appearance of thin coloured leather.

2.8. Waterproofing with Essential Oil and Wax

Only products of plant origin were used for the waterproofing process to maintain the vegan nature of the leather. Essential oil from *Melaleuca alternifolia* and wax from *Copernicia prunifera* were chosen for this process, as these substances are hydrophobic plant products and are easy to find at public markets in the city of Recife, Pernambuco, Brazil. Immediately after the completion of the reconstitution and drying processes, the reconstituted BC surfaces received a thin layer (applied with a brush) of a mixture of 50% w/w wax and oil previously dissolved and homogenised at 75 °C. The samples were then dried at room temperature for 48 h in a naturally ventilated room.

2.9. Characterisation of Biotechnological Vegan Leathers

2.9.1. Determination of Water Contact Angle and Sorption Index

Rectangular samples, with 10 mm in height and 5 mm in length, of dried vegan leather were used for the analysis. To establish the behaviour of the biomaterials in contact with water, each sample was placed in a holder so that the material's surface was flat. Contact angles were determined with the aid of a goniometer using the sessile drop technique. A digital camera (XT10, Fujifilm, Japan) was used for analysis. A droplet of ~25 µL was placed carefully on the upper surface of the leather, and the contact angle was recorded after 1.0 s of spreading [19]. To determine the sorption index (s), the droplet was observed for 10 min until complete water absorption, and the average time was calculated [20].

2.9.2. Swelling Ratio

Rectangular, with 10 mm in height and 5 mm in length, of dried vegan leather samples were weighed and immersed in a 100-mL distilled water bath at 25 °C for 24 h. The samples were then removed from the bath. The excess water was carefully removed from the leather surface with tissue paper, and the samples were weighed immediately. Swelling ratios were determined from the change in weight before and after swelling and expressed as:

$$SR(\%) = \frac{\text{Swollen weight} - \text{Initial weight}}{\text{Initial weight}} \times 100 \qquad (2)$$

2.9.3. Scanning Electron Microscopy (SEM)

For the SEM analysis, dried vegan leather samples were mounted on a copper stub using double adhesive carbon conductive tape and coated with gold for 30 s (SC-701 Quick Coater, Tokyo, Japan). The SEM photographs were obtained using a scanning electron microscope (MIRA3 LM, Tescan, Warrendale, PA, USA) operating at 10.0 kV at room temperature.

2.9.4. Thermogravimetric Analysis (TGA)

The thermal stability of the samples was determined using TGA. Approximately 8 mg of each sample was heated from 30 to 600 °C at a rate of 10 °C/min in a nitrogen atmosphere with a flowrate of 20 mL/min to avoid the oxidative degradation of the samples. The Mettler Toledo TGA 2 Star System was used for this analysis.

2.9.5. Flexibility

To test flexibility, the vegan leather samples were folded by hand 100 times along the same line. The classification of flexibility was based on the number of folds until failure: poor (<20), fair (20–49), good (50–99), and excellent ($\geq$100) [21].

2.9.6. Mechanical Test

Tensile strength (N) and maximum deformation (%) according to time (s) were determined based on Rethwisch and William (2016) [22] for the characterisation of the mechanical properties of the BC. Samples of the dried vegan leathers were cut into rectangular strips (7.5 cm × 3 cm). The mean leather thickness was 0.10 cm. The tensile strength test was performed at room temperature at a velocity of 5 m/min and a static load of 0.5 N using a universal testing machine (EMIC DL–500MF, Brazil), following the ASTM D882 method.

3. Results and Discussion

3.1. BC Yield and Water Retention Capacity

The mean yield of the hydrated BC was 422.12 ± 15.26 g de cellulose/L of fermentation medium. Regarding the dried BC films, the mean yield was 10.07 ± 1.97 g/L with a production time of 14 days. This yield is considered satisfactory when compared to yields reported in previous studies with the same 14-day production time, such as 4.56 g/L in the study by Salari et al. [23] and 6.18 g/L in the study by Ul-Islam et al. [24].

The purification step with NaOH favoured an even colour as well as the removal of metabolites and possible residues from the culture medium adhered to the surface of the biocellulose. Figure 1 displays the purified BC film.

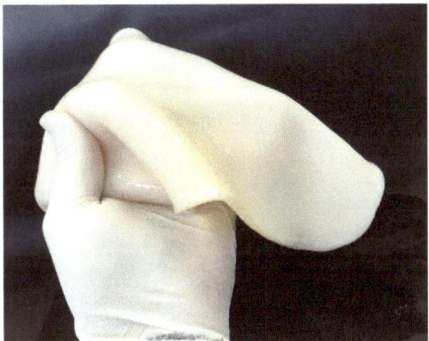

Figure 1. Bacterial cellulose after the purification process.

The results displayed in Table 1 confirm the high WRC (%) of the BC (>97%). Costa et al. [25] and Nascimento et al. [26] describe similar results. The WRC is a fundamental characteristic of efficiency in the incorporation and fixation of hydrophilic dyes.

Table 1. Bacterial cellulose production yield and water retention capacity.

BC 14 Days	Yield (g/L) Mean ± Standard Deviation	WRC (%) Mean ± Standard Deviation
Wet weight	422.12 ± 15.26	97.62 ± 0.39
Dry weight	10.07 ± 1.97	

3.2. BC Dyeing, Dye Fixation, and Waterproofing with Essential Oil and Wax

After the purification step, the membranes were submitted to the dyeing process with the pigments obtained from the plant extracts. The dyed samples were then shredded, followed by reconstitution during the drying process. The dried vegan leather samples were submitted to the waterproofing process with a mixture of essential oil and plant wax. The BC samples used to determine the effectiveness of the methods are listed in Table 2.

Table 2. Bacterial cellulose samples and abbreviations.

Sample	Abbreviation
Pure bacterial cellulose	BC
Waterproofed bacterial cellulose	BC-W
Reconstituted bacterial cellulose	BR
Water-proofed reconstituted bacterial cellulose	BR-W
Reconstituted bacterial cellulose dyed with onion and pomegranate	BRA
Water-proofed reconstituted bacterial cellulose dyed with onion and pomegranate	BRA-W
Reconstituted bacterial cellulose dyed with eucalyptus	BRE
Water-proofed reconstituted bacterial cellulose dyed with eucalyptus	BRE-W

The dyeing and fixation process resulted in the different colours for each sample, corresponding to different tonalities of the plant extracts used *in natura* (Figure 2). All experiments resulted in good visual quality and even pigmentation, corresponding to a varied swatch of colours that could be used in fashion products. The successful fixation of the dyes in the fibres is in agreement with results reported in the literature, such as studies by Verma et al. [27] (dyeing with onion), Maulik et al. [28] (dyeing with eucalyptus) and Tian et al. [29] (dyeing with pomegranate).

Figure 2. Appearance of cellulose leathers before and after dyeing/processing (**a**) Pure Bacterial Cellulose (BC); (**b**) Waterproofed Bacterial Cellulose (BC-W); (**c**) Reconstituted Bacterial Cellulose (BR); (**d**) Waterproofed Reconstituted Bacterial Cellulose (BR-W); (**e**) Reconstituted Bacterial Cellulose Dyed With Onion and Pomegranate (BRA); (**f**) Waterproofed Reconstituted Bacterial Cellulose Dyed With Onion and Pomegranate (BRA-W); (**g**) Reconstituted Bacterial Cellulose Dyed with Eucalyptus (BRE); (**h**) Waterproofed Reconstituted Bacterial Cellulose Dyed with Eucalyptus (BRE-W).

The waterproofing of the surface of the samples also influenced both the visual and physical aspects (Figure 2b,d,f,h), making the material less opaque and adding shine. Moreover, the samples that underwent this process had a more pleasant texture in terms of softness to the touch, losing the initial characteristics of roughness and dehydration.

Thus, the development of a completely vegan microbial cellulose-based textile material submitted to shredding and reconstitution was successful, generating a fabric similar to leather that was more uniform than the post-fermentation BC, with a constant thickness of 0.10 cm. Moreover, the process did not generate any waste. Even the cellulose remnants that did not have the ideal size and thickness for drying and the creation of pieces could be shredded together to form a mass that could be moulded and dried into the desired shape.

Chan et al. [30] also described the non-generation of waste during the production of BC pieces. The researchers used different approaches and novel cultivation techniques, employing recipients with pre-established dimensions so that the cellulose could grow into the desired shape, thereby avoiding cutting and waste and facilitating the creation of future pieces. The authors proved that the organic material can be reused after its fermentation, shredding, and reconstitution and can be grown into any shape of the desired apparel without cuts and with no generation of waste material.

3.3. Water Contact Angle, Swelling Ratio, and Sorption Index

The wettability of a fabric is important to sensorial comfort and is determined by the time required for the fabric to absorb a drop of water. The difference in the contact angle of the water droplet is recorded over time. This time-lapse is denominated by the sorption index (s) [31].

The swelling of the fabric offers the possibility of retaining sweat by the material during its use, which can be beneficial from a practical standpoint [20]. It can also indicate the possibility of the material functioning as a protective moisture barrier. The swelling capacity is expressed as the swelling coefficient and depends on the type and quantity of water retention agents in the fabric. This factor is crucial to enabling the control of the desired properties of the fabric in accordance with its application [20]. Table 3 displays the water contact angle, swelling ratio, and sorption index of the different BC samples.

Table 3. Water contact angle, swelling ratio, and sorption index of bacterial cellulose samples. BC: Pure Bacterial Cellulose; BC-W: Water-proofed bacterial cellulose; BR: Reconstituted bacterial cellulose; BR-W: Water-proofed reconstituted bacterial cellulose; BRA: Reconstituted bacterial cellulose dyed with onion and pomegranate; BRA-W: Water-proofed reconstituted bacterial cellulose dyed with onion and pomegranate BRE: Reconstituted bacterial cellulose dyed with eucalyptus; BRE-W: Water-proofed reconstituted bacterial cellulose dyed with eucalyptus.

Sample	Water Contact Angle (°)	Swelling Ratio (%)	Sorption Index (s)
BC	43.26	57.92 ± 3.12	76.32 ± 3.12
BC-W	67.64	3.81 ± 0.46	600 >
BR	38.58	33.32 ± 4.32	331.22 ± 12.34
BR-W	80.72	16.77 ± 2.33	600 >
BRA	40.60	34.64 ± 1.47	293.32 ± 26.62
BRA-W	76.32	15.85 ± 2.86	600 >
BRE	52.22	31.78 ± 4.74	312.43 ± 17.29
BRE-W	83.96	18.11 ± 1.12	600 >

The obtained water contact angles results (Table 3) are in agreement with studies involving the BC's surface modification with the aim of increasing the water droplets contact angle and, consequently causing a decrease on the hydrophilicity of the biocellulose [31,32].

The vegetal wax was used as a water-repellent agent. It was responsible for creating a layer on the surface of the samples that waterproofed them. As indicated by the sorption index, even after the 600 s time limit was exceeded, the samples that had the protective layer were not able to absorb the liquid. Another factor that indicates the success in this application was the contact angle change between the surface of the samples and the water droplets, indicating that the additional layer reduced the hydrophilicity of the outer layer of the material. Bashari et al. [33] state that the fine particles of *Copernicia prunifera* wax can be used as a natural moisture-repellent agent, therefore responsible for reducing the surface energy of tissues and producing a nano roughness that makes surfaces repellent to moisture.

By observing the swelling ratio results, it was possible to notice a significant value reduction in all the waterproofed samples when compared to the samples without the waterproofing process. Even after being submerged for 24 h, some samples showed a decrease of more than half of the initial swelling ratio value, thus, confirming that the waterproofing process is effective in reducing the swelling ratio (%).

3.4. Scanning Electron Microscopy

SEM was used to investigate the morphology of the surface of the samples before and after shredding, dyeing, and waterproofing. The BR sample (Figure 3c) exhibited a uniform surface, as seen in the BC (Figure 3a), indicating that the reconstitution of the BC was successful. Comparing BR (Figure 3c) to BRA and BRE (Figure 3e,g), the surface of the samples exhibited a new covering, which was related to the dyeing process and pigmentation of the biotechnological vegan leather.

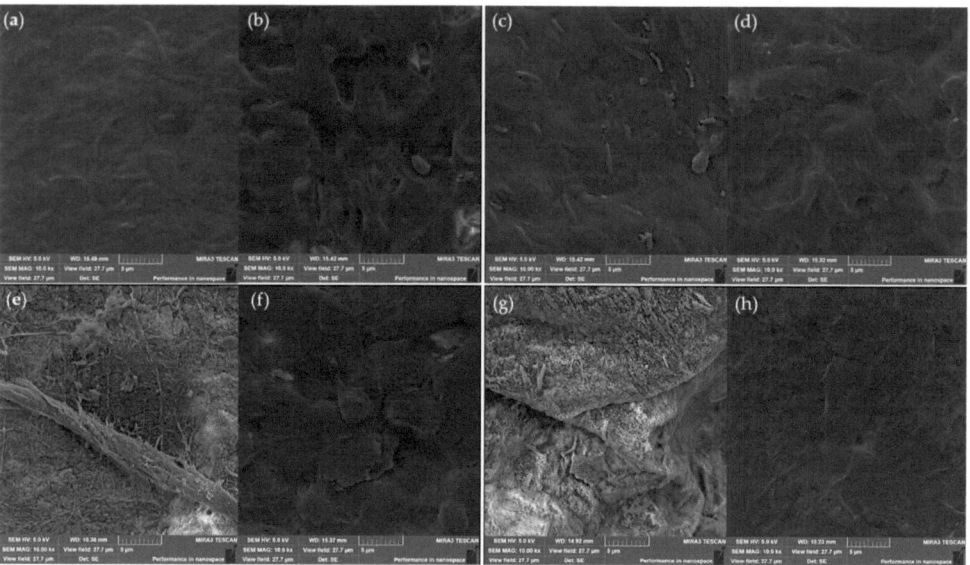

Figure 3. Scanning electron microscopy of all samples: (**a**) Pure Bacterial Cellulose (BC); (**b**) Waterproofed Bacterial Cellulose (BC-W); (**c**) Reconstituted Bacterial Cellulose (BR); (**d**) Waterproofed Reconstituted Bacterial Cellulose (BR-W); (**e**) Reconstituted Bacterial Cellulose Dyed With Onion and Pomegranate (BRA); (**f**) Waterproofed Reconstituted Bacterial Cellulose Dyed With Onion and Pomegranate (BRA-W); (**g**) Reconstituted Bacterial Cellulose Dyed with Eucalyptus (BRE); (**h**) Waterproofed Reconstituted Bacterial Cellulose Dyed with Eucalyptus (BRE-W).

In comparison to the samples before waterproofing (Figure 3a,c,e,g), those after waterproofing (Figure 3b,d,f,h) revealed the emergence of a uniform layer on the surface. This layer is the mixture of the essential oil from *Melaleuca alternifolia* and the wax from *Copernicia prunifera*, indicating that the waterproofing process was successful, giving the surface an impermeable characteristic that would hinder the absorption of water. This new layer is also responsible for the new characteristics of shine and softness discussed above. Moreover, the additives absorbed by the surface of the microbial cellulose could reduce undesired effects related to contact of the material with water. Thus, the dyeing and waterproofing methods constitute a simple, ecologically friendly way to improve the properties of biotextiles.

3.5. Thermogravimetric Analysis (TGA)

Figure 4 and Table 4 show that the materials developed have somewhat similar profiles. According to Rathinamoorthy et al. [34], the initial decomposition temperature (T_{onset}) is when BC begins to disintegrate. This temperature represents the onset of the breakdown of the thermal stability of the material.

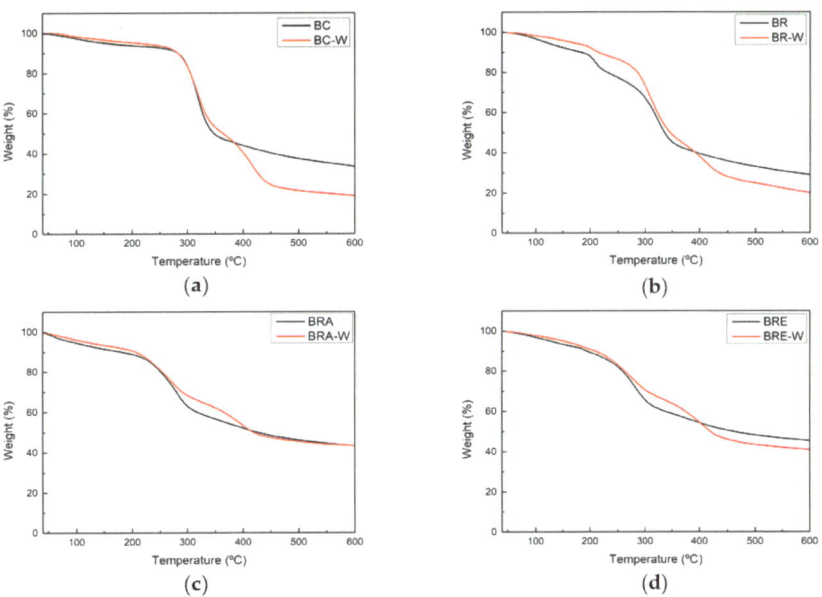

Figure 4. Thermogravimetric graphs of samples: (**a**) Pure (BC) and waterproofed (BC-W) bacterial cellulose; (**b**) Reconstituted (BR) and waterproofed (BR-W) bacterial cellulose; (**c**) Reconstituted (BRA) and waterproofed (BRA-W) bacterial cellulose dyed with onion and pomegranate; (**d**) Reconstituted and (BRE) and waterproofed (BRE-W) bacterial cellulose dyed with eucalyptus.

Table 4. Degradation temperatures of bacterial cellulose samples. BC: Pure Bacterial Cellulose; BC-W: Water-proofed bacterial cellulose; BR: Reconstituted bacterial cellulose; BR-W: Water-proofed reconstituted bacterial cellulose; BRA: Reconstituted bacterial cellulose dyed with onion and pomegranate; BRA-W: Water-proofed reconstituted bacterial cellulose dyed with onion and pomegranate; BRE: Reconstituted bacterial cellulose dyed with eucalyptus; BRE-W: Water-proofed reconstituted bacterial cellulose dyed with eucalyptus.

Samples	Stage 1			Stage 2			Stage 3			Mass Loss at 600 °C (%)
	T_{max}	T_{onset}	T_{endset}	T_{max}	T_{onset}	T_{endset}	T_{max}	T_{onset}	T_{endset}	
BC	301.31	94.72	320.84	339.04	322.04	416.30	-	-	-	33.78
BC-W	289.62	91.22	323.01	341.21	325.78	381.59	447.53	390.87	517.92	20.55
BR	283.23	80.05	326.14	349.77	328.91	423.90	-	-	-	29.24
BR-W	269.13	87.59	327.71	342.66	327.71	384.00	440.65	390.98	503.81	20.30
BRA	236.59	83.30	275.16	312.25	275.64	420.52	-	-	-	42.87
BRA-W	224.05	87.08	298.42	369.34	298.90	404.37	429.19	405.21	504.65	43.00
BRE	245.51	84.35	271.90	309.27	272.26	424.96	-	-	-	45.42
BRE-W	216.25	85.22	293.00	372.07	300.47	406.47	437.52	406.68	495.32	40.31

Two main stages in the loss of mass were found in the samples without waterproofing (BC, BR, BRA, and BRE), whereas three stages were found in the samples with waterproofing (BC-W, BR-W, BRA-W, and BRE-W). The first degradation stage in all samples occurred between 80 and 330 °C, with a mean loss of 33.31 ± 7.41% of mass related to the evaporation of the water adsorbed to the biocellulose and that in the composition of the pigment on the surface of the dyed samples.

The second stage occurred at around 270 to 370 °C, leading to a mean final mass of 47.14 ± 5.77%. This loss was probably related to the degradation of the cellulose, with its de-polymerisation, dehydration, and decomposition of glucose units, as well as the subsequent formation of carbon residues [35], as the main pyrolysis stage of cellulose occurs in a temperature range of 300 to 380 °C [36].

The third stage occurred between 390 and 518 °C and only in the samples with wax in their composition (BC-W, BR-W, BRA-W, BRE-W). Thus, this additional stage was likely related to the waterproofing substance in the samples. At the end of this last degradation stage, the mean residual mass was 34.43 ± 10.12%

The results are compatible with findings described in the literature. In one study, cotton fabrics without waterproofing also had two thermogravimetric phases, the first of which was up to 300 °C related to the degradation of the amorphous regions of the cellulose polymer and with a considerable reduction in mass. The second was related to the crystalline regions of the material, with a maximum temperature of 430 °C [36].

Comparing the results of the intact samples (BC and BC-W) and reconstituted samples (BR, BR-W, BRA, BRA-W, BRE, and BRE-W), the degradation temperatures were an average of 42.21 °C lower for the samples without additives, and an average of 53.14 °C lower for the waterproofed samples in the first stage when compared to the T_{max} of the intact samples. This indicates that the shredding and reconstitution of the cellulose exerted a direct impact on the first degradation phase. The behaviour was inverted in the second stage, as the reconstituted sample without additives (BR) had slightly higher T_{max} (349.77 °C) and residual mass (62.05%) compared to the intact BC (339.04 °C; 56.97%). This is an interesting point and suggests that the BR without dyeing and waterproofing exhibits greater thermal stability during the degradation of the cellulose and formation of carbon residues.

The presence of the dye also diminished the thermal stability in the first and second degradation stages, as T_{max} was reduced by 40 °C in the BRA and BRE samples compared to the BR in both stages. Different results were found concerning waterproofed samples. The waterproofing process led to a reduction in T_{max} ranging from 11.69 to 29.26 °C in the first stage, indicating that the degradation of the wax from *Copernicia prunifera* also occurred in this temperature range. According to Pan et al. [37], the thermal decomposition of this wax in its natural form occurs in the range of 270 to 320 °C. Moreover, the degradation of the low molecular-weight components of the essential oil from *Melaleuca alternifolia* occurs around 170 to 240 °C [38]. In the second stage, waterproofing increased the T_{max} of the BRA-W and BRE-W samples by 57.09 °C and 62.80 °C, respectively, indicating a possible interaction between the waterproofing agent and layer of dye, thereby enhancing the thermal stability of these materials.

3.6. Flexibility and Mechanical Tests

Flexibility is considered one of the critical aspects of the usability and durability of textile products. Textile materials need a surface structure with enough rigidity for wearability and adequate flexibility to be comfortable [39].

The BC samples exhibited excellent flexibility, remaining intact after being folded by hand more than 100 times at the same point. However, some of the samples with a layer of wax on the surface exhibited cracking of this waterproofing film. Thus, further studies are needed to improve the application of the waterproofing agent.

The functioning of a fabric involves its performance during use and is directly linked to its mechanical properties. The results of the mechanical tests (Figure 5 and Table 5) demonstrated that the addition of the plant-based waterproofing agent was capable of enhancing tensile strength by 3.56 to 30.79% and increasing-albeit subtly-the maximum deformation capacity of all samples. These aspects are beneficial, as the elasticity of textiles is of considerable importance and enables pieces to have specific characteristics, such as lightness, less volume, and a tendency to form fewer wrinkles.

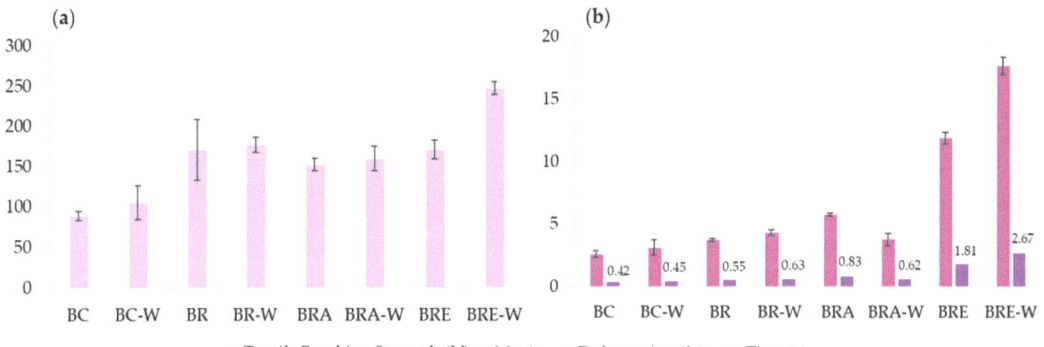

Figure 5. Graphs of results of mechanical tests: (**a**) Tensile strength (N); (**b**) Maximum deformation (%) according to time (s). BC: Pure Bacterial Cellulose; BC-W: Water-proofed bacterial cellulose; BR: Reconstituted bacterial cellulose; BR-W: Water-proofed reconstituted bacterial cellulose; BRA: Reconstituted bacterial cellulose dyed with onion and pomegranate; BRA-W: Water-proofed reconstituted bacterial cellulose dyed with onion and pomegranate; BRE: Reconstituted bacterial cellulose dyed with eucalyptus; BRE-W: Water-proofed reconstituted bacterial cellulose dyed with eucalyptus.

Table 5. Tensile strength and maximum deformation as a function of time of bacterial cellulose samples. BC: Pure Bacterial Cellulose; BC-W: Water-proofed bacterial cellulose; BR: Reconstituted bacterial cellulose; BR-W: Water-proofed reconstituted bacterial cellulose; BRA: Reconstituted bacterial cellulose dyed with onion and pomegranate; BRA-W: Water-proofed reconstituted bacterial cellulose dyed with onion and pomegranate; BRE: Reconstituted bacterial cellulose dyed with eucalyptus; BRE-W: Water-proofed reconstituted bacterial cellulose dyed with eucalyptus.

Sample	Tensile Strength (N)	Maximum Deformation (%)	Time (s)
BC	89.04 ± 11.04	2.61 ± 0.51	0.41 ± 0.05
BC-W	105.11 ± 42.02	3.12 ± 1.19	0.45 ± 0.15
BR	171.01 ± 76.11	3.73 ± 0.23	0.55 ± 0.04
BR-W	177.33 ± 18.55	4.27 ± 0.46	0.63 ± 0.04
BRA	152.53 ± 15.52	5.73 ± 0.23	0.83 ± 0.04
BRA-W	160.02 ± 30.85	3.77 ± 0.96	0.62 ± 0.11
BRE	171.07 ± 23.51	11.87 ± 0.93	1.81 ± 0.14
BRE-W	247.21 ± 16.52	17.63 ± 1.38	2.67 ± 0.18

The samples dyed with eucalyptus (BRE and BRE-W) exhibited excellent results regarding maximum deformation. With the addition of the wax, the deformation time of the BRE-W sample was improved by 147.51% compared to the non-waterproofed sample (BRE).

The samples submitted to the reconstitution process also exhibited improved mechanical properties, as demonstrated by comparing the BC and BR results. The shredding and reconstitution of the sample increased maximum deformation and tensile strength nearly doubled, increasing by 81.97 N. This indicates that the reconstitution process not only improves the visual appearance of the leather and standardisation of production, it also enhances the mechanical properties of leather of a biotechnological origin.

As already recognized by researchers, BC is naturally an exclusive combination of properties such as high polymerization degree, high surface area, high flexibility, high tensile strength, and high water-holding capacity [40]. Based on the results, it is safe to assume that the materials proposed have similar or better tensile strength and deformation properties compared to data described in the literature [17,20]. Thus, the findings indicate the successful development of a waterproof textile material made from reconstituted biocellulose.

4. Conclusions

The use of biotechnological materials, such as BC, indicates new possibilities for the market of sustainable fashion, which is of interest to both manufacturing companies and conscientious consumers. The growing interest in such materials will contribute to improvements in the manufacturing process, making it easier and less expensive to produce sustainable materials that aggregate scientific/technological value and can assist in the protection of the environment. The natural dyes used in the present study were selected due to their low toxicity, making them less likely to provoke allergic reactions. The process described in this work led to the even dyeing of BC leather. Further studies are needed to improve the waterproofing process so that the leather can be folded without losing the layer that protects it from moisture. However, the present results prove the potential of reconstituted BC as a vegan alternative to animal leather.

Author Contributions: Conceptualization, L.A.S., C.J.G.d.S.J. and A.K.L.d.H.C.; methodology, C.J.G.d.S.J., J.D.P.d.A., A.D.M.d.M., A.K.L.d.H.C., M.A.H.; L.J.C.d.N.M., G.M.V.; K.K.d.O.S.S., A.F.d.S.C., L.A.S.; validation, L.A.S., C.J.G.d.S.J., J.D.P.d.A., A.D.M.d.M., A.K.L.d.H.C. and A.F.d.S.C.; formal analysis, C.J.G.d.S.J., M.A.H.; L.J.C.d.N.M., G.M.V.; K.K.d.O.S.S. investigation, C.J.G.d.S.J., J.D.P.d.A., A.D.M.d.M., A.K.L.d.H.C.; writing—original draft preparation, C.J.G.d.S.J., J.D.P.d.A., A.D.M.d.M., A.K.L.d.H.C.; writing—review and editing, C.J.G.d.S.J., J.D.P.d.A., A.D.M.d.M., A.K.L.d.H.C., H.A.d.N.; and A.F.d.S.C.; visualization, C.J.G.d.S.J., A.F.d.S.C., and L.A.S.; supervision, L.A.S.; project administration, L.A.S.; funding acquisition, L.A.S. All authors have read and agreed to the published version of the manuscript.

Funding: This study was funded by the Brazilian fostering agencies *Fundação de Apoio à Ciência e Tecnologia do Estado de Pernambuco* (FACEPE [State of Pernambuco Support to Science and Technology Foundation]) (Grant n. APQ 0378-3.06/14), *Conselho Nacional de Desenvolvimento Científico e Tecnológico* (CNPq [National Council of Scientific and Technological Development]) (Grant n. 405026/2018-8) and *Coordenação de Aperfeiçoamento de Pessoal de Nível Superior* (CAPES [Coordination for the Advancement of Higher Educational Personnel]) (Finance Code 001).

Institutional Review Board Statement: Not applicable.

Informed Consent Statement: Not applicable.

Acknowledgments: The authors are grateful to the Rede Nordeste de Biotecnologia (RENORBIO), Universidade Federal Rural de Pernambuco (UFRPE), Escola UNICAP Icam Tech, Universidade Católica de Pernambuco (UNICAP), Centro de Tecnologia, Departamento de Engenharia Têxtil, Universidade Federal do Rio Grande do Norte (UFRN), Centro de Tecnologia e Geociências, Departamento de Engenharia Química, Universidade Federal de Pernambuco (UFPE) and Instituto Avançado de Tecnologia e Inovação (IATI), Brazil.

Conflicts of Interest: The authors declare no conflict of interest.

References

1. Islam, S. Sustainable raw materials. *Sustain. Technol. Fash. Text.* **2020**, *15*, 343–357. [CrossRef]
2. Galdino, C.J.S.; Maia, A.D.M.; Meira, H.M.; Souza, T.C.; Amorim, J.D.P.; Almeida, F.C.G.; Costa, A.F.S.; Sarubbo, L.A. Use of a bacterial cellulose filter for the removal of oil from wastewater. *Process Biochem.* **2020**, *91*, 288–296. [CrossRef]
3. Rathinamoorthy, R.; Kiruba, T. Bacterial cellulose-A potential material for sustainable eco-friendly fashion products. *J. Nat. Fibers* **2020**, *1*, 1–13. [CrossRef]
4. Fernandes, M.; Souto, A.P.; Dourado, F.; Gama, M. Application of Bacterial Cellulose in the Textile and Shoe Industry: Development of biocomposites. *Polysaccharides* **2021**, *2*, 566–581. [CrossRef]
5. Tabasum, S.; Mahmood, K.; Parveen, B.; Hussain, M. A novel water borne green textile polyurethane dispersions finishes from cotton (Gossypium arboreum) seed oil based polyol used in modification of cellulosic fabrics. *Carbohydr. Polym. Technol. Appl.* **2021**, *2*, 100170. [CrossRef]
6. Costa, A.F.S.; Amorim, J.D.P.; Almeida, F.C.G.; Lima, I.D.; de Paiva, S.C.; Rocha, M.A.V.; Vinhas, G.M.; Sarubbo, L.A. Dyeing of bacterial cellulose films using plant-based natural dyes. *Int. J. Biol. Macromol.* **2019**, *121*, 580–587. [CrossRef]
7. Albuquerque, R.M.B.; Meira, H.M.; Silva, I.D.; Silva, C.J.G.; Almeida, F.C.G.; Amorim, J.D.P.; Vinhas, G.M.; Costa, A.F.S.; Sarubbo, L.A. Production of a bacterial cellulose/poly(3-hydroxybutyrate) blend activated with clove essential oil for food packaging. *Polym. Polym. Compos.* **2020**, *29*, 259–270. [CrossRef]

8. Amorim, J.D.P.; Nascimento, H.A.; Silva Junior, C.J.G.; Medeiros, A.D.M.; Silva, I.D.L.; Costa, A.F.S.; Vinhas, G.M.; Sarubbo, L.A. Obtainment of bacterial cellulose with added propolis extract for cosmetic applications. *Polym. Eng. Sci.* **2021**, *62*, 565–575. [CrossRef]
9. Qasim, U.; Osman, A.I.; Al-Muhtaseb, A.H.; Farrell, C.; Al-Abri, M.; Ali, M.; Vo, D.-V.N.; Jamil, F.; Rooney, D.W. Renewable cellulosic nanocomposites for food packaging to avoid fossil fuel plastic pollution: A review. *Environ. Chem. Lett.* **2020**, *19*, 613–641. [CrossRef]
10. Medeiros, A.D.M.; Silva Junior, C.J.G.; Amorim, J.D.P.; Nascimento, H.A.; Converti, A.; Costa, A.F.S.; Sarubbo, L.A. Biocellulose for Treatment of Wastewaters Generated by Energy Consuming Industries: A review. *Energies* **2021**, *14*, 5066. [CrossRef]
11. Ververis, C.; Georghiou, K.; Christodoulakis, N.; Santas, P.; Santas, R. Fiber dimensions, lignin and cellulose content of various plant materials and their suitability for paper production. *Ind. Crops Prod.* **2004**, *19*, 245–254. [CrossRef]
12. Amorim, J.D.P.; Souza, K.C.; Duarte, C.R.; Duarte, I.S.; Ribeiro, F.A.S.; Silva, G.S.; Farias, P.M.A.; Stingl, A.; Costa, A.F.S.; Vinhas, G.M. Plant and bacterial nanocellulose: Production, properties and applications in medicine, food, cosmetics, electronics and engineering: A review. *Environ. Chem Lett.* **2020**, *18*, 851–869. [CrossRef]
13. Gelin, K.; Bodin, A.; Gatenholm, P.; Mihranyan, A.; Edwards, K.; Strømme, M. Characterization of water in bacterial cellulose using dielectric spectroscopy and electron microscopy. *Polymer* **2007**, *48*, 7623–7631. [CrossRef]
14. Ng, F.M.C.; Wang, P.W. Natural Self-grown Fashion from Bacterial Cellulose: A paradigm shift design approach in fashion creation. *Des. J.* **2016**, *19*, 837–855. [CrossRef]
15. Ng, M.C.F.; Wang, W. A Study of the Receptivity to Bacterial Cellulosic Pellicle for Fashion. *Res. J. Text. Appar.* **2015**, *19*, 65–69. [CrossRef]
16. Galdino, C.J.S.; Medeiros, A.D.M.; Amorim, J.D.P.; Nascimento, H.A.; Henrique, M.A.; Costa, A.F.S.; Sarubbo, L.A. The Future of Sustainable Fashion: Bacterial cellulose biotextile naturally dyed. *Chem. Eng. Trans.* **2021**, *86*, 1333–1338. [CrossRef]
17. Domskiene, J.; Sederaviciute, F.; Simonaityte, J. Kombucha bacterial cellulose for sustainable fashion. *Int. J. Cloth. Sci.* **2019**, *31*, 644–652. [CrossRef]
18. Villarreal-Soto, S.A.; Beaufort, S.; Bouajila, J.; Souchard, J.-P.; Taillandier, P. Understanding Kombucha Tea Fermentation: A review. *J. Food Sci.* **2018**, *83*, 580–588. [CrossRef]
19. Marin, E.; Rojas, J. Preparation and characterization of crosslinked poly (vinyl) alcohol films with waterproof properties. *Int. J. Pharm. Pharm. Sci.* **2015**, *7*, 242–248.
20. Kamiński, K.; Jarosz, M.; Grudzień, J.; Pawlik, J.; Zastawnik, F.; Pandyra, P.; Kołodziejczyk, A.M. Hydrogel bacterial cellulose: A path to improved materials for new eco-friendly textiles. *Cellulose* **2020**, *27*, 5353–5365. [CrossRef]
21. Chen, G.; Zhang, B.; Zhao, J.; Chen, H. Improved process for the production of cellulose sulfate using sulfuric acid/ethanol solution. *Carbohydr. Polym.* **2013**, *95*, 332–337. [CrossRef] [PubMed]
22. Rethwisch, D.G.J.; William, D.C. *Ciência e Engenharia de Materiais: Uma Introdução*, 9th ed.; LTC: Rio de Janeiro, Brazil, 2016.
23. Salari, M.; Khiabani, S.M.; Mokarram, R.R.; Ghanbarzadeh, B.; Kafil, S.H. Preparation and characterization of cellulose nanocrystals from bacterial cellulose produced in sugar beet molasses and cheese whey media. *Int. J. Biol. Macromol.* **2019**, *122*, 280–288. [CrossRef] [PubMed]
24. Ul-Islam, M.; Ullah, M.W.; Khan, S.; Park, J.K. Production of bacterial cellulose from alternative cheap and waste resources: A step for cost reduction with positive environmental aspects. *Korean J. Chem. Eng.* **2020**, *37*, 925–937. [CrossRef]
25. Costa, A.F.S.; Almeida, F.C.G.; Vinhas, G.M.; Sarubbo, L.A. Production of bacterial cellulose by *Gluconacetobacter hansenii* using corn steep liquor as nutrient sources. *Front. Microbiol.* **2017**, *8*, 2017. [CrossRef]
26. Nascimento, H.A.; Amorim, J.D.P.; Silva, C.J.G.J.; Medeiros, A.D.M.; Costa, A.F.S.; Napoleão, D.C.; Vinhas, G.M.; Sarubbo, L.A. Influence of gamma irradiation on the properties of bacterial cellulose produced with concord grape and red cabbage extracts. *Curr. Res. Biotechnol.* **2022**, *4*, 119–128. [CrossRef]
27. Verma, M.; Gahlot, N.; Singh, S.S.J.; Rose, N.M. UV protection and antibacterial treatment of cellulosic fibre (cotton) using chitosan and onion skin dye. *Carbohydr. Polym.* **2021**, *257*, 117612. [CrossRef]
28. Maulik, R.S.; Chakraborty, L.; Pandit, P. Evaluation of Cellulosic and Protein Fibers for Coloring and Functional Finishing Properties Using Simultaneous Method with Eucalyptus Bark Extract as a Natural Dye. *Fibers Polym.* **2021**, *22*, 711–719. [CrossRef]
29. Tian, Y.; Liu, Q.; Zhang, Y.; Hou, X.; Zhang, Y. Obtaining colored patterns on polyamide fabric with laser pretreatment and pomegranate peel dyeing. *Text. Res. J.* **2022**, *1*, 405175221076038. [CrossRef]
30. Chan, C.K.; Shin, J.; Jiang, S.X.K. Development of Tailor-Shaped Bacterial Cellulose Textile Cultivation Techniques for Zero-Waste Design. *Cloth. Text. Res. J.* **2017**, *36*, 33–44. [CrossRef]
31. Classen, E. Comfort testing of textiles. *Adv. Charact. Test. Text.* **2018**, *3*, 59–69. [CrossRef]
32. Wang, F.; Zhao, X.; Wahid, F.; Zhao, X.; Qin, X.; Bai, H.; Xie, Y.-Y.; Zhong, C.; Jia, S. Sustainable, superhydrophobic membranes based on bacterial cellulose for gravity-driven oil/water separation. *Carbohydr. Polym.* **2021**, *253*, 117220. [CrossRef]
33. Bashari, A.; Amir, H.S.K.; Salamatipour, N. Bioinspired and green water repellent finishing of textiles using carnauba wax and layer-by-layer technique. *J. Text. Inst.* **2019**, *111*, 1148–1158. [CrossRef]
34. Rathinamoorthy, R.; Aarthi, T.; Shree, C.A.A.; Haridharani, P.; Shruthi, V.; Vaishnikka, R.L. Development and Characterization of Self-assembled Bacterial Cellulose Nonwoven Film. *J. Nat. Fibers* **2019**, *18*, 1857–1870. [CrossRef]

35. Xu, K.; Li, Q.; Xie, L.; Shi, Z.; Su, G.; Harper, D.; Tang, Z.; Zhou, J.; Du, G.; Wang, S. Novel flexible, strong, thermal-stable, and high-barrier switchgrass-based lignin-containing cellulose nanofibrils/chitosan biocomposites for food packaging. *Ind. Crop. Prod.* **2022**, *179*, 114661. [CrossRef]
36. Zhu, P.; Sui, S.; Wang, B.; Sun, K.; Sun, G. A study of pyrolysis and pyrolysis products of flame-retardant cotton fabrics by DSC, TGA, and PY–GC–MS. *J Anal. Appl. Pyrol.* **2004**, *71*, 645–655. [CrossRef]
37. Pan, Y.F.; Xiao, H.N.; Xu, J.X.; Zhao, Y. Cellulose Fibers Modified with Starch-Based Microcapsules for Green Packaging. *Adv. Mater. Res.* **2013**, *781–784*, 2734–2737. [CrossRef]
38. Silveira, M.P.; Silva, H.C.; Pimentel, I.C.; Poitevin, C.G.; Stuart, A.K.C.; Carpiné, D.; Jorge, L.M.M.; Jorge, R.M.M. Development of active cassava starch cellulose nanofiber-based films incorporated with natural antimicrobial tea tree essential oil. *J. Appl. Polym. Sci.* **2019**, *137*, 48726. [CrossRef]
39. Ismar, E.; Bahadir, K.S.; Kalaoglu, F.; Koncar, V. Futuristic Clothes: Electronic Textiles and Wearable Technologies. *Glob. Chall.* **2020**, *4*, 1900092. [CrossRef]
40. Gorgieva, S.; Trček, J. Bacterial Cellulose: Production, Modification and Perspectives in Biomedical Applications. *Nanomaterials* **2019**, *9*, 1352. [CrossRef]

Article

An In Vitro and In Vivo Study of the Efficacy and Toxicity of Plant-Extract-Derived Silver Nanoparticles

Anjana S. Desai [1,†], Akanksha Singh [2,†], Zehra Edis [3,4,*], Samir Haj Bloukh [4,5], Prasanna Shah [6], Brajesh Pandey [1], Namita Agrawal [2,*] and Neeru Bhagat [1,*]

1. Department of Applied Science, Symbiosis Institute of Technology, Symbiosis International (Deemed University), Pune 412115, India; desaianjana89@gmail.com (A.S.D.); bpandey@gmail.com (B.P.)
2. Department of Zoology, University of Delhi, New Delhi 110007, India; akankshasingh81293@gmail.com
3. Department of Pharmaceutical Sciences, College of Pharmacy and Health Science, Ajman University, Ajman P.O. Box 346, United Arab Emirates
4. Center of Medical and Bio-allied Health Sciences Research, Ajman University, Ajman P.O. Box 346, United Arab Emirates; s.bloukh@ajman.ac.ae
5. Department of Clinical Sciences, College of Pharmacy and Health Science, Ajman University, Ajman P.O. Box 346, United Arab Emirates
6. Department of Physics, Acropolis Institute of Technology and Research, Indore 453771, India; psnimc@gmail.com
* Correspondence: z.edis@ajman.ac.ae (Z.E.); nagrawaluci@gmail.com (N.A.); neerubhagat@hotmail.com (N.B.); Tel.: +971-5-6694-7751 (Z.E.)
† These authors contributed equally to this work.

Abstract: Silver nanoparticles (AgNPs) display unique plasmonic and antimicrobial properties, enabling them to be helpful in various industrial and consumer products. However, previous studies showed that the commercially acquired silver nanoparticles exhibit toxicity even in small doses. Hence, it was imperative to determine suitable synthesis techniques that are the most economical and least toxic to the environment and biological entities. Silver nanoparticles were synthesized using plant extracts and their physico-chemical properties were studied. A time-dependent in vitro study using HEK-293 cells and a dose-dependent in vivo study using a *Drosophila* model helped us to determine the correct synthesis routes. Through biological analyses, we found that silver nanoparticles' cytotoxicity and wound-healing capacity depended on size, shape, and colloidal stability. Interestingly, we observed that out of all the synthesized AgNPs, the ones derived from the turmeric extract displayed excellent wound-healing capacity in the in vitro study. Furthermore, the same NPs exhibited the least toxic effects in an in vivo study of ingestion of these NPs enriched food in *Drosophila*, which showed no climbing disability in flies, even at a very high dose (250 mg/L) for 10 days. We propose that stabilizing agents played a superior role in establishing the bio-interaction of nanoparticles. Our study reported here verified that turmeric-extract-derived AgNPs displayed biocompatibility while exhibiting the least cytotoxicity.

Keywords: silver nanoparticles; characterizations; in vitro wound healing assay; in vivo *Drosophila* model; aloe vera; turmeric; biocompatibility; cytotoxicity

1. Introduction

Engineered nanomaterials (ENMs) are widely incorporated into numerous technologies and consumer products owing to their remarkable plasmonic and optical properties. Physico-chemical properties, such as shape, size, surface charge, agglomeration, dispersity, and colloidal stability, influence the behavior of metal and metal oxides to a great extent. These nanoparticles are rapidly finding applications in diverse areas, including textile, cosmetics, agriculture, food packaging, pharmaceutical, chemical, engineering, and medical applications [1–6]. Metal and metal oxide nanoparticles, such as silver, gold, zinc oxide, and copper oxide, encompass diverse therapeutic applications, such as targeted drug release and antibacterial and antimicrobial activities, and are thus considered ideal candidates for wound dressings [7–9] and wound healing [10,11]. Metal nanoparticles prepared using plant extracts exhibit good antimicrobial activity, possibly due to the eco-friendly nature of the reactants as compared with chemically synthesized particles [12–14]. However, the intrinsic toxicity of metal NPs remains a disadvantage, which hinders their application in wound healing and hence demands further improvement in either the synthesis process or growth [15]. Among these metal NPs, silver nanoparticles (AgNPs) have a crucial role in wound healing due to their inherent antibacterial and anti-inflammatory properties [16]. Physical and chemical methods are mainly being used to synthesize metal nanoparticles, but the intrinsic toxicity of the nanoparticles synthesized by these methods is a matter of concern properties [17,18]. Cuts and burns cause damage to the epidermis, resulting in a wound. Open, unhealed wounds become sites for pathogen activity, causing severe infection and resulting in permanent damage or death [8]. Thus, it becomes imperative to heal the wound quickly to carry out the body's normal functioning without microbial infection. The healing process helps minimize the scars and permits hard scabs to form [9]. Silver nanoparticles were discovered to be hazardous to cells, inhibiting cell growth and multiplication and triggering cell death depending on concentrations and exposure time [13,18]. It is evident that synthesis techniques play a major role in the preparation of less toxic nanoparticles. The biogenic mode of nanoparticle (NP) synthesis is gaining momentum due to the wide range of advantages it offers in economic viability, environmental sustainability, safety, and easy access to source materials [17]. The biogenic synthesis of AgNPs was achieved, and the antimicrobial activity of these NPs was demonstrated against many pathogens. In recent years, researchers used numerous plant extracts, such as *Moringa olifera*, coriander, ginger, turmeric, aloe vera, and neem [19,20]. Out of these, turmeric and aloe vera gained popularity because of their anti-inflammatory, anti-oxidative, and anti-neoplastic properties. Botanically, aloe vera is known as *Aloe barbadensis miller*; it is a perennial, xerophytic, succulent, shrubby or arborescent, pea-green hue plant that belongs to the *Asphodelaceae* (*Liliaceae*) family. There are 75 biologically active components, viz., vitamins (vitamin A, C, E, B12), enzymes (alkaline, alkaline phosphatase, amylase, bradykinase, carboxypeptidase, catalase, cellulose, lipase, and peroxidase), minerals (calcium, chromium, copper, selenium, magnesium, manganese, potassium, sodium and zinc), sugars (monosaccharides, polysaccharides), anthraquinones (laxatives), fatty acids (steroids, cholesterol, campesterol, β-sisosterol, and lupeol), hormones (auxins and gibberellins) and some amino acids [21,22]. The active ingredient present in turmeric is curcumin, botanically known as *Curcuma longa*, and is a member of the ginger family (*Zingiberaceae*). The mineral composition of turmeric is 0.63% phosphorous, 0.46% potassium, 0.20% calcium, and 0.05% iron. It was observed that elemental compositions of nitrogen (N), phosphorus (P), and potassium (K) are found to greater extents in saplings [23,24]. Researchers used various kinds of mammalian cell lines for in vitro study, but we chose human embryonic kidney (HEK-293) cells due to their easy growth and higher reproducibility. The scratch assay was conducted to study the toxicity in these cell lines. Amongst the different vertebrate models used for in vivo toxicity studies, *Drosophila* has gained wide attention because of its cost-effectiveness, shorter life cycle, ease of genetic manipulation, and ease of growth in laboratory conditions [25]. Developmental assays were done on the *Drosophila* model to study the behavioral pattern through toxic drug exposure. The aim of the present study was to increase the efficacy of wound-healing properties of nanosilver by using edible plants that are known in Ayurveda for their anti-bacterial, regenerative, and therapeutic properties. To achieve this, we used wholesome turmeric, and aloe vera leaves to synthesize silver nanoparticles because of the abovementioned properties of turmeric and aloe vera. The silver nanoparticles, synthesized with different reducing agents, were used to investigate the in vitro cytotoxicity in HEK-293 cell lines and the in vivo toxicity in a *Drosophila* model.

2. Materials and Methods

2.1. Preparation of the Biosynthesized AgNPs

Silver nitrate (AgNO$_3$) solution from Sigma-Aldrich with a purity of 99.999% was used to prepare all silver nanoparticles. Four samples of AgNPs were synthesized using two different routes, viz., the chemical route (named AG-C), and using plant extracts, i.e., the biological route (AG-B). The AG-C sample was prepared by mixing 300 mL of 1 mM silver nitrate solution into 50 mL of 0.5 mM polyvinyl pyrrolidone (PVP) solution. The resulting solution was kept at 80 °C, and 0.5 M hydrazine hydrate solution was added drop by drop into it. This mixture was then kept at 80 °C for 1 h, resulting in the precursor for sample AG-C.

Two different plant extracts were used to synthesize the samples via biological routes. The relevant part of the plants was cleaned, crushed, and boiled in Milli-Q water and filtered to obtain the extracts described in detail elsewhere [26]. For the turmeric extract, dry turmeric roots bought from the market were used, and for the aloe vera extract, stems of the aloe vera plants grown in our pots were used. A 300 mL aqueous solution of 1 mM AgNO$_3$ was mixed with different plant extracts, viz., aloe vera (A), turmeric (T), and a 1:1 mixture of aloe vera and turmeric (AT), to get a total of 500 mL of each AG-B precursor solution. The sample thus formed using aloe vera was named AG-BA and the sample formed using turmeric was named AG-BT. Similarly, the sample prepared using both aloe vera and turmeric extracts (1:1) was named AG-BAT.

The abovementioned precursors were initially kept at 80 °C for 1 h, then left at room temperature for a day to settle down in colloidal form and change color. The color change indicated the formation of nanoparticles. These colloids were then centrifuged at 5000 rpm for 20 min, filtered, and dried overnight in an oven at 50 °C to obtain powder samples.

2.2. Analytical Methods

The crystallographic structure of the prepared samples was studied using an X-ray diffraction (XRD) technique. A Bruker D8 Advance X-ray Diffractometer with Cu-Kα radiation (λ = 1.506 Å) was used to record X-ray diffraction (XRD) patterns. X-ray photoelectron spectroscopy (XPS) was done using XPS-SPECS GmbH, Germany, (Al K$_\alpha$ (1486.6 eV) X-rays). A field-emission scanning electron microscope (FESEM) was used to study the surface morphology. An FEI Nova NanoSEM 450 scanning electron microscope attached with Bruker X Flash 6130 with excellent energy resolution (123 eV for Mn Kα and 45 eV for Cu Kα) was used for the energy-dispersive X-ray spectroscopy (EDS). The transmission electron microscopy (TEM) images of silver nanoparticles were obtained with a TECHNAI20G2 transmission electron microscope.

All the AgNPs solutions were independently analyzed for their hydrodynamic diameter and polydispersity index on a Malvern Instrument Zetasizer Nano-ZS, Malvern (Malvern-Aimil Instruments private limited, New Delhi, India). The zeta potentials of the nanoparticles were also ascertained using the same instrument.

2.3. Biological Methods

The wound-healing property and cytotoxicity (in vitro) of the prepared AgNPs were studied using human embryonic kidney (HEK-293) cells. HEK-293 cells were cultured in a Petri dish and nourished using 10% fetal bovine serum along with Dulbecco modified eagle medium (DMEM). Inoculation was done with 200 µL of diluted (1:20) various AgNPs. The changes in the cells were captured using an optical microscope and the area covered by the cells was measured using the Image-J software. Details of the tests carried out are included in the Results section.

The toxicities of the AgNPs were tested on *Drosophila* fly stocks. All the assays were performed on wild-type *Drosophila* (Oregon-R), obtained from Bloomington *Drosophila* Stock Centre (BDSC), and conditions were maintained at 24.5 ± 0.5 °C, 60% humidity, and a 12 h light cycle.

A stock of 5% (w/v) AgNPs (all types) suspension was sonicated (QSONICA Sonicators) for 30 min at an amplitude of 30 to obtain a homogenous suspension. The final concentrations of 25 mg/L, 50 mg/L, and 250 mg/L were obtained by vigorously mixing different AgNPs in partially cooled fly food. The nanoparticles and the dose used were AG-C (25 mg/L, 50 mg/L), AG-BT (25 mg/L, 50 mg/L, 250 mg/L), AG-BA (25 mg/L, 50 mg/L), and AG-BAT (25 mg/L, 50 mg/L). Further details of various tests and studies are included in the Results section.

2.4. Statistical Analysis

All the plotted values of in vivo represent mean + SEM (standard error of the mean), and error bars represent the positive SEM value. Statistical analysis was carried out for all the assays using a two-tailed independent Student's *t*-test for pairwise comparisons. The significance level was ascribed at 0.05. Regarding *p*-values, *** $p < 0.001$, ** $p = 0.001$–0.01, and * $p = 0.01$–0.05.

3. Results

3.1. Physicochemical Characterization

3.1.1. XRD and EDS

XRD and EDS were utilized to study the morphology and composition of the samples.

Figure 1 shows the X-ray diffraction patterns of all the synthesized silver nanoparticles. The X-ray diffraction (XRD) pattern of silver nanoparticles prepared using PVP (AG-C) showed a pure phase of silver.

XRD patterns of nanoparticles synthesized with turmeric extract (AG-BT) and both aloe vera and turmeric extracts (AG-BAT) displayed oxide peaks corresponding to Ag_4O_4 (α) (silver I, III) (also known as Ag_2O_2) [27].

The silver nanoparticles prepared using aloe vera extract (AG-BA) and using both aloe vera and turmeric extracts (AG-BAT) showed the presence of silver oxide phases, such as $Ag_3O_4(\gamma)$ and $Ag_2O(\beta)$ [28,29]. All these peaks were matched with JCPDS data and indexed. A few other peaks were also noticed in this diffractogram. These might have been due to the in situ formation of some unstable oxides of silver. The average particle/crystallite size as calculated by Scherer's formula was 19 nm for AG-C, 12 nm for AG-BT, 21 nm for AG-BA, and 14 nm for AG-BAT. The Scherrer formula is the standard formula used for calculating the crystallite size and is given by

$$\tau = \frac{K\lambda}{\beta\cos\theta} \tag{1}$$

where τ is the crystallite size, K is the dimensionless shape factor with a value of 0.9, λ is the wavelength of the incident X-rays (1.506 Å), β is the full width at half maximum of the strongest peak, and θ is the Bragg's angle at which the peak is situated.

One can also observe there was a slight shift in the peak positions toward higher angles in AG-BAT. This indicated stress in the sample, which could have been introduced due to the formation of different silver oxides of different sizes.

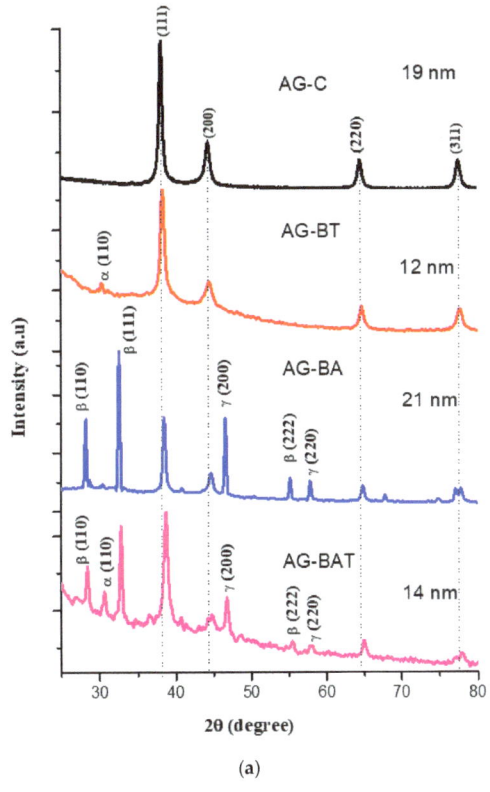

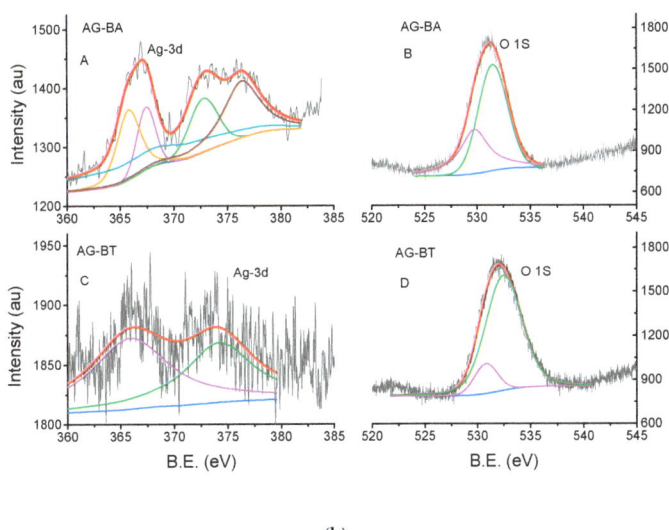

Figure 1. (**a**) X-ray diffraction patterns of different AgNPs samples. Dotted lines are drawn to show peaks corresponding to pure Ag. Various peaks in the pattern correspond to different phases of silver oxide, viz., α-Ag_4O_4, β-Ag_2O, and γ-Ag_3O_4. (**b**) XPS binding energy spectra of samples AG-BA and AG-BT: (**A**,**C**) silver Ag 3d spectra and (**B**,**D**) oxygen with a high resolution.

Some oxides of silver are quite elusive in XRD. Nevertheless, their presence was indicative in other tests. Hence, to correctly identify and claim their presence in our samples, we performed XPS. XPS gives the oxidation states of elements by giving their binding energy values (BE). Pure silver shows narrow peaks corresponding to Ag($3d_{5/2}$) at 368.5 eV and Ag($3d_{3/2}$) at 374.5 eV, and the oxygen O 1s peak is observed at 530.5 eV. It is known that when silver forms an oxide, it lowers the Ag(3d) binding energies [30] and increases the O 1s binding energy. Oxides also lead to the widening of the peaks (i.e., FWHM) [31]. Figure 1b represents the fitted XPS data of our samples AG-BA and AG-BT. In the sample AG-BA, we observed the fitted peaks O 1s corresponding to 529.8 eV and 531.46 eV and the Ag peaks at 376.3 eV, 372.7 eV, 367.4 eV, 366.8 eV, and 365.7 eV. As per the literature, these peaks correspond to metallic silver (376.3 eV, 372.7 eV), Ag (I)—367.4 eV, Ag (III)—366.8 eV, and Ag (II)—365.7 eV oxidation states. The O 1s BE values corresponding to 529.8 eV indicated the presence of Ag/Ag_2O and that between 530–532 eV showed the presence of superoxides, carbonates, or hydroxyl ions [32–34]. No peaks of carbonates were detected in the XRD; hence, their presence in our samples was ruled out by us. The sample AG-BA showed more toxicity; hence, it is safe to say that this sample contained more superoxide than hydroxyl ions. The presence of these binding energy peaks was indicative of the presence of AgNPs, Ag_2O, and Ag_3O_4 in the prepared sample AG-BA. In the sample AG-BT, we observed O 1s peaks corresponding to BE 530.2 eV (AgO or silver (I, III)) and 532.38 eV. The peak at 532.38 eV may have been because of the presence of superoxide, leading to increased toxicity or, as per some studies, may reflect the presence of contamination [35], which in our case could have been an outcome of biological synthesis. The Ag 3d binding energy values observed in the sample were 374.08 eV and 366.1 eV corresponding to Ag and Ag (III) [36]. These observations confirmed the presence of mainly AgNPs and a small amount of Ag_4O_4 (silver (I, III)) in this sample. These silver oxides came up in our samples due to the process of synthesis, which involved interaction with a biological environment. This interaction led to the oxidation of silver, which would normally be absent in chemically capped nanoparticles [24,37].

The energy-dispersive X-ray spectroscopic (EDS) technique was used to determine the composition of various elements in the synthesized AgNPs. All samples showed some amount of elemental oxygen. The XRD pattern of AG-C was not able to reflect the formation of oxides. There were oxide peaks observed in the XRD pattern of samples prepared in the presence of the aloe vera extract, but a very small fraction of oxide peaks was observed in the silver prepared using the turmeric extract. It is interesting to note that the elemental analysis gave almost equal amounts of oxygen in the samples prepared in the presence of turmeric. This indicated that the presence of turmeric extract restricted the formation of oxides and trapped a lot of free oxygen in the silver nanoparticles. The EDS studies, combined with the XRD, confirmed silver oxides in the samples prepared using the aloe vera extract. The presence of elemental silver and elemental oxide in each sample can be seen in the EDS images (Figure 2) and the amounts are presented in Table 1.

Table 1. EDS results showing the wt.% of elemental silver, elemental oxygen, and trace elements (N, Si, etc.) in the synthesized samples.

Sample Name	Silver	Oxygen	Trace Elements
AG-C	74.9	17.6	7.5
AG-BT	17.7	59.4	22.9
AG-BA	90.7	8.5	0.8
AG-BAT	15.5	77	7.5

It was evident from the EDS results that the elemental silver in AG-BA and AG-C was much higher than that in AG-BT and AG-BAT (Figure 2).

3.1.2. Field-Emission Scanning Electron Microscope (FESEM) Results

The surfaces of the chemically synthesized silver nanoparticles (AG-C) showed monodispersed spherical granular particles, with some patches of agglomeration (Figure 3).

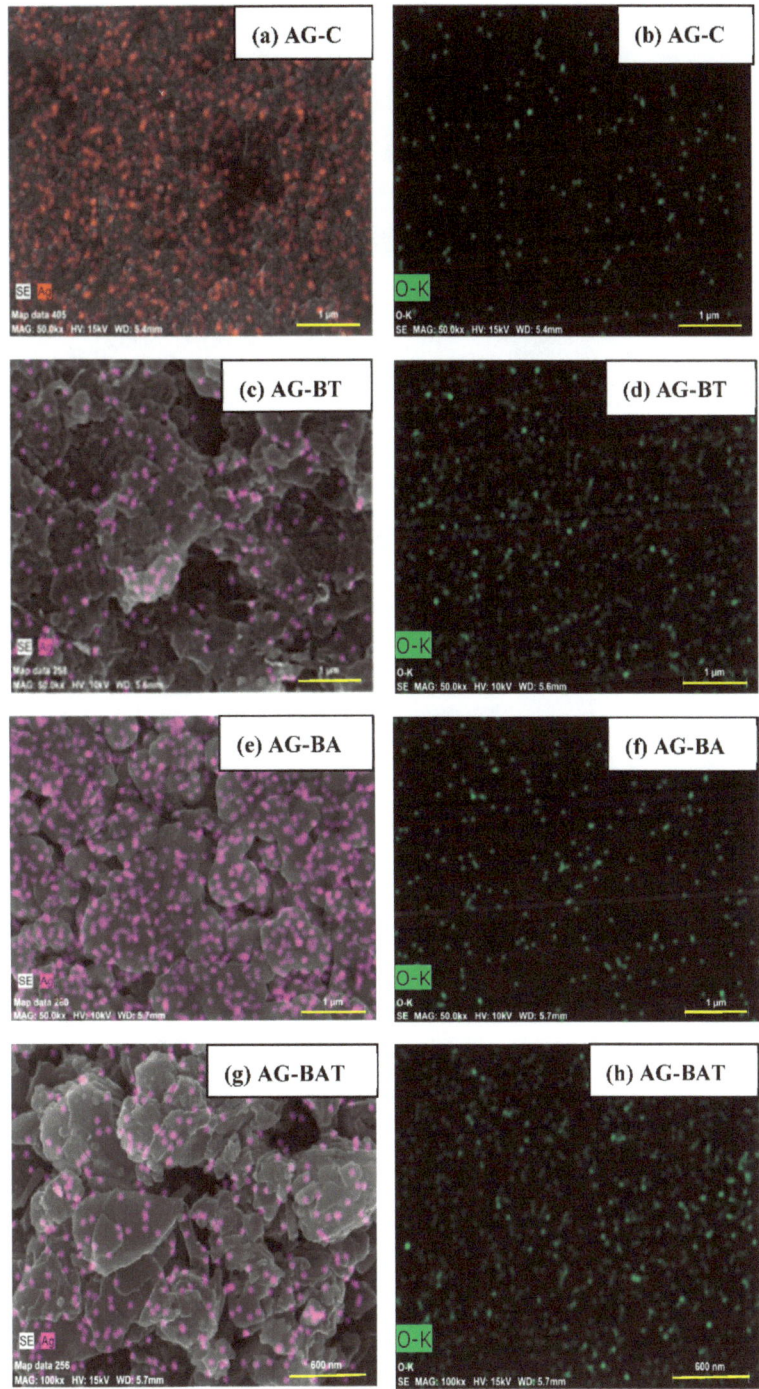

Figure 2. EDS images of AgNPs showing: elemental silver in the (**a**) chemically synthesized silver, (**c**) synthesis achieved using turmeric extract, (**e**) synthesis achieved using aloe vera extract, and (**g**) synthesis achieved using aloe vera and turmeric extracts in equal proportion; elemental oxygen (**b**,**d**,**f**,**h**) in each sample, respectively.

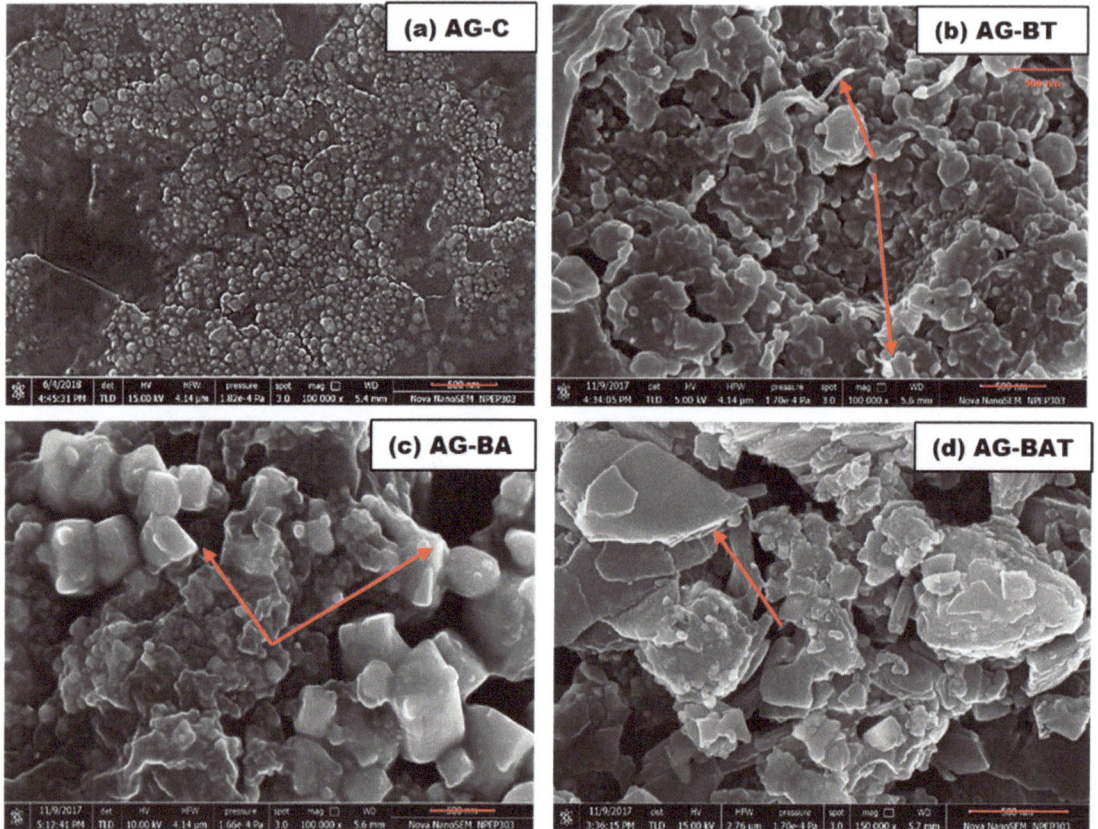

Figure 3. FESEM micrographs of differentially synthesized AgNPs: (**a**) AG-C, (**b**) AG-BT, (**c**) AG-BA, and (**d**) AG-BAT. The arrows indicate some oxygen-rich regions. The bar shown under each image depicts 500 nm.

A predominant arrowhead or pointed rod-like structures mixed with flakes and clusters were formed when turmeric extract (AG-BT) was used (Figure 3b).

On the other hand, the AgNPs prepared using aloe vera extract (AG-BA) showed cubic pillar-like monoclinic structures (Figure 3c). It is also evident from the FESEM images that this structure was monoclinic (related to Ag_3O_4) and some areas had a low melting point. Almost all the corners and edges were melted due to the energy of the electron beam. When both extracts were used (AG-BAT), the surface morphology was completely different as compared with the other samples and had layered flakes (2D structures) (Figure 3d). Shiny regions observed in the FESEM micrographs indicated the presence of oxide-rich regions. XRD analysis gave an indication of the type of oxides formed in the three samples synthesized using plant extract. There were some common silver oxides that can be seen in AG-BA and AG-BAT. In the sample AG-BAT, the shining edges might have been because of the presence of Ag_3O_4. Higher magnification was achieved due to higher electron energy. Silver oxides other than Ag_3O_4 had lower melting points, and hence, tended to melt at lower temperatures obtained using high electron energy. Since Ag_3O_4 has a relatively high melting point (1605 °C), it does not melt easily and gave shiny edges in the FESEM images. It is interesting to note that the silver nanoparticles synthesized in the presence of turmeric extracts led to flake-like structures. The release of oxygen is easier in flake-like structures and may be useful in applications where free oxygen is helpful in healing.

3.1.3. TEM and SAED

Transmission electron micrographs with the corresponding SAED patterns of all the synthesized silver nanoparticles are shown in Figure 4.

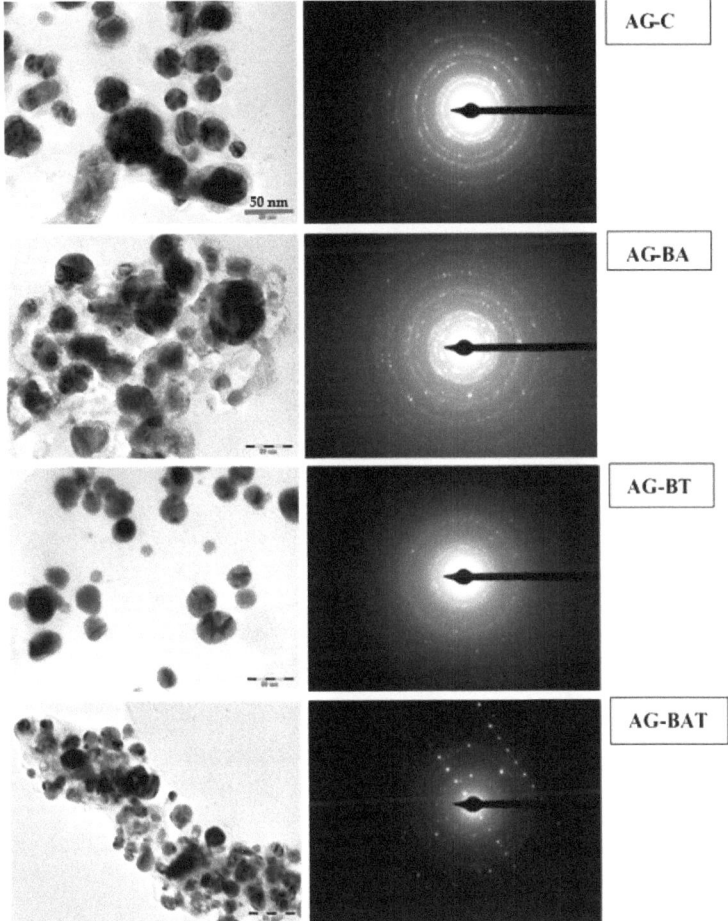

Figure 4. TEM micrographs and SAED patterns of silver nanoparticles. All TEM micrographs are of equal magnification.

The Ag-C sample showed mainly capped and crystalline spherical nanoparticles [38].

The structure of the silver nanoparticles was quite distinct when synthesized using turmeric extracts. Core–shell-type spherical nanoparticles with an almost uniform distribution were observed by Chircov et al. in their study of iron oxide and were reported with the help of SAED patterns as well [39]. It was also observed that when these NPs are used for healing purposes (scratch test), this led to better healing properties than others. One reason could be due to the oxygen trapped in the core–shells, and when they interact with live tissue, they release O_2, which helps with the growth of cells.

The shapes of the nanoparticles were irregular and fussed when synthesis was done in presence of aloe vera. When both turmeric and aloe vera were used we obtained non-uniform (also evident by PDI > 0.7 in Table 2) hollow spherical particles [40]. It is interesting to note that only by changing the medium of synthesis; nanoparticles with different shapes were produced.

Table 2. DLS and zeta potential measurements of nanoparticles.

Sample	Hydrodynamic Diameter (nm)	Polydispersity Index (PDI)	Zeta Potential (mV)
AG-C (PVP AgNPs)	124.4	0.425	−4.84
AG-BT (turmeric AgNPs)	494.5	0.589	−28.9
AG-BAT (aloe vera–turmeric AgNPs)	778.9	0.724	−23.7
AG-BA (aloe vera AgNPs)	463.2	0.449	−15.35

3.1.4. Dynamic Light Scattering (DLS) and Zeta Potential Measurements

Dynamic light scattering (DLS) measures the Brownian motion of particles in a mixture and gives the size distribution based on intensity fluctuations.

It is important to note that DLS measures the size distribution of clustered particles and, therefore, shows the tendency of particles to agglomerate in the colloidal form. The average hydrodynamic diameters of AG-C, AG-BT, AG-BAT, and AG-BA as determined by DLS were 124.4 nm, 494.5 nm, 778.9 nm, and 463.2 nm, respectively (Table 2).

Apart from this, all the nanoparticles assessed displayed a polydispersity index below 1 and, therefore, highlighted their monodispersity and lower tendency to agglomerate in the solution (Table 2).

AG-BT displayed good colloidal stability and exhibited a zeta potential of −28.9 mV. AG-BAT and AG-BA displayed moderate colloidal stability, exhibiting zeta potentials of −23.7 mV and −15.35 mV, respectively [41].

Contrarily, AG-C was highly unstable in the colloidal solution, displaying a zeta potential of −4.84 mV. It was evident from our study that the use of turmeric extract for the synthesis of NPs led to greater colloidal stability and uniformity in the size of nanoparticles (monodispersity).

3.2. Biological Studies

3.2.1. In Vitro Studies

The human embryonic kidney (HEK-293) cells were seeded in nine-well culture plates (seed density (3–6) $\times 10^5$ cells/cm^3). In order to produce a scratch in a confluent monolayer, the 200 µL sterile pipette tip was used. Scratches were made one at a time with constant pressure so that the same scratch width was maintained throughout the plates. The cells were washed extensively with phosphate-buffered saline (PBS) to remove the detached cells and debris. The prepared samples, along with the A and T extracts, were inoculated in nine-well culture plates. The images were observed and captured with the help of an inverted optical microscope with an attached camera. All the Petri dishes were observed for 2 days at the same time after inoculation. These studies were done thrice and the average of three sets of the assay was considered for evaluation.

The 'wound healing test' or 'scratch test' is the most common laboratory test performed to study the 2D cell migration, cell-to-cell interaction, cell proliferation, etc. Many commercially available creams and ointments used for healing surface wounds contain silver. Hence it is important to study the effect of silver nanoparticles in cell migration (tissue migration), as the wounds heal mainly via this process. Some nanoparticles are also known to cause cell death via apoptosis [42].

Hence, studying an in vitro 'scratch test' is imperative and was achieved by inoculating the medium with the various synthesized nanoparticles. The wound-healing capability of AgNPs was assessed upon the inoculation of the scratched cell culture with different samples for AgNPs for 2 days. On day 1 post-inoculation, wound healing (marked by the growth of cells near the scratch) was higher in all the dishes containing AgNPs than in the control dish (Figure 5A (actual pictures of the cells), Figure 5B (graphical depiction of the area covered by cell growth)).

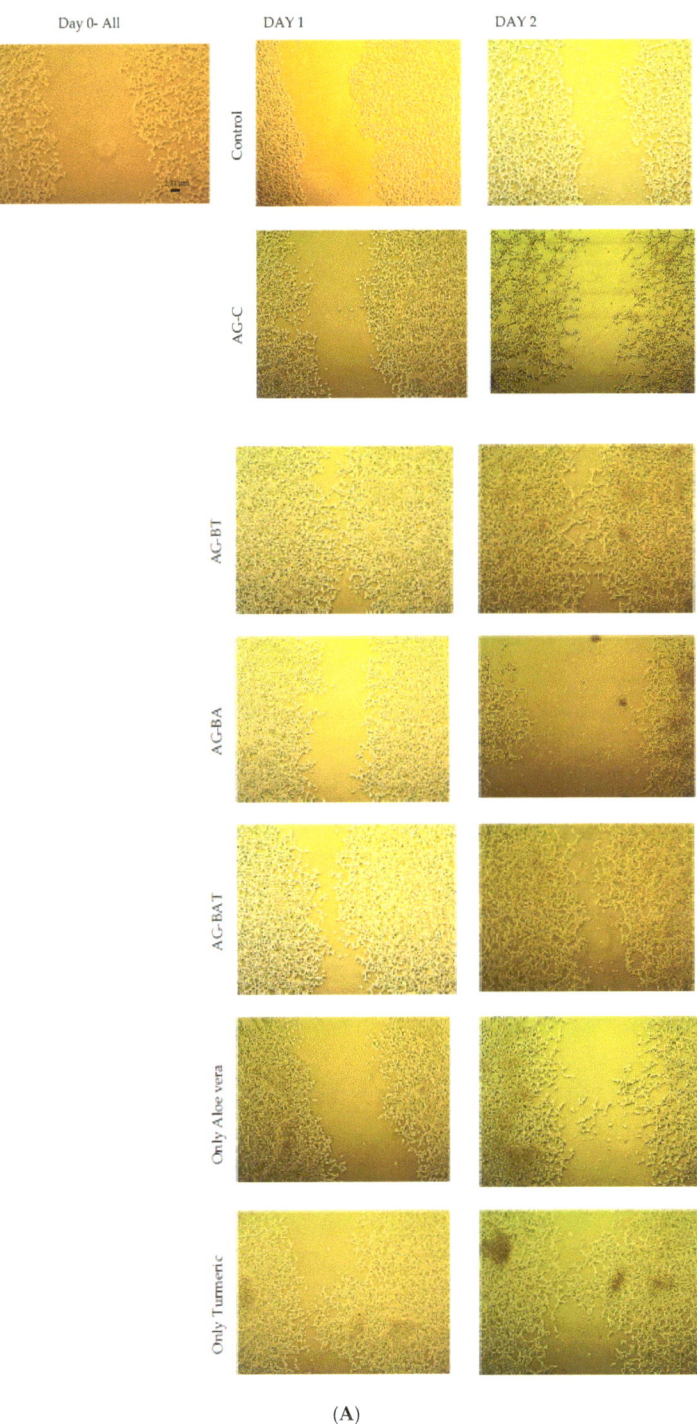

(A)

Figure 5. *Cont.*

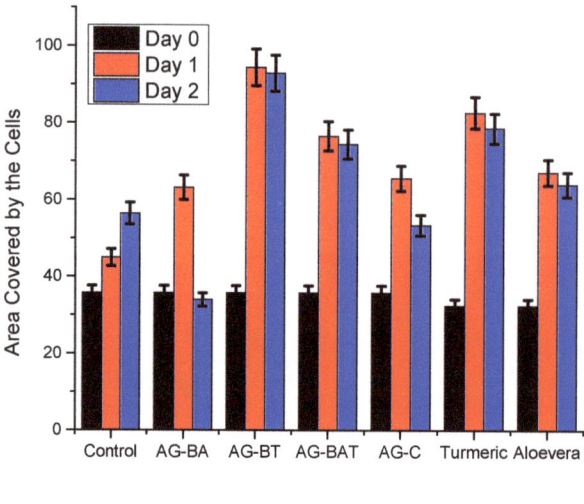

(B)

Figure 5. (**A**) Images from the scratch test analysis; (**B**) quantitative representation of cell growth measured on day 1 and day 2 using all the synthesized samples, as well as the plant extracts, in comparison with the control. Error: ±5%.

On day 2, however, cells inoculated with AG-C and AG-BA showed extensive death (Figure 5A). Cell cultures inoculated with AG-BT and AG-BAT exhibited almost full wound closure, with significant cell proliferation (Figure 5A). From the figures, the effect of just the plant extracts on cell growth can also be observed. It is evident that though the turmeric extract alone also enhanced the cell growth, it was to a lesser extent as compared with AG-BT. This shows that the presence of the AgNPs played an important role in wound healing capacity and is explained in detail later in the Discussion section. On the other hand, the aloe vera extract alone showed higher healing potential as compared with AG-BA.

To ascertain the role of aloe vera and turmeric extracts alone on cell proliferation, we also undertook the scratch test using just the extracts. Our results showed that the aloe vera extract, in its current concentration, was on par with AG-C on day 1 and better than the control and AG-BA in terms of its wound healing properties. However, the extract showed ineffectiveness in further cell growth on day 2 but did not show toxicity either. Similar observations were made for the turmeric extract, yet better than the aloe vera extract.

Though the aloe vera extract, turmeric extract, AG-BAT, and AG-BT all showed reduced or negligible cytotoxicity, amongst all samples tested, AG-BT fared best and enhanced the wound-healing capacity of cells with accelerated growth, while AG-C and AG-BA proved to be cytotoxic.

3.2.2. In Vivo Studies

Analysis of Larval Development and Adult Eclosion

The 4–5 days old parental wild-type flies were allowed to lay eggs in a chamber overnight, and 75 eggs were transferred to each vial containing media supplemented with and without different doses of various nanoparticles. Pupation success was evaluated by counting the number of viable pupae formed out of the 75 eggs transferred to a vial.

Similarly, the adult eclosion percentage from each condition was estimated by scoring the total number of F1 flies that eclosed from the vials of that condition.

Five vials per condition were scored for pupation and eclosion assays [25,43].

Climbing Assay

A climbing assay was performed on the eclosed F1 flies from and aged upon AgNP-treated fly food on days 10 and 30. Climbing ability was assessed by monitoring the vertical climbing of flies in an empty glass tube (diameter 2.2 cm) with marked gradations.

A group of 10 flies in two replicates per condition were transferred to a glass tube and acclimatized, and then we tapped to the bottom of the tube and allowed them to climb.

The number of flies passing the 10 cm mark in 15 s was recorded for each condition. Three trials per replicate condition were allowed and noted [25].

Pupation and Eclosion Rates of Flies Exposed to AgNPs

To ascertain the effect of ingestion of various nanoparticles on developmental viability at pupal and adult stages, we comprehensively analyzed the pupation rate and eclosion success in *Drosophila*. The larvae were fed with AG-C (25 mg/L, $p < 0.001$; 50 mg/L, $p < 0.001$), AG-BA (25 mg/L, $p = 0.008$), AG-BT (50 mg/L, $p < 0.001$; 250 mg/L, $p = 0.002$), and AG-BAT (25 mg/L, $p < 0.001$; 50 mg/L, $p = 0.007$).

The percentage of larvae successfully pupating upon ingestion of the AgNPs was substantially lower than the larvae reared on standard food. However, it is worth noting that the pupation percentage in larvae reared with AG-BT (25 mg/L, $p = 0.013$) was comparable to the control (Figure 6A).

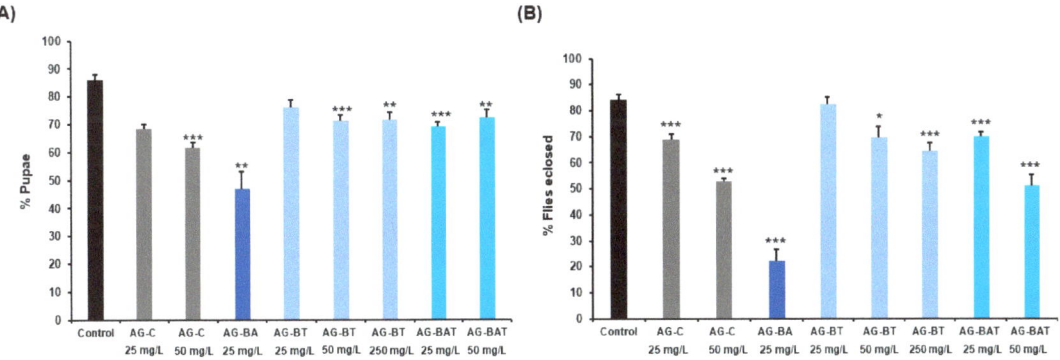

Figure 6. (**A**) Percentage of pupae formation; (**B**) adult eclosion assessed in each condition. Values represent mean ± SEM. p-value: *** $p < 0.001$; ** $p = 0.001$–0.01; * $p = 0.01$–0.05. Statistical analysis was done by Student's t-test.

Significant lethality witnessed in most AgNP-treated conditions highlighted the extent of toxicity caused by nanoparticle ingestion from early developmental stages. The toxicity observed was the lowest in larvae that ingested 25 mg/L AG-BT, indicating that the AgNPs synthesized using turmeric extract were potentially safer than nanoparticles derived chemically or with other plant extracts.

Larvae reared on 50 mg/L AG-BA (aloe-derived AgNPs) died very early in development, did not pupate, and hence, are not presented in the graph.

Eclosion observed for the same doses of AgNPs showed significantly lower flies eclosing from AG-C (25 mg/L), AG-BA (25 mg/L), AG-BT (50 mg/L, 250 mg/L), and AG-BAT (25 mg/L, 50 mg/L) as compared with the control condition. Flies eclosing from AG-BT (25 mg/L) displayed the same eclosion percentage as the control ($p = 0.595$) (Figure 6B).

Conclusively, our results indicated that 25 mg/L of turmeric-derived AgNPs (AG-BT) was found to be compatible with developmental survival.

Cuticular Melanization of Flies Exposed to AgNPs

Cuticular pigmentation is a critical parameter of an insect's life, as it influences various physiological and behavioral aspects, including lifespan, immunity, stress tolerance, and courtship. Loss of cuticular melanization is a well-known manifestation of AgNPs exposure in *Drosophila* [44–46]. Importantly, it was observed that the extent of loss of pigmentation in flies upon NP ingestion is directly correlated with decreased longevity, decreased locomotor activity, and overall toxicity.

To assess phenotypic aberrations in flies ingesting differentially synthesized silver nanoparticles, we monitored cuticular melanization in the pupal and adult stages.

We observed a similar degree of demelanization in pupae and adults reared upon equal doses of AG-C and AG-BAT (i.e., 25 mg/L AG-C = 25 mg/L AG-BAT and 50 mg/L AG-C = 50 mg/L AG-BAT) (Figures 7 and 8).

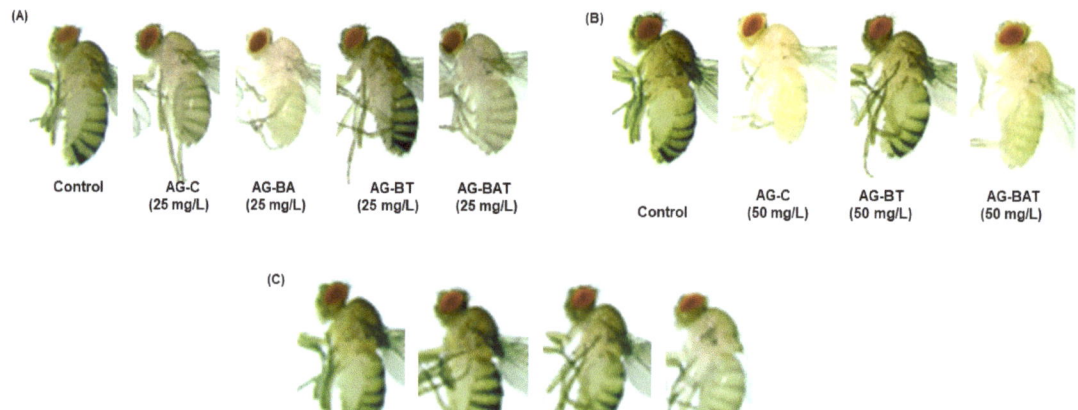

Figure 7. Cuticular melanization of the pupae formed upon exposure to (**A**) nanoparticles at 25 mg/L, (**B**) nanoparticles at 50 mg/L, and (**C**) turmeric-derived nanoparticles at different concentrations.

Figure 8. Dose-related effect of nanoparticle ingestion on cuticular melanization of adults: (**A**) 25 mg/L fed nanoparticles, (**B**) 50 mg/L ingested nanoparticles, and (**C**) ingested AG-BT nanoparticles at different concentrations.

While pupae and flies reared upon 25 mg/L AG-BA displayed the maximum loss of pigmentation as compared with those reared on other nanoparticles at the same dose, larvae reared upon 50 mg/L AG-BA did not pupate and hence could not be assessed for any further assays. Remarkably, pupae and adult flies reared on an AG-BT-enriched diet did not lose cuticular melanization at 25 mg/L and 50 mg/L and displayed loss of melanization only at a very high dose of 250 mg/L (Figures 7C and 8C).

It is generally accepted that the extent of AgNP-mediated loss of melanization is proportional to the extent of system dysfunctions observed in *Drosophila* [44,45].

These results suggest that AG-BT exerted the least toxicity on cuticular pigmentation of all the assessed nanoparticles and did not interfere with the melanin synthesis pathway up to the concentration of 50 mg/L.

Impact of AgNPs on Degree of Climbing of Flies

Further, to explore the impact of ingestion of various AgNPs supplemented food on *Drosophila*, we monitored the crucial vertical climbing behavior and survival of adults for 30 days. The climbing assay is widely used in neurodegenerative fly models and the degree of climbing defect directly correlates to the degree of neuronal atrophy (or toxicity).

As the ingestion of AgNPs was reported to cause severe impedance in climbing ability at different concentrations, we evaluated the extent of climbing disability in various experimental conditions with aging. On day 10, the climbing ability was significantly reduced in flies reared upon 50 mg/L of AG-C and AG-BAT while remaining comparable to the control in all other feeding conditions (Figure 9A).

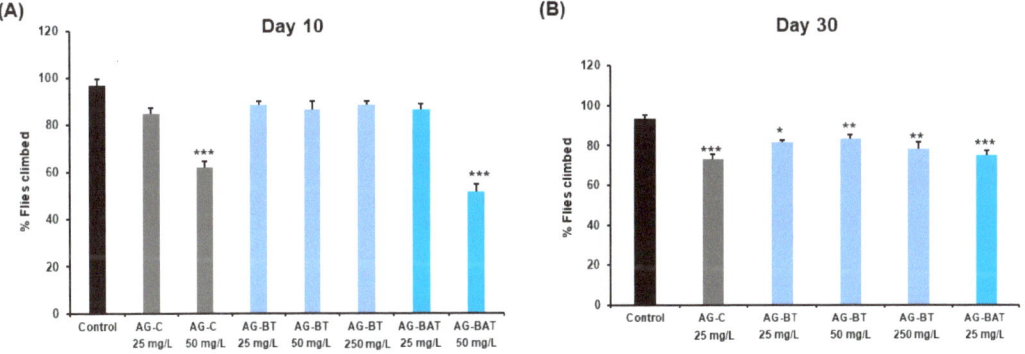

Figure 9. Assessment of the climbing ability of flies upon exposure to different nanoparticles. The climbing ability (**A**) of 10-day-old flies was compromised at 50 mg/L concentrated ingestion of AG-C and AG-BAT nanoparticles, and (**B**) was affected in all the surviving conditions at day 30 to various degrees. Values represent mean + SEM. *p*-value: *** $p < 0.001$; ** $p = 0.001$–0.01; * $p = 0.01$–0.05. Statistical analysis was done using Student's *t*-test.

By day 30, the vertical climbing ability declined in all the adults exposed to different doses and types of AgNPs with maximum impairment seen upon aging on AG-C and AG-BAT at 25 mg/L and least impedance observed in AG-BT 25 mg/L raised adults (Figure 9B).

Notably, flies fed with and aged upon 50 mg/L AG-C and AG-BAT did not survive up to day 30 and hence could not be monitored for climbing ability.

Conclusively, while on day 10, no climbing disability was observed in flies reared upon AG-BT-enriched food, even at very high dose ingestion (250 mg/L), the climbing ability was reduced by day 30 in a dose-dependent manner in all the treated conditions.

4. Discussion

Despite a plethora of studies depicting the characteristics, assimilation, behavior, and metabolism of nanoparticles in living systems, ambiguity regarding their cytotoxicity persists and presents challenges regarding their application. It was been earlier reported by

Olga et al. [47] that the smaller the size of the AgNPs, the greater the cytotoxicity; moreover, Ag nanoparticles, being neutral, show less cytotoxicity as compared with Ag ions. Further, Hamouda et al. reported that some biosynthesized AgNPs are less cytotoxic on certain human cell lines, such as MCF-7 and HCT-116 [48].

To further explore the toxic potency, we comprehensively analyzed the physicochemical characteristics of chemically synthesized and biosynthesized AgNPs and tested them in vitro and in vivo.

Much of the cytotoxicity encompassing AgNPs was shown to result from its conversion to Ag+ ions intracellularly.

It was shown earlier that the biogenic nanoparticles exhibit less toxicity in vitro. Cytotoxicity of nanoparticles is mainly attributed to the generation of reactive oxygen species (ROS), which further goes on to activate inflammatory and apoptotic pathways [49]. ROS plays an important physiological role in cell proliferation, repair, and signaling at low levels. However, when the levels of ROS increase beyond a threshold, the cells undergo oxidative stress, which eventually leads to cell death.

The interaction of ROS (H_2O_2) with silver nanoparticles, explained by Asharani et al. [50], is as follows:

$$2Ag + H_2O_2 + 2H^+ \rightarrow 2Ag^+ + 2H_2O \quad (2)$$

and according to Di He et al. [51]:

$$Ag\,(NP) + H_2O_2 \rightarrow Ag^+ + O_2^{\bullet -} + H_2O \quad (3)$$

From the scratch test, similar behavior was observed for AG-C and AG-BA. The common observation was that both these samples initially (day 1) seemed to help in cell growth and then later (day 2), caused the death of the cells.

In comparison, the amount of cell death was greater with AG-BA. The initial growth and subsequent death were observed because cell viability depends on the concentration of nanoparticles and time. The initial growth of the cells indicated that the ROS produced by the interaction of the cell organelle with the Ag+ ions were low in concentration. Low levels of ROS are known to aid the growth of cells [52]. However, as time progresses, the amount of ROS produced increases leading to cell death by apoptosis. In accordance with the earlier studies [16,49], higher toxicity in AG-BA is due to its larger size, a higher concentration of elemental silver, and the existence of significant amounts of hazardous oxide (Ag_2O). Similarly, these oxides may also interact with the SOD (H_2O_2) (SOD—superoxide dismutase) inside the cell to produce Ag+ ions.

From the above discussion, it is clear that Ag+ ions are the prime agents for producing highly toxic ROS, both extracellular and intracellular in the presence of Ag_2O.

In contrast, the non-toxicity and enhanced cell viability shown in samples inoculated with AG-BT and lower toxicity observed for AG-BAT could be attributed to the lower concentration of elemental silver [16,53–56], smaller size [52,53,57], spherical shape, and higher colloidal stability (zeta potential report) of these AgNPs [16,52,53]. A lower amount of AgNPs leads to low levels of H_2O_2 (ROS) and smaller AgNPs increase extracellular superoxide dismutase (SOD_3). This helps in cell growth and cell proliferation and decreases apoptosis. Earlier studies also suggested that increased oxygen levels in cell culture led to lower oxidative stress and lower ROS production [58]. It was also evident from the pointed rod-like structure seen in the FESEM (Figure 3) that this was rich in oxygen (the bright tip), which might easily release oxygen and increase the concentration of dissolved oxygen in the cell culture. This oxygen-releasing theory was further verified by the core–shell/hollow structures of AG-BT and AG-BAT, as seen in the TEM images.

The core–shell/hollow structure might release trapped oxygen while interacting with the cellular environment. Though most silver oxides were reported to dissolve in the cell environment, Ag_4O_4 (the oxide present in AG-BT and AG-BAT) is stable, and hence does not dissolve to give Ag+. The absence of Ag+ ions in AG-BT reduces the risk of cytotoxicity.

The effect of Ag$_4$O$_4$ is important, as its small presence in the sample AG-BAT seemed to influence its behavior and overcome the effect of Ag$_2$O, thus making it less toxic compared with AG-BA. AG-C may also owe its increased toxicity to its instability, which is clear from our DLS studies. In contrast, the zeta potential data confirmed the high colloidal stability for AG-BT in water.

Therefore, our in vitro results highlighted the wound-healing and cell proliferative effects of turmeric-derived AgNPs, which may have largely been due to the ability of turmeric to quench excess ROS and its compatible physicochemical properties with the biological system.

We found a substantial impact of AgNPs ingestion on larval development, as was evidenced by a significant decrease in the formation of viable pupae in the treated conditions. Out of all the nanoparticles assessed, aloe-derived AgNPs ingestion caused the most detrimental impact on larval development and pupal formation, as the larvae died at a moderate dose (50 mg/L) exposure and did not pupate, while those reared at 25 mg/L displayed a significant reduction in pupation and died within 2 days of eclosion. This inherent toxicity associated with aloe-derived AgNPs may have been due to the presence of two silver oxide phases in the NPS, as ascertained using XRD, XPS, and EDS. The eclosion percentage of flies was also significantly reduced upon exposure to different AgNPs. The AgNPs are known to cause endocrine disruption in various model organisms [59] and may act through disruption of ecdysone and juvenile hormone (JH) to impede developmental progress. Induction of oxidative stress and apoptosis was also evidenced upon exposure to AgNPs in larval and adult tissues [46,60]. At the same time, phytoconstituents of turmeric, especially curcumin, possess well-known anti-oxidative activity. Therefore, turmeric-derived AgNPs might provide their positive effects during development by alleviating oxidative stress.

We observed a direct relationship between pigmentation loss in pupae and adults with the concentration of chemically derived, aloe-vera-derived, and aloe–turmeric-derived AgNPs ingested. However, ingestion of turmeric-derived AgNPs induced pigmentation loss only at a very high dose (i.e., 250 mg/L). These results indicate a superior biological response to the ingestion of turmeric-based AgNPs than that of the ingestion of chemically, aloe- or aloe–turmeric-derived AgNPs.

We found that the ingestion of AgNPs incorporated food throughout aging led to a decline in the climbing ability of flies, which was dependent on the type and concentration of AgNPs and the age of the flies.

At an early age, only moderate doses of chemical AgNPs and aloe vera–turmeric AgNPs lowered the climbing ability. Later, the deterioration in climbing ability became evident with all the types and doses of AgNPs. Therefore, it can be inferred that the negative effects of nanoparticle exposure on the motor deficit could be manifested progressively with aging, and maybe because of in vivo processing of NPs and their accumulation over a period that needs to be assessed in greater detail.

These results suggested a progressively aggravating system-wide toxicity, leading to the death of most flies at high-dose exposures of AG-C and AG-BAT, and hence, highlighted the extensive toxicity associated with these NPs.

However, in the present study, we showed that AgNPs tended to reduce the biotoxicity when synthesized from scratch with the turmeric extract as the precursor. The effect of curcumin added along with the silver-nanoparticle-rich diet in *Drosophila* was reported earlier [25].

To further understand whether turmeric alone is enough to lead to increased efficacy and lower toxicity in the systems, a separate in vitro test was done by inoculating the HEK-293 culture with only the turmeric extract of the same concentration as the other nanoparticles.

It was observed that using the same concentration of only turmeric extract led to the death of cells. Hence, we concluded that only when turmeric extract is used to synthesize the AgNPs, incorporated with the AgNPs, or is used in dilution, does it lose its toxic effect and helps with cell growth. This, too, is an important finding, as many ethnicities traditionally use turmeric directly for wound healing. The mechanism is yet to be studied.

This study also showed that not all biologically derived AgNPs are safe for biological applications. Their effects depend on specific properties of the phytochemicals that can influence their interaction with silver.

The mechanism of how the plant extracts impart unique properties to nanoparticles is still not fully understood. Furthermore, the mechanism of how nanoparticles interact with cells (in both animals and plants) is also not well understood. Moreover, the NPs that work well for animal/mammalian cells might not do so for plants. Hence an extensive study is needed in this field to have a universally good nanoparticle. We are working on finding these answers in our next project.

5. Conclusions

Our studies showed that the use of plant extract during the synthesis process is a good option for a low cost, environmentally friendly, and easy method of preparation of silver nanoparticles. The incorporation of turmeric extract in the synthesis process slowed the growth of NPS, whereas aloe vera enhanced it. The presence of aloe vera extract in the synthesis led to the formation of various oxides of silver. From in vitro studies, we could conclude that the AgNPs synthesized using turmeric extract showed faster and more sustained cell growth as compared with other AgNPs. These NPs also showed no cytotoxicity.

Our studies also showed that though some plant extracts are known for their healing properties (turmeric and aloe vera in our case), the presence of the nanoparticles can either enhance (AG-BT) or undermine (AG-BA) their efficacy. Hence, the use of the right plant extract is essential and there needs to be an extensive study to aid with choosing these extracts.

We also provided strong evidence of a healthy biological response observed upon in vivo exposure to turmeric-derived AgNPs in a dose-dependent manner. Turmeric AgNPs hence may be shown to be a safer alternative to the extensively used chemically synthesized AgNPs. With further detailed toxicity analysis and standardization of synthesis, turmeric-derived AgNPs can be commercialized.

Author Contributions: Conceptualization, N.B. and B.P.; methodology, A.S.D. and N.B.; software, A.S.D., N.B. and S.H.B.; validation, N.B., Z.E. and S.H.B.; formal analysis, A.S.D., N.B., A.S. and N.A.; investigation, A.S.D., N.B., A.S. and N.A.; resources, N.B., Z.E. and S.H.B.; data curation, A.S.D., N.B., A.S. and N.A.; writing—original draft preparation, A.S.D., N.B., A.S. and B.P.; writing—review and editing, A.S.D., N.B., A.S., B.P., N.A., P.S., S.H.B. and Z.E.; visualization, A.S.D., N.B., A.S. and N.A.; supervision, N.B., B.P., P.S. and N.A.; project administration, N.B.; funding acquisition, N.B., S.H.B. and Z.E. All authors have read and agreed to the published version of the manuscript.

Funding: This project received technical support and research fellowship from the Consortium for Scientific Research (CSR)—DAEF Indore (CSR-IC/BL-19/CRS-116/2018-19/1390). Some equipment was also funded by Research Support Fund provided to Neeru Bhagat by SIU. This work was kindly supported by the Deanship of Research and Graduate Studies, Ajman University, Ajman, United Arab Emirates (Project ID No: Ref. No. 2021-IRG-PH-2).

Institutional Review Board Statement: Not applicable.

Informed Consent Statement: Not applicable.

Data Availability Statement: Not applicable.

Acknowledgments: The authors are grateful to Joyita Banerjee and Swagata Roy of the Symbiosis School of Biomedical Sciences for allowing us to carry out the scratch assay in their laboratory. Anjana S. Desai is grateful to UGC-DAEF, CSR, Indore, for allowing us to carry out most of the investigations of the samples. Akanksha Singh would like to acknowledge CSIR India for research fellowship. The authors would also like to convey their gratitude to Jayant Yadav, JRF, Symbiosis Institute of Technology, for helping us with the XPS fitting.

Conflicts of Interest: The authors declare no conflict of interest.

References

1. Rujido-Santos, I.; Herbello-Hermelo, P.; Barciela-Alonso, M.C.; Bermejo-Barrera, P.; Moreda-Pineiro, A. Metal Content in Textile and (Nano)Textile Products. *Int. J. Environ. Res. Public Health* **2022**, *19*, 944. [CrossRef] [PubMed]
2. Chaudhary, R.; Nawaz, K.; Khan, A.K.; Hano, C.; Abbasi, B.H.; Anjum, S. An Overview of the Algae-Mediated Biosynthesis of Nanoparticles and Their Biomedical Applications. *Biomolecules* **2020**, *10*, 1498. [CrossRef] [PubMed]
3. Hoang, N.H.; Le Thanh, T.; Sangpueak, R.; Treekoon, J.; Saengchan, C.; Thepbandit, W.; Papathoti, N.K.; Kamkaew, A.; Buensanteai, N. Chitosan Nanoparticles-Based Ionic Gelation Method: A Promising Candidate for Plant Disease Management. *Polymers* **2022**, *14*, 662. [CrossRef] [PubMed]
4. Khezerlou, A.; Tavassoli, M.; Alizadeh Sani, M.; Mohammadi, K.; Ehsani, A.; McClements, D.J. Application of Nanotechnology to Improve the Performance of Biodegradable Biopolymer-Based Packaging Materials. *Polymers* **2021**, *13*, 4399. [CrossRef]
5. Yilmaz, H.; Culha, M. A Drug Stability Study Using Surface-Enhanced Raman Scattering on Silver Nanoparticles. *Appl. Sci.* **2022**, *12*, 1807. [CrossRef]
6. Balos, S.; Dramicanin, M.; Janjatovic, P.; Kulundzic, N.; Zabunov, I.; Pilic, B.; Klobcar, D. Influence of Metallic Oxide Nanoparticles on the Mechanical Properties of an A-TIG Welded 304L Austenitic Stainless Steel. *Materials* **2020**, *13*, 4513. [CrossRef]
7. Gobi, R.; Ravichandiran, P.; Babu, R.S.; Yoo, D.J. Biopolymer and Synthetic Polymer-Based Nanocomposites in Wound Dressing Applications: A Review. *Polymers* **2021**, *13*, 1962. [CrossRef]
8. Han, S.K. Basics of wound healing. In *Innovations and Advances in Wound Healing*; Springer: Berlin/Heidelberg, Germany, 2016; pp. 1–37.
9. Naskar, A.; Kim, K.S. Recent Advances in Nanomaterial-Based Wound-Healing Therapeutics. *Pharmaceutics* **2020**, *12*, 499. [CrossRef]
10. Razzaq, A.; Khan, Z.U.; Saeed, A.; Shah, K.A.; Khan, N.U.; Menaa, B.; Iqbal, H.; Menaa, F. Development of Cephradine-Loaded Gelatin/Polyvinyl Alcohol Electrospun Nanofibers for Effective Diabetic Wound Healing: In-Vitro and In-Vivo Assessments. *Pharmaceutics* **2021**, *13*, 349. [CrossRef]
11. Khan, B.A.; Ullah, S.; Khan, M.K.; Uzair, B.; Menaa, F.; Braga, V.A. Fabrication, Physical Characterizations, and In Vitro, In Vivo Evaluation of Ginger Extract-Loaded Gelatin/Poly(Vinyl Alcohol) Hydrogel Films Against Burn Wound Healing in Animal Model. *AAPS PharmSciTech* **2020**, *21*, 323. [CrossRef]
12. Shao, J.; Wang, B.; Li, J.; Jansen, J.A.; Walboomers, X.F.; Yang, F. Antibacterial effect and wound healing ability of silver nanoparticles incorporation into chitosan-based nanofibrous membranes. *Mater. Sci. Eng. C Mater. Biol. Appl.* **2019**, *98*, 1053–1063. [CrossRef] [PubMed]
13. Ahmad, N.; Bhatnagar, S.; Ali, S.S.; Dutta, R. Phytofabrication of bioinduced silver nanoparticles for biomedical applications. *Int. J. Nanomed.* **2015**, *10*, 7019–7030. [CrossRef]
14. Dutta, T.; Chowdhury, S.K.; Ghosh, N.N.; Chattopadhyay, A.P.; Das, M.; Mandal, V. Green synthesis of antimicrobial silver nanoparticles using fruit extract of Glycosmis pentaphylla and its theoretical explanations. *J. Mol. Struct.* **2022**, *1247*, 131361. [CrossRef]
15. Chenthamara, D.; Subramaniam, S.; Ramakrishnan, S.G.; Krishnaswamy, S.; Essa, M.M.; Lin, F.H.; Qoronfleh, M.W. Therapeutic efficacy of nanoparticles and routes of administration. *Biomater. Res.* **2019**, *23*, 20. [CrossRef]
16. Kim, K.S.; Lee, D.; Song, C.G.; Kang, P.M. Reactive oxygen species-activated nanomaterials as theranostic agents. *Nanomedicine* **2015**, *10*, 2709–2723. [CrossRef]
17. Saif, S.; Tahir, A.; Chen, Y. Green Synthesis of Iron Nanoparticles and Their Environmental Applications and Implications. *Nanomaterials* **2016**, *6*, 209. [CrossRef]
18. Chung, I.M.; Park, I.; Seung-Hyun, K.; Thiruvengadam, M.; Rajakumar, G. Plant-Mediated Synthesis of Silver Nanoparticles: Their Characteristic Properties and Therapeutic Applications. *Nanoscale Res. Lett.* **2016**, *11*, 40. [CrossRef]
19. Makarov, V.V.; Love, A.J.; Sinitsyna, O.V.; Makarova, S.S.; Yaminsky, I.V.; Taliansky, M.E.; Kalinina, N.O. "Green" nanotechnologies: Synthesis of metal nanoparticles using plants. *ActaNaturae* **2014**, *6*, 20. [CrossRef]
20. Sood, K.; Kaur, J.; Singh, H.; Kumar Arya, S.; Khatri, M. Comparative toxicity evaluation of graphene oxide (GO) and zinc oxide (ZnO) nanoparticles on Drosophila melanogaster. *Toxicol. Rep.* **2019**, *6*, 768–781. [CrossRef]
21. Surjushe, A.; Vasani, R.; Saple, D. Aloe vera: A short review. *Indian J. Dermatol.* **2008**, *53*, 163. [CrossRef]
22. Hęś, M.; Dziedzic, K.; Górecka, D.; Jędrusek-Golińska, A.; Gujska, E. Aloe vera (L.) Webb.: Natural sources of antioxidants–a review. *Plant Foods Hum. Nutr.* **2019**, *74*, 255–265. [CrossRef] [PubMed]
23. Mughal, M.H. Turmeric polyphenols: A comprehensive review. *Integr. Food Nutr. Metab.* **2019**, *6*, 1–6.
24. Kocaadam, B.; Şanlier, N. Curcumin, an active component of turmeric (*Curcuma longa*), and its effects on health. *Crit. Rev. Food Sci. Nutr.* **2017**, *57*, 2889–2895. [CrossRef] [PubMed]
25. Singh, A.; Raj, A.; Padmanabhan, A.; Shah, P.; Agrawal, N. Combating silver nanoparticle-mediated toxicity in Drosophila melanogaster with curcumin. *J. Appl. Toxicol. JAT* **2021**, *41*, 1188–1199. [CrossRef] [PubMed]
26. Neeru Bhagat, B.P. Synthesis of Antibacterial Oxide of Copper for Potential Application as Antifouling Agent. *Curr. Nanosci.* **2021**, *17*, 1–7.
27. Jalal, M.R.; Hojjati, H.; Jalal, J.R.; Ebrahimi, S. Synthesis of Ag_2O_2 microparticles combined with Ag_2CO_3 nanoparticles by a pin-to-solution electrical discharge in atmospheric air. *Mater. Lett.* **2019**, *251*, 218–221. [CrossRef]

28. Gao, X.-Y.; Wang, S.-Y.; Li, J.; Zheng, Y.-X.; Zhang, R.-J.; Zhou, P.; Yang, Y.-M.; Chen, L.-Y. Study of structure and optical properties of silver oxide films by ellipsometry, XRD and XPS methods. *Thin Solid Film.* **2004**, *455*, 438–442. [CrossRef]
29. Hammad, A.; Abdel-Wahab, M.; Alshahrie, A. Structural and morphological properties of sputtered silver oxide thin films: The effect of thin film thickness. *Dig. J. Nanomater. Bios.* **2016**, *11*, 1245–1252.
30. Bielmann, M.; Schwaller, P.; Ruffieux, P.; Gröning, O.; Schlapbach, L.; Gröning, P. AgO investigated by photoelectron spectroscopy: Evidence for mixed valence. *Phys. Rev. B* **2002**, *65*, 235431. [CrossRef]
31. Kumar-Krishnan, S.; Prokhorov, E.; Hernández-Iturriaga, M.; Mota-Morales, J.D.; Vázquez-Lepe, M.; Kovalenko, Y.; Sanchez, I.C.; Luna-Bárcenas, G. Chitosan/silver nanocomposites: Synergistic antibacterial action of silver nanoparticles and silver ions. *Eur. Polym. J.* **2015**, *67*, 242–251. [CrossRef]
32. Tsendzughul, N.T.; Ogwu, A.A. Physicochemical aspects of the mechanisms of rapid antimicrobial contact-killing by sputtered silver oxide thin films under visible light. *ACS Omega* **2019**, *4*, 16847–16859. [CrossRef] [PubMed]
33. Kaspar, T.C.; Droubay, T.; Chambers, S.A.; Bagus, P.S. Spectroscopic evidence for Ag (III) in highly oxidized silver films by X-ray photoelectron spectroscopy. *J. Phys. Chem. C* **2010**, *114*, 21562–21571. [CrossRef]
34. Lützenkirchen-Hecht, D. Electrochemically grown silver oxide (Ag2O) by XPS. *Surf. Sci. Spectra* **2011**, *18*, 96–101. [CrossRef]
35. Murray, B.; Newberg, J.; Walter, E.; Li, Q.; Hemminger, J.; Penner, R. Reversible resistance modulation in mesoscopic silver wires induced by exposure to amine vapor. *Anal. Chem.* **2005**, *77*, 5205–5214. [CrossRef] [PubMed]
36. Al-Sarraj, A.; Saoud, K.M.; Elmel, A.; Mansour, S.; Haik, Y. Optoelectronic properties of highly porous silver oxide thin film. *SN Appl. Sci.* **2021**, *3*, 1–13. [CrossRef]
37. de Jesús Ruíz-Baltazar, Á.; Reyes-López, S.Y.; de Lourdes Mondragón-Sánchez, M.; Estevez, M.; Hernández-Martinez, A.R.; Pérez, R. Biosynthesis of Ag nanoparticles using Cynara cardunculus leaf extract: Evaluation of their antibacterial and electrochemical activity. *Results Phys.* **2018**, *11*, 1142–1149. [CrossRef]
38. Kora, A.J.; Rastogi, L. Enhancement of antibacterial activity of capped silver nanoparticles in combination with antibiotics, on model gram-negative and gram-positive bacteria. *Bioinorg. Chem. Appl.* **2013**, *2013*, 871097. [CrossRef]
39. Chircov, C.; Matei, M.-F.; Neacșu, I.A.; Vasile, B.S.; Oprea, O.-C.; Croitoru, A.-M.; Trușcă, R.-D.; Andronescu, E.; Sorescu, I.; Bărbuceanu, F. Iron oxide–silica core–shell nanoparticles functionalized with essential oils for antimicrobial therapies. *Antibiotics* **2021**, *10*, 1138. [CrossRef]
40. Begum, S.; Jones, I.P.; Jiao, C.; Lynch, D.E.; Preece, J.A. Characterisation of hollow Russian doll microspheres. *J. Mater. Sci.* **2010**, *45*, 3697–3706. [CrossRef]
41. Mukherjee, S.; Chowdhury, D.; Kotcherlakota, R.; Patra, S. Potential theranostics application of bio-synthesized silver nanoparticles (4-in-1 system). *Theranostics* **2014**, *4*, 316. [CrossRef]
42. Vinken, M.; Blaauboer, B.J. In vitro testing of basal cytotoxicity: Establishment of an adverse outcome pathway from chemical insult to cell death. *Toxicol. In Vitro* **2017**, *39*, 104–110. [CrossRef] [PubMed]
43. Raj, A.; Shah, P.; Agrawal, N. Dose-dependent effect of silver nanoparticles (AgNPs) on fertility and survival of Drosophila: An in-vivo study. *PLoS ONE* **2017**, *12*, e0178051. [CrossRef] [PubMed]
44. Armstrong, N.; Ramamoorthy, M.; Lyon, D.; Jones, K.; Duttaroy, A. Mechanism of silver nanoparticles action on insect pigmentation reveals intervention of copper homeostasis. *PLoS ONE* **2013**, *8*, e53186.
45. Silver Key, S.C.; Reaves, D.; Turner, F.; Bang, J.J. Impacts of Silver Nanoparticle Ingestion on Pigmentation and Developmental Progression in Drosophila. *Atlas J. Biol.* **2011**, *1*, 52–61. [CrossRef]
46. Raj, A.; Shah, P.; Agrawal, N. Sedentary behavior and altered metabolic activity by AgNPs ingestion in Drosophila melanogaster. *Sci. Rep.* **2017**, *7*, 15617. [CrossRef]
47. Dlugosz, O.; Szostak, K.; Staron, A.; Pulit-Prociak, J.; Banach, M. Methods for Reducing the Toxicity of Metal and Metal Oxide NPs as Biomedicine. *Materials* **2020**, *13*, 279. [CrossRef]
48. Hamouda, R.A.; Hussein, M.H.; Abo-Elmagd, R.A.; Bawazir, S.S. Synthesis and biological characterization of silver nanoparticles derived from the cyanobacterium Oscillatoria limnetica. *Sci. Rep.* **2019**, *9*, 1–17. [CrossRef]
49. Kim, J.-H.; Ma, J.; Jo, S.; Lee, S.; Kim, C.S. Enhancement of Antibacterial Performance of Silver Nanowire Transparent Film by Post-Heat Treatment. *Nanomaterials* **2020**, *10*, 938. [CrossRef]
50. AshaRani, P.; Low Kah Mun, G.; Hande, M.P.; Valiyaveettil, S. Cytotoxicity and genotoxicity of silver nanoparticles in human cells. *ACS Nano* **2009**, *3*, 279–290. [CrossRef]
51. He, D.; Garg, S.; Waite, T.D. H_2O_2-mediated oxidation of zero-valent silver and resultant interactions among silver nanoparticles, silver ions, and reactive oxygen species. *Langmuir* **2012**, *28*, 10266–10275. [CrossRef]
52. Redza-Dutordoir, M.; Averill-Bates, D.A. Activation of apoptosis signalling pathways by reactive oxygen species. *Biochim. Biophys. Acta (BBA)-Mol. Cell Res.* **2016**, *1863*, 2977–2992. [CrossRef] [PubMed]
53. Akter, M.; Sikder, M.T.; Rahman, M.M.; Ullah, A.; Hossain, K.F.B.; Banik, S.; Hosokawa, T.; Saito, T.; Kurasaki, M. A systematic review on silver nanoparticles-induced cytotoxicity: Physicochemical properties and perspectives. *J. Adv. Res.* **2018**, *9*, 1–16. [CrossRef] [PubMed]
54. Lennicke, C.; Rahn, J.; Lichtenfels, R.; Wessjohann, L.A.; Seliger, B. Hydrogen peroxide–production, fate and role in redox signaling of tumor cells. *Cell Commun. Signal.* **2015**, *13*, 1–19. [CrossRef] [PubMed]
55. Canaparo, R.; Foglietta, F.; Limongi, T.; Serpe, L. Biomedical Applications of Reactive Oxygen Species Generation by Metal Nanoparticles. *Materials* **2020**, *14*, 53. [CrossRef]

56. Yun'an Qing, L.C.; Li, R.; Liu, G.; Zhang, Y.; Tang, X.; Wang, J.; Liu, H.; Qin, Y. Potential antibacterial mechanism of silver nanoparticles and the optimization of orthopedic implants by advanced modification technologies. *Int. J. Nanomed.* **2018**, *13*, 3311. [CrossRef]
57. Almeer, R.S.; Ali, D.; Alarifi, S.; Alkahtani, S.; Almansour, M. Green Platinum Nanoparticles Interaction with HEK293 Cells: Cellular Toxicity, Apoptosis, and Genetic Damage. *Dose-Response* **2018**, *16*, 1559325818807382. [CrossRef]
58. Ast, T.; Mootha, V.K. Oxygen and mammalian cell culture: Are we repeating the experiment of Dr. Ox? *Nat. Metab.* **2019**, *1*, 858–860. [CrossRef]
59. Iavicoli, I.; Fontana, L.; Leso, V.; Bergamaschi, A. The effects of nanomaterials as endocrine disruptors. *Int. J. Mol. Sci.* **2013**, *14*, 16732–16801. [CrossRef]
60. Mao, B.H.; Chen, Z.Y.; Wang, Y.J.; Yan, S.J. Silver nanoparticles have lethal and sublethal adverse effects on development and longevity by inducing ROS-mediated stress responses. *Sci. Rep.* **2018**, *8*, 2445. [CrossRef]

Article

Development and Characterization of the Biodegradable Film Derived from Eggshell and Cornstarch

Joseph Merillyn Vonnie [1], Kobun Rovina [1,*], Rasnarisa Awatif Azhar [1], Nurul Huda [1], Kana Husna Erna [1], Wen Xia Ling Felicia [1], Md Nasir Nur'Aqilah [1] and Nur Fatihah Abdul Halid [2]

[1] Faculty of Food Science and Nutrition, Universiti Malaysia Sabah, Kota Kinabalu 88400, Sabah, Malaysia; vonnie.merillyn@gmail.com (J.M.V.); rasnarisa@gmail.com (R.A.A.); drnurulhuda@ums.edu.my (N.H.); mn1911017t@student.ums.edu.my (K.H.E.); felicialingling.97@gmail.com (W.X.L.F.); aqilah98nash@gmail.com (M.N.N.)

[2] Borneo Marine Research Institute, Universiti Malaysia Sabah, Kota Kinabalu 88400, Sabah, Malaysia; fatihahhalid@ums.edu.my

* Correspondence: rovinaruby@ums.edu.my; Tel.: +60-88-320000 (ext. 8713); Fax: +60-88-320993

Abstract: In the current study, cornstarch (CS) and eggshell powder (ESP) were combined using a casting technique to develop a biodegradable film that was further morphologically and physicochemically characterized using standard methods. Scanning electron microscopy (SEM) and transmission electron microscopy (TEM) were used to characterize the morphology of the ESP/CS film, and the surface of the film was found to have a smooth structure with no cracks, a spherical and porous irregular shape, and visible phase separation, which explains their large surface area. In addition, the energy dispersive X-ray (EDX) analysis indicated that the ESP particles were made of calcium carbonate and the ESP contained carbon in the graphite form. Fourier Transform Infrared Spectroscopy indicated the presence of carbonated minerals in the ESP/CS film which shows that ESP/CS film might serve as a promising adsorbent. Due to the inductive effect of the O–C–O bond on calcium carbonate in the eggshell, it was discovered that the ESP/CS film significantly improves physical properties, moisture content, swelling power, water solubility, and water absorption compared to the control CS film. The enhancement of the physicochemical properties of the ESP/CS film was principally due to the intra and intermolecular interactions between ESP and CS molecules. As a result, this film can potentially be used as a synergistic adsorbent for various target analytes.

Keywords: biodegradable film; starch; calcium carbonate; eggshell; physicochemical

1. Introduction

Eggshell (ES) is a byproduct of poultry production and kitchen waste. Most country to dispose of these waste products primarily in landfills without additional treatment, which has serious environmental consequences [1]. Malaysia produces approximately 642,600 tons of eggs per year and is expected to generate 70,686 tons of ES waste per year [2]. This is due to the fact that eggs are a common ingredient in a variety of products, including cakes, fast food, and everyday meals. However, statistics show that egg production in Malaysia has increased dramatically to 773 thousand tonnes, with an expected annual growth rate of 3.6% by 2020. In this case, ES waste can be used as a source of calcium carbonate ($CaCO_3$), which accounts for approximately 95% of a shell mass and 5% of organic materials such as sulfated polysaccharides, collagen, and other proteins [3]. Furthermore, $CaCO_3$ has been used to remove manganese, cadmium, lead, zinc, nickel, and chromium [4]; additionally, $CaCO_3$ is known for having good morphology and porosity and removing more than 90% of metal ions when the pH, contact time, agitation speed, and adsorbent dosage are optimized. Previously, Cree and Rutter [5], as well as Betancourt and Cree [6], have conducted morphological studies to compare mineral limestone and eggshell powder (ESP) under a scanning electron microscope at the same magnification. ESP revealed the presence of pores, while these were absent in the limestone. This implies that eggshells have bigger pores than limestone, which allows them to serve as an adsorbent.

One of the greatest threats to human life is environmental pollution caused by synthetic polymers, plastics, and non-degradable packaging materials entering the environment [7]. Thin film-based biopolymers, such as starch, have become increasingly important in the manufacturing sector, particularly in the food industry. Biodegradable films are typically composed of polypeptides, lipids, polysaccharides, or a combination of these materials. The development of biodegradable films made from renewable natural resources has been investigated as a good alternative to reduce the negative environmental impacts caused by waste disposal as they can easily be decomposed by the environment [8,9]. Corn starch is a natural biopolymer that has been widely employed in the development of environmentally friendly packaging materials [10]. Moreover, corn is very rich in starch, which is one of the foremost copious and biodegradable biopolymers and is commonly utilized within the food, pharmaceutical, textile, biomass energy, and chemical industries [11]. However, due to its continuous matrix formation capacity and high amylose content, starch has a higher film-forming capability when compared with other polysaccharides [12,13]. Starch has numerous benefits in the development of biodegradable films, including low cost, renewability, appropriate physical, chemical, and functional properties, and a diverse source base [14,15]. Furthermore, starch is a natural carbohydrate polymer derived from plants, such as cereals, roots, and tubers. Several novel bio-based materials have been investigated as a potential alternative to resolve environmental issues, including wheat [16], corn [17], potato [18], cassava [19], sugar palm starch [20], and pea [21]. ESP/CS film was employed as a potential polymeric adsorbent for the adsorption and removal of analytes.

The previous research indicates that ES has the potential to be a good sorbent due to its high carbon and calcium concentrations, as well as its high porosity and good accessibility of functional groups. Recently, Wang et al. [22] studied the application of granular bentonite—eggshell composites (BEP) for the removal of heavy metal (Pb). BEP had an elimination ratio of over 99.90% and an adsorption capacity of over 40 mg/g of Pb, respectively. The Elovich kinetic model described the Pb adsorption process well and suggested removal processes, including ion exchange, electrostatic adsorption, and complexation, between BEP and Pb. Isa et al. [23] proposed the use of two natural agro-wastes composed of eggshells and sugarcane bagasse as potential adsorbents in the removal of Pb (II) and Cd (II) from aqueous solutions. The maximum concentrations of Pb and Cd absorbed per gram of biosorbent were determined to be 277.8 and 13.62 mg/g for eggshells and 31.45 and 19.49 mg/g for bagasse, respectively. Antecedently, Mostafavi et al. [24] designed and fabricated nanocomposite-based polyurethane filters with excellent physical and chemical adsorption capabilities for particular contaminants and the ability to alleviate or minimize pollutants from wastewater. This filter could be a promising device for the removal of heavy metals from water.

In this study, there is a focus on developing and characterizing the eggshell powder for the development of biodegradable films containing cornstarch since it is abundant, cost-effective, easily processed, and demonstrates strong potential as an adsorbent. The surface morphology, functional groups, and hydrophobic characteristics of composite film were studied. This research will serve to remediate and alleviate the side effects of contaminants in various types of food and wastewater in the industrial sectors.

2. Materials and Methods

2.1. Materials

Raw and fresh eggshells were collected from a local restaurant in Sabah, Malaysia. Cornstarch powder (purity 100%) and potassium bromide (KBr, purity $\geq$ 99%) were purchased from Sigma Aldrich, St. Louis, MO, USA.

2.2. Preparation of Eggshell Powder (ESP)

Fresh ES was collected and rinsed several times with standard tap water followed by deionized water to eliminate contaminants and interfering materials on the surface of the ES. Then, the washed ES was air dried for 30 min and dried for 5 h at 50 °C in a hot air oven (Binder, 07-32195). The dried ES was then blended into fine particles using a blender (Panasonic MX-337) and sieved through a 250 m sieve to obtain the finely powdered sample, which was then stored in an airtight container [25,26].

2.3. Preparation of Biodegradable ESP/CS Films

ES solution was prepared by grinding 0.6 g of ESP and dispersing it in 30 mL of distilled water. The mixtures were continuously stirred at room temperature for 2 h to thoroughly wet the ESP particles before being filtered through Whatman filter paper. Following that, 0.6 g of CS powder was dissolved in 10 mL of distilled water and ES solution to produce control films of CS film solution and ESP/CS film solution, respectively. The solutions were then stirred with a magnetic stirrer at 100 °C until gelatinized. Each film-forming solution was poured into 90 mm internal diameter Petri dishes and allowed to form films at room temperature for nearly 24 h.

2.4. Morphological Characterization

The surface morphology and elemental composition of ESP/CS and CS films were analyzed using scanning electron microscopy (SEM, Carl Zeiss Ma 10), energy-dispersive X-ray spectroscopy (EDX), and transmission electron microscopy (TEM, Tecnai G Spirit Biotwin).

2.5. Physical Characterization

2.5.1. Thickness (e) and Density (ρ)

A micrometer with a precision of ±0.001 mm was used to measure the thickness of developed films. The film dimensions were determined by taking samples at five different locations [27]. For the ρ determination, 6.25 cm^2 (area) sample discs from each film were used. The ρ film was then calculated as the ratio of weight (W) to volume (V), where V equals area (A), by multiplying the e of each film as in Equation (1).

$$\rho = \frac{w}{v} = \frac{w}{A*e} \tag{1}$$

2.5.2. Moisture Content (MC), Water Solubility (WS), Water Absorption (WA) and Swelling Power (SP) of Films

For MC, the films were cut into 2.5 cm × 2.5 cm pieces and weighed (W_i). Then, it was dried at 105 °C for 24 h and weighed as W_f. The MC of the different films was calculated using Equation (2) [28].

$$MC\ (\%) = \frac{W_i - W_f}{W_i} \times 100 \tag{2}$$

For WS, the initial dry weight (W_i) of film was determined by drying it at 105 °C for 24 h. Each sample was then immersed in 50 mL of H$_2$O and kept at 25 °C for 24 h. Subsequently, the films were dried in an oven at 105 °C for 24 h and weighed (W_f). The data obtained was used to calculate the WS for each film using Equation (3).

$$WS\ (\%) = \frac{W_i - W_f}{W_i} \times 100 \tag{3}$$

The WA of the films was determined according to Wu et al. [25]. The films (2.5 cm × 2.5 cm) dried at 105 °C for 24 h, and the initial weight (W_i) was obtained. The films were then immersed in H$_2$O at room temperature for 60 min. The samples were taken out and weighed immediately after wiping off the H₂O surface with filter papers (W_f). The percentages of WA were calculated using Equation (4).

$$WA\ (\%) = \frac{W_f - W_i}{W_i} \times 100 \tag{4}$$

The SP % of the developed films was determined according to the method described by Herniou et al. [29]. Each of the film samples was initially cut into 2 cm diameter ($\varnothing_i$) discs and immersed in 20 mL of H₂O. The containers were then sealed and maintained at room temperature for 24 h. The disc diameters ($\varnothing_f$) of the samples were recorded and calculated using Equation (5).

$$SP\ (\%) = \frac{\varnothing_f - \varnothing_i}{\varnothing_i} \times 100 \tag{5}$$

2.6. Storage Stability

The storage stability of the films was determined in three different conditions by measuring the weight loss (W_L) over 28 days. Three containers consisting of five replicate film samples were stored in the incubator (37 °C), cold room (4 °C), and room temperature. The weights were recorded on day 0, 7, 14, 21, and 28. The W_L was calculated using Equation (6).

$$\text{Weight Loss}\ (W_L) = W_i - W_f \tag{6}$$

2.7. Fourier Transform Infrared (FTIR) Spectroscopy and X-ray Diffraction (XRD) Analysis

CS and ESP/CS films' functional groups were measured using an FTIR spectrometer (Perkin-Elmer Universal Attenuated Total Reflectance spectrometer). The film samples were blended with potassium bromide (KBr) powder and pressed into a pellet. The crystalline structure and particle size of the film samples were observed by X-ray diffraction (XRD) analysis. The XRD patterns were recorded using a diffractometer (PANalytical-Empyrean instrument; Co radiation: 1.54056 A°) and analyzed between 0 and 90° (2 theta). The voltage, current, and amount of time passed were 40 Kv, 40 mA, and 1 s.

2.8. Statistical Analysis

The results are expressed as the mean ± standard deviation (n = 5) to ensure the data's accuracy. The SPSS IBM Statistics 19 software was used to analyze the data. The experimental data were analyzed using one-way ANOVA, with Tukey Post-Hoc test at a 95% confidence interval with $p < 0.05$ to establish significant differences between the samples.

3. Results and Discussion

3.1. Morphology Characterization

Figure 1 shows the surface of the CS and ESP/CS films, as well as their composition, which was determined by scanning electron microscopy (SEM). $CaCO_3$ is the primary component of ES and is composed of three polymorphs: vaterite, aragonite, and calcite. These polymorphs are temperature-dependent, with different temperatures producing different $CaCO_3$ structures with different morphology [30]. Figure 1a,b shows the spheroidal-shaped crystallites of vaterite, as well as the smallest crystallites, which are about 1–2 μm in size [31]. The porous surface textures of the ESP/CS films support the adsorbent by increasing the surface area and allowing the adsorption process to occur while also exhibiting a non-adhesive appearance and agglomeration formation. Due to its high adsorption capacity, raw ES's porous external and internal layers made it an appealing bioadsorbent. Furthermore, the external surface of the ES is primarily composed of $CaCO_3$, whereas the inner layer is made up of uncalcified fibrous membranes composed primarily of organic compounds [32].

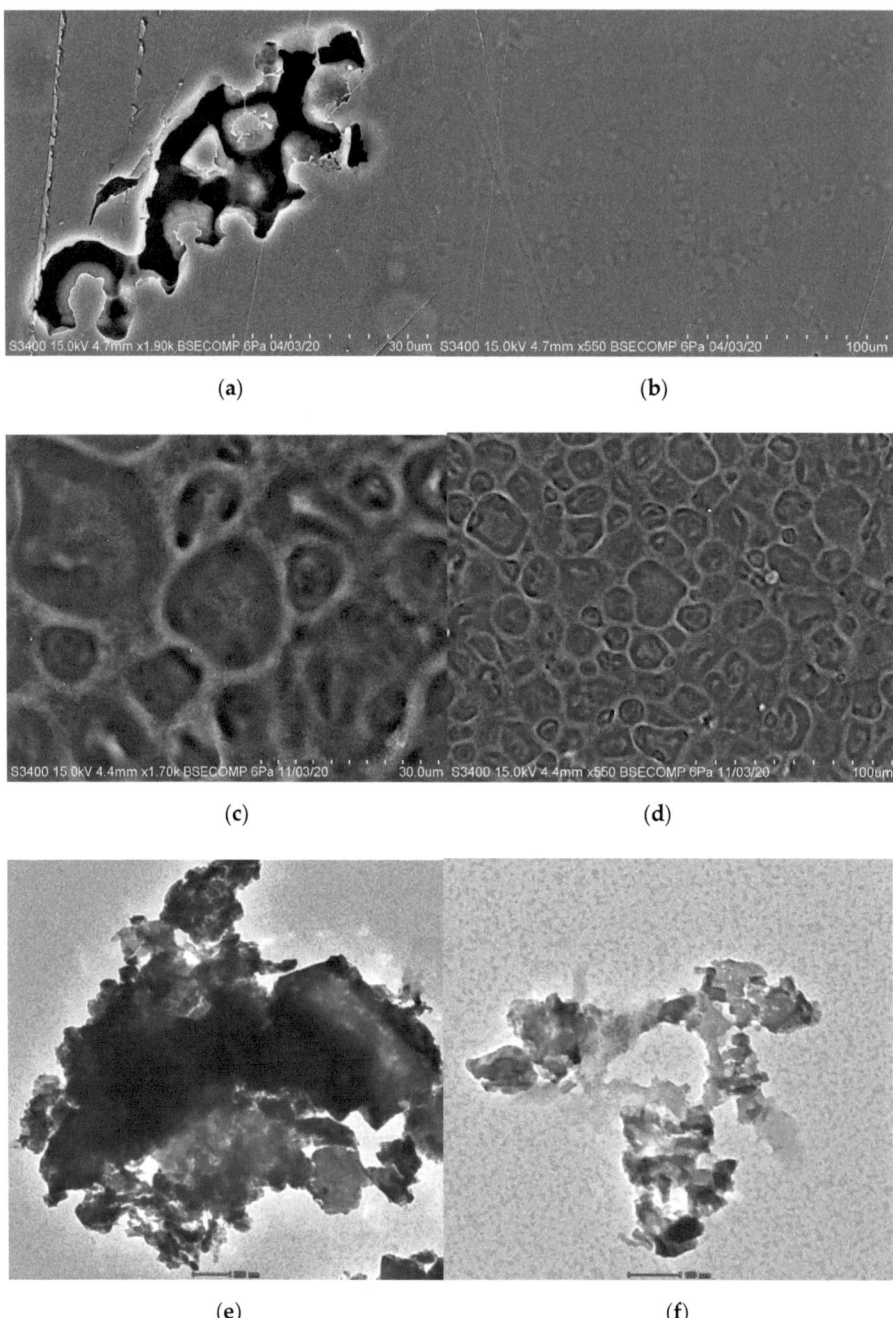

Figure 1. *Cont.*

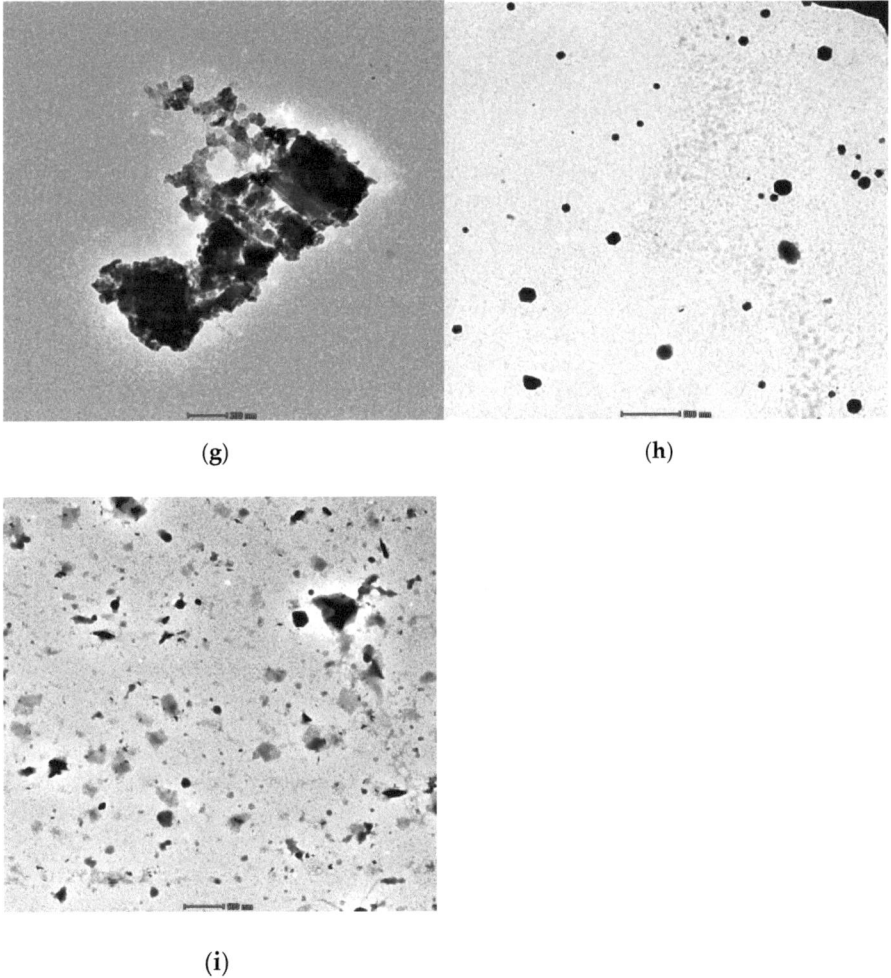

(i)

Figure 1. The SEM images of biodegradable (**a**,**b**) ESP/CS and (**c**,**d**) CS film with different magnifications (×30 µm, ×100 µm) and TEM images of (**e**–**g**) ESP/CS and (**h**,**i**) CS film with different magnifications (×100 µm, ×200 µm, ×500 µm).

The surfaces of the ESP/CS films, as shown in Figure 1b, have a homogeneous and smooth structure with no cracks, visible phase separation, and embedded ESP in the starch matrix. According to Sun et al. [33], phase separation could destabilize the interface adhesion between the nanofiller and the matrix, lowering the tensile strength of films with increased nanoparticle content. As a result of the organic components in ESP, the ESP particles are spread uniformly in the film matrix, resulting in improved adhesion to the starch matrix [34]. The microstructure of the ESP reveals that the particle size and shape varies; however, they are composed of porous irregularly-shaped particles. According to King'Ori [35], the microstructure reveals relatively uniform distributions of ESP particles in the matrix. The distribution of ESP particles is influenced by the ESP particles' excellent wettability by molten metal and good interfacial bonding between the particles and matrix material. In the microstructures of the composites, the ESP was well retained and distributed along the grain boundaries.

The granular shape of CS was round and polygonal, as shown in Figure 1c,d, but the surface became rough as many pores formed on the starch granules, resulting in a compact structure [34]. Some of the pores even extend from the surface to the interior of the starch granules, as shown in Figure 1d, and the starch granules have large interval cavities. Due to these structural properties, a large specific surface area is possible. As a result, it is possible to conclude that these macropores can improve the adsorption of adsorbates and can be used as adsorbents in a variety of applications [36,37]. Furthermore, starch granules with various sizes or apparent densities had inherent crystallinity variations, which influenced the granule's functional properties.

The EDX of the ESP particles demonstrates that the particles contain calcium (Ca), oxygen (O), copper (Cu), gold (Au), nitrogen (N), and silica (Si), with a presence of carbon (C) (61.83%) in the ESP particles. According to King'Ori [35], ES contains approximately 98.2% of calcium carbonate, 0.9% of magnesium, and 0.9% of phosphorus, which showed that the ESP particles are made of $CaCO_3$ and contain carbon in the graphite. These analyses are comparable to other authors' reinforcement analyses. The microstructure of the unreinforced Au (16.42%) and Cu (1.87%) also contained in this sample led to the preparation of a sample in which all specimens (external surfaces) were coated with a thin layer of gold under a high vacuum. According to the EDX analysis of the composite material, there is a possibility of a chemical reaction between aluminium (Al) melt and the ESP particle, which resulted in the release of Si (0.57%), N (11.94%), O (19.77%), C, and Ca (0.12%) [38].

Next, TEM was used to determine the characteristics of the various materials, including the morphology, shape, and particle size of ESP/CS and CS. The TEM images in Figure 1e,f show that the particles have irregular shapes, which explains their large surface area [39]. In addition, irregular shapes in ES have been observed, which may be due to the calcite shape of the ESP [40]. According to Minakshi et al. [39], bright-field TEM imaging of the ES revealed dense and angular calcite crystals with particle sizes ranging from 30 to 500 nm. The bright field can be seen mostly in Figure 1f and in some parts of Figure 1e. The porous and nanocrystalline nature can be seen in the high-resolution TEM image (Figure 1g). This occurred because the nucleation and growth of CaO crystals from the $CaCO_3$ parent contributed to a roughened surface texture on the particles [39].

In contrast, high-resolution TEM measurements revealed that the CS particles were small, had good uniformity, and an almost perfect spherical shape, with a diameter ranging from 53.1 nm–83.1 nm (Figure 1h). The surface of the CS film without any ESP was smooth. As shown in Figure 1h, the spherical particles and the particle size distribution were relatively homogeneous and no agglomeration occurred. The particles of CS in Figure 1i consist of both large and small irregular bodies entangled by thin filaments which have no encapsulated structures, and more dispersion could be observed. However, Ponsanti et al. [41] intensely observed that CS has more dispersion and a smaller particle size than other starches, including cassava starch and sago starch.

3.2. Film Thickness and Density Measurement

After the addition of ESP, the thickness of the composite film changed. The thickness of the CS film was 0.018 ± 0.021 mm, whereas the thickness of the ESP/CS film was increased to 0.026 ± 0.018 mm due to the presence of $CaCO_3$ particles in the ESP, which strengthened the interfacial interaction between ESP and the CS film matrix, hence reducing matrix chain mobility and increasing the macroscopic thickness of the ESP/CS composite film. At the same time, incorporation of ESP significantly increased the density of ESP/CS film due to the establishment of a continuous network facilitated by the incorporation of ESP, which minimized the effect of water [34].

3.3. Moisture Content (MC)

The CS film had a higher MC value (48.533 ± 18.213) than the ESP/CS film (38.781 ± 16.139), which was due to the greater number of available OH groups in the starch film. Furthermore, this led to the outstanding ability of CS film to absorb moisture from the environment, and this finding was compatible with the FTIR spectra analysis. The lower *MC* value of the ESP/CS film was due to the cross-linking between ES ($CaCO_3$) and CS, which reduced the polarity and the interaction between hydrophilic groups and water [42]. The objective of determining the film's moisture content was to ascertain its superior removal capacity. The moisture content of a film indicates the percentage of water capacity present. The lower the moisture content, the greater the adsorption efficiency due to lesser water molecules binding to the active sites of the film matrix and allowing target analytes to bind onto the active sites [43]. Therefore, the ESP/CS film showed promise as an adsorbent due to its lower moisture content.

3.4. Water Solubility (WS)

The ESP/CS film (39.022 ± 12.251) displayed a lower solubility than the CS film (46.632 ± 9.109), as shown in Table 1. The intramolecular interaction and arrangement of the ESP and CS granules can be attributed to these outcomes. CS in ESP solution can form highly cross-linked systems, preventing water molecules from penetrating the ESP/CS films and dissolving ESP fiber proteins and starch granules [13]. Consequently, the capacity of the film to absorb water was reduced [44] due to the greater amount of ESP filler loaded onto the film matrix [45]. The ESP/CS film showed a moderate decrease in WS during the gelatinization process due to the interaction with ES molecules where the degradation of ES and starch co-occurred, wherein the hydroxyl groups (-OH) of degraded starch interacted with the functional groups of ES, resulting in the formation of a cross-linking network [13]. The reduced number of free -OH groups made the ESP/CS film less attractive to water molecules, which also helped to slow down the release rate of free polymer chains from collagen and starch to water, and eventually resulted in the lower solubility of ESP/CS films compared to CS films [13,46].

Table 1. The physicochemical values of CS and ESP/CS film.

Parameters	CS Film	ESP/CS Film
Thickness	0.018 ± 0.021 [a]	0.026 ± 0.018 [b]
Density	0.535 ± 0.386 [c]	0.617 ± 0.210 [d]
MC (%)	48.533 ± 18.213	38.781 ± 16.139 [e]
WS (%)	46.632 ± 9.109	39.022 ± 12.251
WA (%)	87.072 ± 2.758	87.700 ± 3.374
SP (%)	8.000 ± 2.739	7.000 ± 2.739

Different lowercase letters in the same column indicate a statistically significant difference ($p < 0.05$).

3.5. Water Adsorption (WA)

The presence of free hydroxyl groups (–OH) in ESP/CS film can result in high sorption of target analytes. Table 1 shows the *WA* value for ESP/CS and CS films. It showed that both films absorbed water very well during the 60 min immersion process to reach the saturation level. The addition of ESP into CS can increase the *WA* of ESP/CS film (87.700 ± 3.374) compared to CS film (87.072 ± 2.758). These results indicated that a higher content of filler and the presence of high levels of $CaCO_3$ in the ESP, thus increasing the *WA* of ESP/CS film. In addition, the high *WA* value in the ESP/CS film with higher ESP loading was due to the increased number of voids between ESP and CS matrices [47].

3.6. Swelling Power (SP)

Based on Table 1, the *SP* value of the ESP/CS film was reduced compared to the CS film with values of 7.000 ± 2.739 and 8.000 ± 2.739, respectively. A decreasing *SP* value of the ESP/CS film indicated better resistance to swelling, probably due to the presence of more vital associative forces maintaining the granule structure [48]. However, the rise in the *SP* of the CS films is attributed to the hydrogen bond in water molecules to the exposed hydroxyl groups of amylose and amylopectin.

3.7. Storage Stability

The storage stability test applied to the films aimed to provide evidence on how the quality of the films varies with time under the influence of environmental factors, including temperature, humidity, and light. The storage stability for both films was analyzed for 28 days at three different places with different temperatures, as shown in Figure 2. The storage time and temperature affected water activity. From the obtained result, the trends for both biodegradable films started to decrease after day 7, which was attributed to the decreasing water-holding capacity of the films and the moisture loss [49].

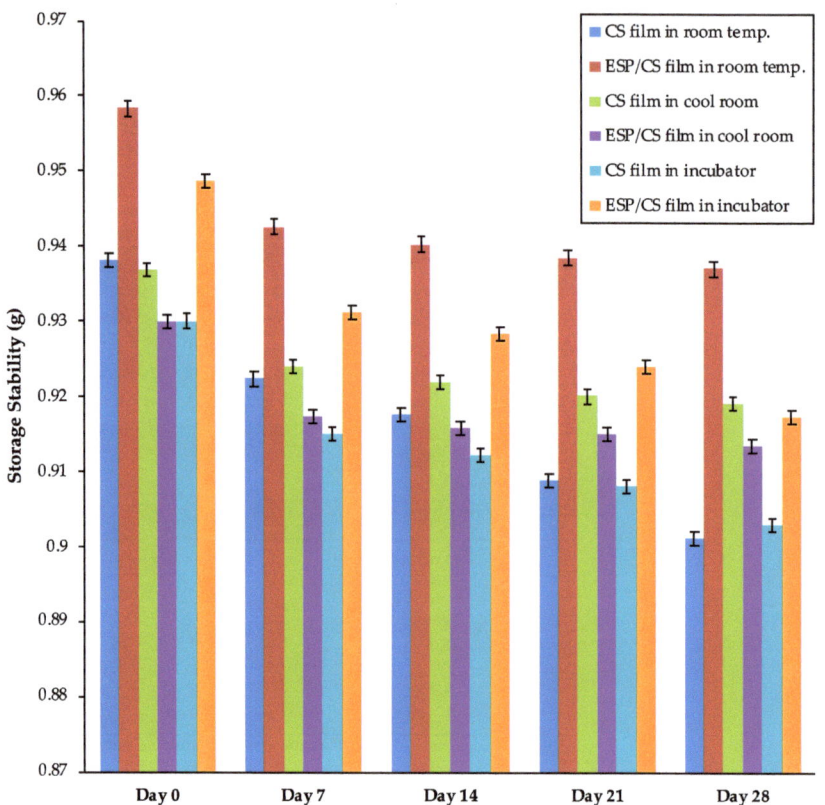

Figure 2. The storage stability of films for day 0, 7, 14, 21, and 28.

Based on Figure 2, the storage stability for ESP/CS film kept at room temperature (25 °C) and in an incubator (37 °C) was higher relative to those kept in a cool room (4 °C). Previously, Jiang et al. [34] reported that composite films consisting of CS and ESP have excellent thermal stability due to the presence of ESP which affects the matrix's heat conductivity and prevents the starch films from degrading thermally. Moreover, it was reported that CS and ESP show good compatibility. For example, ESP was tightly dispersed in the matrix, resulting in increased thermal stability and an extended degradation rate of the composite [50]. However, the storage stability of the ESP/CS film was lower in refrigerated conditions due to the slower active reaction of the composite film [51]. Also, the ESP/CS film provided a significantly higher quality of film at room temperature when compared with other conditions; the ESP/CS film was 0.9370 g, while the CS film was 0.9011 g after 28 days. This scenario happened because ESP/CS film had a lower peroxide value and a higher induction period than CS film, indicating greater stability [52]. The peroxide values increased faster at room temperature than at lower temperatures (cool room) during the storage time, and the highest storage stability for the ESP/CS film stored at room temperature was exposure to daylight. Thus, higher peroxide values and a longer induction period reduced the decomposition of the ESP/CS film and provided greater storage stability.

3.8. Fourier Transform Infrared (FTIR) and X-ray Diffraction (XRD) Analysis

The results of the stretching and bending vibrations of the functional groups in the ESP/CS, which is responsible for the adsorption of the adsorbate molecules, are shown in Figure 3. Figure 3a shows the IR spectral analysis for ESP/CS with a distinct peak at 654.34 cm^{-1}, 712.13 cm^{-1}, 872.42 cm^{-1}, and 1400.03 cm^{-1}. An intense peak and a strong absorption band of the ES particle was observed at 1400.03 cm^{-1}, which can be strongly associated with the presence of carbonate minerals within the ES matrix [53]. There are also two observable sharp bands at 712.13 and 872.42 cm^{-1}, which were associated with the asymmetric stretching of the C-H band assigned and the in-plane deformation and out-plane deformation, indicating the existence of $CaCO_3$ [53]. Equally significant, the prominent absorption peaks of carbonates at 1400.03 cm^{-1} support Onwubu et al.'s [1] findings. They claimed that carbonate-based materials commonly detected the broad stretching frequency of the C=O bond in carbonate ions.

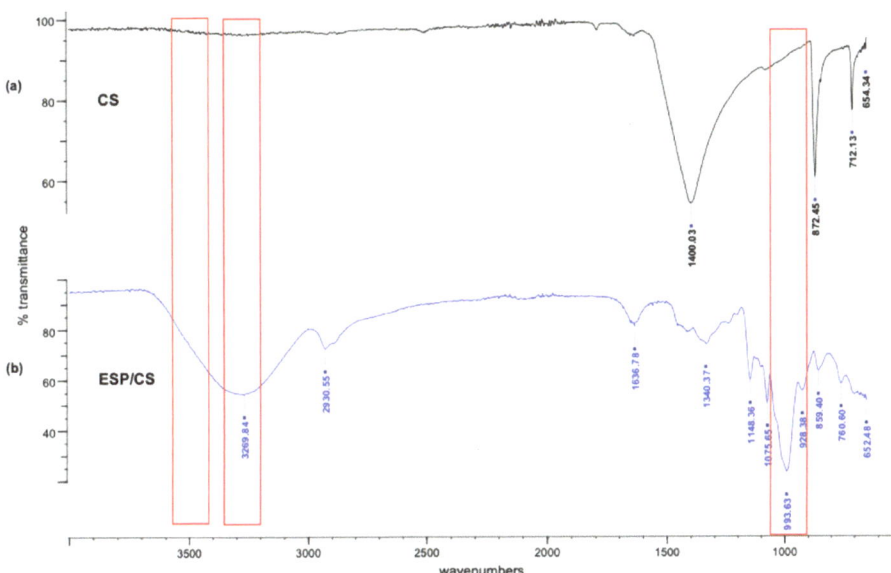

Figure 3. FTIR spectra of (**a**) ESP and (**b**) CS films.

Figure 3b shows the characteristic peaks of CS at 3269.84 cm^{-1} (O–H stretching), 2930.55 cm^{-1} (CH_2 group), 1636.78 cm^{-1} (CH–O–H group), and 1075.65 cm^{-1} (C–O stretching) [53]. The absorption band occurred in CS at approximately 652.48 cm^{-1}, reflecting the typical stretching vibrations of C-C skeletons. The peak and strong absorption band at 3269.84 cm^{-1} were attributed to the presence of functional groups containing O-H and N-H stretching of starch and carboxylic acid (-COOH) groups at 1636.78 cm^{-1} and 1075.65 cm^{-1} wavelengths, respectively, which has a significant effect on the process of adsorption and the efficiency of adsorbing target analytes via ion exchange [54,55]. Moreover, a strong band observed at 1636.78 cm^{-1} was attributed to the water adsorbed in the amorphous amylose region of starch [56]. These results indicated an increase in the number of hydrogen bonds between the starch molecules and the ESP particles, which led to the vibrational frequency of the pyranoid ring skeleton decreasing and the absorption bands shifting to a lower wavenumber. It showed that the addition of ESP promoted the formation of hydrogen bonds and reduced the number of active sites for water adsorption, which decreased the free space in the film network and strengthened the compact structure of composite films [57]. Thus, the water and oxygen resistance abilities of the ESP/CS films improved.

The crystalline structure of starch granules consists of the ordered crystalline region and the disordered amorphous region. X-ray diffraction (XRD) patterns were used to evaluate the amorphous crystalline structure characterized by sharp peaks related to crystalline diffraction and an amorphous zone [58]. The wide-angle XRD patterns of the ESP/CS and CS films with the presence and absence of the particles in between $10°(2\theta)$ and $50°(2\theta)$ are shown in Figure 4. A new strong characteristic reflection appearing at $2\theta = 29.4°$ can be seen from the XRD patterns of the ESP/CS films, which led to the semi-crystalline structure of the composite films. In addition, the greater intensity of the diffraction peaks at $2\theta = 29.4°$ may be related to the amount of ESP, attributable to the calcite crystal and agglomerates of ESP. Also, the peak broadening is caused by small crystal size. This indicated that the ESP was uniformly dispersed in the CS matrix and built a strong interaction with CS [34]. However, the XRD pattern shown in Figure 4 was similar to the commercial calcium carbonate pattern exhibited by previous research by Ji et al. [59], where it showed that the main component of eggshell powder was 95% $CaCO_3$. The peak of the ESP/CS film at 8.5° disappeared in the nanocomposite films, which led to the excellent compatibility between the CS and $CaCO_3$ particles. As expected, they modified the peak intensity of the films, where the ESP/CS film produced slightly higher peak intensities than the CS film, resulting in the greater strength of CS films.

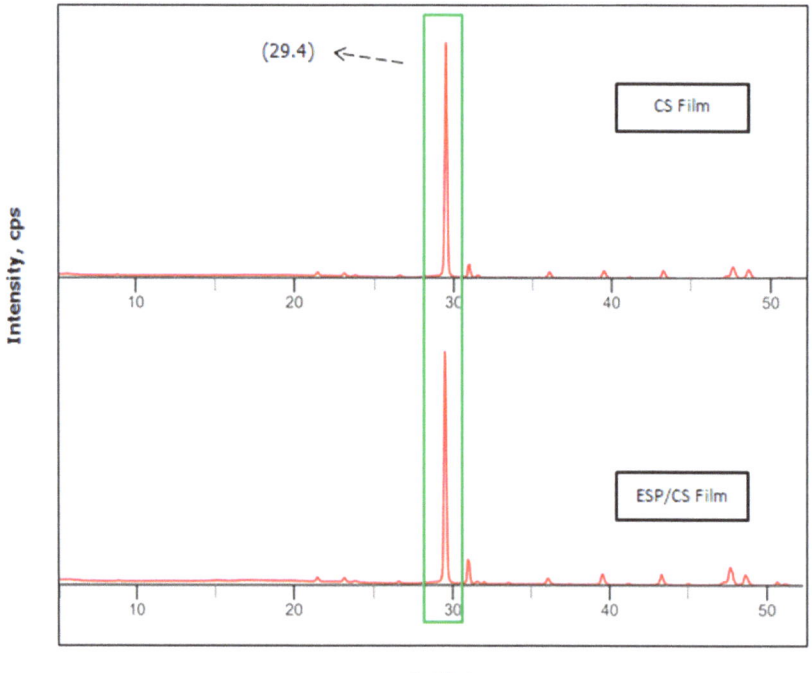

Figure 4. The wide-angle XRD patterns of the CS and ESP/CS films.

4. Conclusions

In this research, the ESP/CS film was prepared by a solution-casting method. The ESP particles were uniformly dispersed and embedded in the film matrix according to SEM analysis. In addition, SEM analysis of ESP/CS films showed a homogeneous and smooth structure with no cracks and visible phase separation. The image from TEM revealed irregular shapes of particles, which explains their larger surface area. The integration of ESP enhanced the thickness, density, water absorption, and swelling power in the film when compared to CS film. Moisture content, water solubility, and swelling power of ESP is lower than CS. The storage stability of ESP is higher in room temperature and in an incubator. Furthermore, FTIR analysis revealed that the addition of ESP aided the formation of hydrogen bonds and strengthened the compact structure of composite films, as well as the potential to produce a strong absorption band, which can be strongly associated with the presence of carbonate minerals. XRD analysis showed the semi-crystalline structure of the composite film and that the peak intensity of the ESP/CS film is higher than the CS film, which proves that ESP built strong interactions with CS. Finally, as a biofiller, ESP can improve the mechanical strength, hardness, and barrier properties of films without using cross-linking agents. This ESP/CS film is suitable to be used for the adsorption of target analytes, such as contaminants that are present in food products.

Author Contributions: J.M.V. and R.A.A. drafted the manuscript and collected the data. K.R. developed the framework, finalized, and supervised the research. J.M.V., R.A.A., N.H., K.H.E., M.N.N., W.X.L.F. and N.F.A.H. conceived the study and participated in its design and coordination. All authors have read and agreed to the published version of the manuscript.

Funding: This research was funded by the Universiti Malaysia Sabah, grant number PHD0024-2019.

Institutional Review Board Statement: Not applicable.

Informed Consent Statement: Not applicable.

Data Availability Statement: Not applicable.

Acknowledgments: The authors would like to thank all the researchers involved in this project.

Conflicts of Interest: The authors declare no conflict of interest.

Nomenclature

%	Percent
°C	Degrees Celsius
°	Degree
ø	Theta
μm	Micrometer
e	Thickness
ρ	Density
cm	Centimeter
g	Gram
mA	Milliampere
m	Meter
mm	Millimeter
nm	Nanometer
Kv	Kilovolt

Abbreviations

ANOVA	Analysis of variance
BEP	Bentonite–eggshell composites
EDX	Energy dispersive X-ray
ES	Eggshell
ESP	Eggshell powder
FTIR	Fourier Transform Infrared spectroscopy
MC	Moisture content
SEM	Scanning electron microscopy
SP	Swelling power
TEM	Transmission electron microscopy
WA	Water adsorption
WS	Water solubility

References

1. Onwubu, S.C.; Vahed, A.; Singh, S.; Kanny, K. Physicochemical Characterization of a Dental Eggshell Powder Abrasive Material. *J. Appl. Biomater. Funct. Mater.* **2017**, *15*, e341–e346. [CrossRef] [PubMed]
2. Sethupathi, S.; Kai, Y.C.; Kong, L.L.; Munusamy, Y.; Bashir, M.J.K.; Iberahim, N. Preliminary study of sulfur dioxide removal using calcined egg shell. *Malays. J. Anal. Sci.* **2017**, *21*, 719–725. [CrossRef]
3. Liu, Z.; Song, L.; Zhang, F.; He, W.; Linhardt, R. Characteristics of global organic matrix in normal and pimpled chicken eggshells. *Poult. Sci.* **2017**, *96*, 3775–3784. [CrossRef] [PubMed]
4. Jacob, J.J.; Varalakshmi, R.; Gargi, S.; Jayasri, M.A.; Suthindhiran, K. Removal of Cr (III) and Ni (II) from tannery effluent using calcium carbonate coated bacterial magnetosomes. *NPJ Clean Water* **2018**, *1*, 1–10. [CrossRef]
5. Cree, D.; Rutter, A. Sustainable Bio-Inspired Limestone Eggshell Powder for Potential Industrialized Applications. *ACS Sustain. Chem. Eng.* **2015**, *3*, 941–949. [CrossRef]
6. Betancourt, N.G.; Cree, D.E. Mechanical Properties of Poly (lactic acid) Composites Reinforced with $CaCO_3$ Eggshell Based Fillers. *MRS Adv.* **2017**, *2*, 2545–2550. [CrossRef]
7. Pirsa, S.; Sharifi, K.A. A review of the applications of bioproteins in the preparation of biodegradable films and polymers. *J. Chem. Lett.* **2020**, *1*, 47–58. [CrossRef]
8. Jabraili, A.; Pirsa, S.; Pirouzifard, M.K.; Amiri, S. Biodegradable Nanocomposite Film Based on Gluten/Silica/Calcium Chloride: Physicochemical Properties and Bioactive Compounds Extraction Capacity. *J. Polym. Environ.* **2021**, *29*, 2557–2571. [CrossRef]
9. Hosseini, S.N.; Pirsa, S.; Farzi, J. Biodegradable nano composite film based on modified starch-albumin/MgO; antibacterial, antioxidant and structural properties. *Polym. Test.* **2021**, *97*, 107182. [CrossRef]
10. García, A.V.; Álvarez-Pérez, O.B.; Rojas, R.; Aguilar, C.N.; Garrigós, M.C. Impact of Olive Extract Addition on Corn Starch-Based Active Edible Films Properties for Food Packaging Applications. *Foods* **2020**, *9*, 1339. [CrossRef]
11. Abotbina, W.; Sapuan, S.M.; Sultan, M.T.H.; Alkbir, M.F.M.; Ilyas, R.A. Development and characterization of cornstarch-based bioplastics packaging film using a combination of different plasticizers. *Polymers* **2021**, *13*, 3487. [CrossRef] [PubMed]
12. Chen, X.; Guo, L.; Du, X.; Chen, P.; Ji, Y.; Hao, H.; Xu, X. Investigation of glycerol concentration on corn starch morphologies and gelatinization behaviours during heat treatment. *Carbohydr. Polym.* **2017**, *176*, 56–64. [CrossRef] [PubMed]
13. Wang, K.; Wang, W.H.; Ye, R.; Liu, A.J.; Xiao, J.D.; Liu, Y.W.; Zhao, Y.N. Mechanical properties and solubility in water of corn starch-collagen composite films: Effect of starch type and concentrations. *Food Chem.* **2017**, *216*, 209–216. [CrossRef]
14. Sudheesh, C.; Sunooj, K.V.; Sasidharan, A.; Sabu, S.; Basheer, A.; Navaf, M.; Raghavender, C.; Sinha, S.; George, J. Energetic neutral N2 atoms treatment on the kithul (*Caryota urens*) starch biodegradable film: Physico-chemical characterization. *Food Hydrocoll.* **2020**, *103*, 105650. [CrossRef]
15. Seligra, P.G.; Jaramillo, C.M.; Famá, L.; Goyanes, S. Data of thermal degradation and dynamic mechanical properties of starch–glycerol based films with citric acid as crosslinking agent. *Data Briefs* **2016**, *7*, 1331–1334. [CrossRef] [PubMed]
16. Song, X.; Zuo, G.; Chen, F. Effect of essential oil and surfactant on the physical and antimicrobial properties of corn and wheat starch films. *Int. J. Biol. Macromol.* **2018**, *107*, 1302–1309. [CrossRef] [PubMed]
17. Li, L.; Chen, H.; Wang, M.; Lv, X.; Zhao, Y.; Xia, L. Development and characterization of irradiated-corn-starch films. *Carbohydr. Polym.* **2018**, *194*, 395–400. [CrossRef]
18. Malmir, S.; Montero, B.; Rico, M.; Barral, L.; Bouza, R.; Farrag, Y. Effects of poly (3-hydroxybutyrate-co-3-hydroxyvalerate) microparticles on morphological, mechanical, thermal, and barrier properties in thermoplastic potato starch films. *Carbohydr. Polym.* **2018**, *194*, 357–364. [CrossRef]
19. Edhirej, A.; Sapuan, S.M.; Jawaid, M.; Zahari, N.I. Effect of various plasticizers and concentration on the physical, thermal, mechanical, and structural properties of cassava-starch-based films. *Starch-Stärke* **2017**, *69*, 1500366. [CrossRef]
20. Jumaidin, R.; Sapuan, S.; Jawaid, M.; Ishak, M.; Sahari, J. Characteristics of thermoplastic sugar palm Starch/Agar blend: Thermal, tensile, and physical properties. *Int. J. Biol. Macromol.* **2016**, *89*, 575–581. [CrossRef]

21. Saberi, B.; Vuong, Q.; Chockchaisawasdee, S.; Golding, J.; Scarlett, C.J.; Stathopoulos, C. Physical, Barrier, and Antioxidant Properties of Pea Starch-Guar Gum Biocomposite Edible Films by Incorporation of Natural Plant Extracts. *Food Bioproc. Technol.* **2017**, *10*, 2240–2250. [CrossRef]
22. Wang, G.; Liu, N.; Zhang, S.; Zhu, J.; Xiao, H.; Ding, C. Preparation and application of granular bentonite-eggshell composites for heavy metal removal. *J. Porous Mater.* **2022**, *29*, 817–826. [CrossRef]
23. Harripersadth, C.; Musonge, P.; Isa, Y.M.; Morales, M.G.; Sayago, A. The application of eggshells and sugarcane bagasse as potential biomaterials in the removal of heavy metals from aqueous solutions. *S. Afr. J. Chem. Eng.* **2020**, *34*, 142–150. [CrossRef]
24. Mostafavi, S.; Rezaverdinejad, V.; Pirsa, S. Design and fabrication of nanocomposite-based polyurethane filter for improving municipal waste water quality and removing organic pollutants. *Adsorpt. Sci. Technol.* **2019**, *37*, 95–112. [CrossRef]
25. Wu, H.; Xiao, D.; Lu, J.; Li, T.; Jiao, C.; Li, S.; Lu, P.; Zhang, Z. Preparation and Properties of Biocomposite Films Based on Poly(vinyl alcohol) Incorporated with Eggshell Powder as a Biological Filler. *J. Polym. Environ.* **2020**, *28*, 2020–2028. [CrossRef]
26. Shukla, S.K.; Al Mushaiqri, N.R.S.; Al Subhi, H.M.; Yoo, K.; Al Sadeq, H. Low-cost activated carbon production from organic waste and its utilization for wastewater treatment. *Appl. Water Sci.* **2020**, *10*, 62. [CrossRef]
27. Hajizadeh, H.; Peighambardoust, S.J.; Peressini, D. Physical, mechanical, and antibacterial characteristics of bio-nanocomposite films loaded with Ag-modified SiO_2 and TiO_2 nanoparticles. *J. Food Sci.* **2020**, *85*, 1193–1202. [CrossRef]
28. Ploypetchara, T.; Gohtani, S. Change in characteristics of film based on rice starch blended with sucrose, maltose, and trehalose after storage. *J. Food Sci.* **2020**, *85*, 1470–1478. [CrossRef]
29. Herniou–Julien, C.; Mendieta, J.R.; Gutiérrez, T.J. Characterization of biodegradable/non-compostable films made from cellulose acetate/corn starch blends processed under reactive extrusion conditions. *Food Hydrocoll.* **2018**, *89*, 67–79. [CrossRef]
30. Chong, K.-Y.; Chia, C.-H.; Zakaria, S. Polymorphs calcium carbonate on temperature reaction. *AIP Conf. Proc.* **2014**, *1614*, 52–56. [CrossRef]
31. Weiss, C.A.; Torres-Cancel, K.; Moser, R.D.; Allison, P.G.; Gore, E.R.; Chandler, M.Q.; Malone, P.G. Influence of temperature on calcium carbonate polymorph formed from ammonium carbonate and calcium acetate. *J. Nanotech. Smart Mater.* **2014**, *1*, 1–6. [CrossRef]
32. Sankaran, R.; Show, P.L.; Ooi, C.-W.; Ling, T.C.; Shu-Jen, C.; Chen, S.-Y.; Chang, Y.-K. Feasibility assessment of removal of heavy metals and soluble microbial products from aqueous solutions using eggshell wastes. *Clean Technol. Environ. Policy* **2020**, *22*, 773–786. [CrossRef]
33. Sun, Q.; Xi, T.; Li, Y.; Xiong, L. Characterization of Corn Starch Films Reinforced with $CaCO_3$ Nanoparticles. *PLoS ONE* **2014**, *9*, e106727. [CrossRef] [PubMed]
34. Jiang, B.; Li, S.; Wu, Y.; Song, J.; Chen, S.; Li, X.; Sun, H. Preparation and characterization of natural corn starch-based composite films reinforced by eggshell powder. *CyTA J. Food* **2018**, *16*, 1045–1054. [CrossRef]
35. King'ori, A. A Review of the Uses of Poultry Eggshells and Shell Membranes. *Int. J. Poult. Sci.* **2011**, *10*, 908–912. [CrossRef]
36. Zhang, B.; Cui, D.; Liu, M.; Gong, H.; Huang, Y.; Han, F. Corn porous starch: Preparation, characterization and adsorption property. *Int. J. Biol. Macromol.* **2012**, *50*, 250–256. [CrossRef]
37. Gao, J.; Sun, S.-P.; Zhu, W.-P.; Chung, T.-S. Chelating polymer modified P84 nanofiltration (NF) hollow fiber membranes for high efficient heavy metal removal. *Water Res.* **2014**, *63*, 252–261. [CrossRef]
38. Hassan, S.; Aigbodion, V. Effects of eggshell on the microstructures and properties of Al–Cu–Mg/eggshell particulate composites. *J. King Saud Univ. Eng. Sci.* **2015**, *27*, 49–56. [CrossRef]
39. Minakshi, M.; Higley, S.; Baur, C.; Mitchell, D.R.G.; Jones, R.T.; Fichtner, M. Calcined chicken eggshell electrode for battery and supercapacitor applications. *RSC Adv.* **2019**, *9*, 26981–26995. [CrossRef]
40. Onwubu, S.C.; Mdluli, P.S.; Singh, S.; Tlapana, T. A novel application of nano eggshell/titanium dioxide composite on occluding dentine tubules: An in vitro study. *Braz. Oral Res.* **2019**, *33*, e016. [CrossRef]
41. Ponsanti, K.; Tangnorawich, B.; Ngernyuang, N.; Pechyen, C. A flower shape-green synthesis and characterization of silver nanoparticles (AgNPs) with different starch as a reducing agent. *J. Mater. Res. Technol.* **2020**, *9*, 11003–11012. [CrossRef]
42. Yin, P.; Liu, J.; Zhou, W.; Li, P. Preparation and Properties of Corn Starch/Chitin Composite Films Cross-Linked by Maleic Anhydride. *Polymers* **2020**, *12*, 1606. [CrossRef] [PubMed]
43. Jalu, R.G.; Chamada, T.A.; Kasirajan, R. Calcium oxide nanoparticles synthesis from hen eggshells for removal of lead (Pb(II)) from aqueous solution. *Environ. Chall.* **2021**, *4*, 100193. [CrossRef]
44. da Rosa Zavareze, E.; Pinto, V.Z.; Klein, B.; El Halal, S.L.M.; Elias, M.C.; Prentice-Hernández, C.; Dias, A.R.G. Development of oxidised and heat–moisture treated potato starch film. *Food Chem.* **2012**, *132*, 344–350. [CrossRef]
45. Halimatul, M.; Sapuan, S.; Jawaid, M. Water absorption and water solubility properties of sago starch biopolymer composite films filled with sugar palm particles. *Polimery* **2019**, *64*, 596–604. [CrossRef]
46. Ahmad, M.; Hani, N.M.; Nirmal, N.; Fazial, F.F.; Mohtar, N.F.; Romli, S.R. Optical and thermo-mechanical properties of composite films based on fish gelatin/rice flour fabricated by casting technique. *Prog. Org. Coat.* **2015**, *84*, 115–127. [CrossRef]
47. Farahana, R.N.; Supri, A.G.; Teh, P.I. Tensile and water absorption properties of eggshell powder filled recycled high-density polyethylene/ethylene vinyl acetate composites: Effect of 3-aminopropyltriethoxysilane. *Adv. Mat. Res.* **2015**, *5*, 1–9. [CrossRef]
48. Valcárcel-Yamani, B.; Rondan-Sanabria, G.G.; Finardi-Filho, F. The physical, chemical and functional characterization of starches from Andean tubers: Oca (*Oxalis tuberosa* Molina), olluco (*Ullucus tuberosus* Caldas) and mashua (*Tropaeolum tuberosum* Ruiz & Pavón). *Braz. J. Pharm. Sci.* **2013**, *49*, 453–464. [CrossRef]

49. Ket-On, A.; Pongmongkol, N.; Somwangthanaroj, A.; Janjarasskul, T.; Tananuwong, K. Properties and storage stability of whey protein edible film with spice powders. *J. Food Sci. Technol.* **2016**, *53*, 2933–2942. [CrossRef]
50. Ji, M.; Li, F.; Li, J.; Li, J.; Zhang, C.; Sun, K.; Guo, Z. Enhanced mechanical properties, water resistance, thermal stability, and biodegradation of the starch-sisal fibre composites with various fillers. *Mater. Des.* **2021**, *198*, 109373. [CrossRef]
51. Lyu, J.S.; Lee, J.-S.; Han, J. Development of a biodegradable polycaprolactone film incorporated with an antimicrobial agent via an extrusion process. *Sci. Rep.* **2019**, *9*, 1–11. [CrossRef] [PubMed]
52. Rabadán, A.; Álvarez-Ortí, M.; Pardo, J.E.; Alvarruiz, A. Storage stability and composition changes of three cold-pressed nut oils under refrigeration and room temperature conditions. *Food Chem.* **2018**, *259*, 31–35. [CrossRef] [PubMed]
53. Tizo, M.S.; Blanco, L.A.V.; Cagas, A.C.Q.; Cruz, B.R.B.D.; Encoy, J.C.; Gunting, J.V.; Arazo, R.O.; Mabayo, V.I.F. Efficiency of calcium carbonate from eggshells as an adsorbent for cadmium removal in aqueous solution. *Sustain. Environ. Res.* **2018**, *28*, 326–332. [CrossRef]
54. Al-Senani, G.M.; Al-Fawzan, F.F. Adsorption study of heavy metal ions from aqueous solution by nanoparticle of wild herbs. *Egypt. J. Aquat. Res.* **2018**, *44*, 187–194. [CrossRef]
55. Nacke, H.; Gonçalves, D.S.M.; Campagnolo, M.A.; Coelho, G.F.; Schwantes, D.; Dos Santos, M.G.; Briesch, D.L.; Zimmermann, J. Adsorption of Cu (II) and Zn (II) from Water by Jatropha curcas L. as Biosorbent. *Open Chem. J.* **2016**, *14*, 103–117. [CrossRef]
56. Salaheldin, H.I. Optimizing the synthesis conditions of silver nanoparticles using corn starch and their catalytic reduction of 4-nitrophenol. *Adv. Nat. Sci. Nanosci. Nanotechnol.* **2018**, *9*, 025013. [CrossRef]
57. Oleyaei, S.A.; Almasi, H.; Ghanbarzadeh, B.; Moayedi, A.A. Synergistic reinforcing effect of TiO_2 and montmorillonite on potato starch nanocomposite films: Thermal, mechanical and barrier properties. *Carbohydr. Polym.* **2016**, *152*, 253–262. [CrossRef]
58. Li, C.; Zhu, W.; Xue, H.; Chen, Z.; Chen, Y.; Wang, X. Physical and structural properties of peanut protein isolate-gum Arabic films prepared by various glycation time. *Food Hydrocoll.* **2015**, *43*, 322–328. [CrossRef]
59. Ji, G.; Zhu, H.; Qi, C.; Zeng, M. Mechanism of interactions of eggshell microparticles with epoxy resins. *Polym. Eng. Sci.* **2009**, *49*, 1383–1388. [CrossRef]

Review

Alginate-Based Bio-Composites and Their Potential Applications

Khmais Zdiri [1,2,*], Aurélie Cayla [1], Adel Elamri [3], Annaëlle Erard [1] and Fabien Salaun [1]

[1] Laboratoire de Génie et Matériaux Textiles, École Nationale Supérieure des Arts et Industries Textiles, Université de Lille, 59000 Lille, France
[2] Laboratoire de Physique et Mécanique Textiles, École Nationale Supérieure d'Ingénieurs Sud-Alsace, Université de Haute Alsace, EA 4365, 68100 Mulhouse, France
[3] Unité de Recherche Matériaux et Procédés Textiles, École Nationale d'Ingénieurs de Monastir, Université de Monastir, UR17ES33, Monastir 5019, Tunisia
* Correspondence: khmaiszdiri@gmail.com

Abstract: Over the last two decades, bio-polymer fibers have attracted attention for their uses in gene therapy, tissue engineering, wound-healing, and controlled drug delivery. The most commonly used bio-polymers are bio-sourced synthetic polymers such as poly (glycolic acid), poly (lactic acid), poly (e-caprolactone), copolymers of polyglycolide and poly (3-hydroxybutyrate), and natural polymers such as chitosan, soy protein, and alginate. Among all of the bio-polymer fibers, alginate is endowed with its ease of sol–gel transformation, remarkable ion exchange properties, and acid stability. Blending alginate fibers with a wide range of other materials has certainly opened many new opportunities for applications. This paper presents an overview on the modification of alginate fibers with nano-particles, adhesive peptides, and natural or synthetic polymers, in order to enhance their properties. The application of alginate fibers in several areas such as cosmetics, sensors, drug delivery, tissue engineering, and water treatment are investigated. The first section is a brief theoretical background regarding the definition, the source, and the structure of alginate. The second part deals with the physico-chemical, structural, and biological properties of alginate bio-polymers. The third part presents the spinning techniques and the effects of the process and solution parameters on the thermo-mechanical and physico-chemical properties of alginate fibers. Then, the fourth part presents the additives used as fillers in order to improve the properties of alginate fibers. Finally, the last section covers the practical applications of alginate composite fibers.

Keywords: alginate; fiber; thermo-mechanical; physico-chemical; wound-healing

1. Introduction

There is a great variety of bio-polymers, including the family of polysaccharides such as sodium alginate, which are produced from marine products. These bio-polymers constitute an interesting alternative for replacing polymers derived from petrochemicals, as they have important physico-chemical and biological properties [1,2]. Alginate bio-polymers can form a reticulated structure when they link with chloride ions, Na^+, or calcium ions, Ca^{2+} [3]. These properties make these bio-polymers very useful materials in various fields, including water treatment, packaging, textiles, agriculture, pharmaceuticals, electronics, and the biomedical field [4,5].

This bio-polymer has been exploited in several forms. Among these forms, fibers, nano-fibers, micro-spheres, micro-particles, and nano-particles have been studied. Many authors have spun alginate fibers using a variety of methods: electro-spinning, the microfluidic system, and the traditional wet-spinning technique [6]. Wet-spinning is a technique that can be easily scaled up and is commonly used in high-performance fiber industries to create materials such as aramid fibers. This spinning technique is a higher throughput, more viable and cost-effective technique than the electro-spinning method [7].

However, there are limitations for the use of calcium alginate fibers. Indeed, alginate fibers have a low mechanical strength and poor thermal stability [8]. Additionally, the contact of alginate with the physiological environment makes this bio-polymer inappropriate for uses related to load-bearing body parts [9]. The incorporation of adhesive peptide and natural or synthetic polymers into alginate fibers gives them their desired properties and allows them to be used in innovative processes. Indeed, alginate fiber composites have shown a strong development potential, particularly in biological, medical, electronics, and food packaging applications [10].

In the present review, the structure and specific properties of alginate are described, followed by the discussion of the different strategies of alginate spinning and the impact of the process and solution parameters on the properties of these fibers. A comprehensive study of literature works on the incorporation of a wide range of materials into alginate fibers is carried out. Additionally, the exploitation and exploration of bio-composite fibers of alginate in several applications are discussed. This review can be used as a reference for researchers and industrials in their potential works.

2. Definition, Source, and Structure of Alginate

Alginate is a relatively abundant natural polysaccharide, since it is the structural component of brown algae. It is mainly generated from *Laminaria hyperborea*, *Laminaria digitata*, *Laminaria japonica*, *Ascophyllum*, and *Macrocystis pyrifera* [11].

Nowadays, this polysaccharide is used in many sectors, thinks to its unique colloidal properties, which means that it can be used as a thickener, stabilizer, film-forming agent, gelling agent, etc. [12].

The extraction of alginate from algae is based on the water solubility of this polymer. Alginic acid is insoluble in water, whereas sodium Na^+ or potassium K^+ salts are soluble. It is therefore necessary to convert all alginate salts (calcium and potassium) into sodium alginate salts using sodium carbonate. The main steps of the alginate extraction process are [13,14]:

- ✓ Pre-treatment: the algae are washed several times with water and then rinsed with distilled water in order to remove any impurities. The algae are then dried and finely crushed.
- ✓ Purification: the seaweed powder is treated with a dilute solution of acid, capable of dissolving sugars. This causes the formation of alginic acid.
- ✓ Extraction: the alginic acid is redissolved in a slightly basic solution of sodium carbonate $NaHCO_3$ (concentration 1.5%) at temperatures over a range of 50–90 °C for 1–2 h. This converts the alginic acid into sodium alginate.
- ✓ Recovery: The sodium alginate, which is not soluble in a mixture of alcohol and water, can be separated from the system. Indeed, the sodium alginate solution is then filtered and an addition of ethanol allows for the precipitation of the alginate. The process ends with a drying and grinding step to obtain a powder with the appropriate particle size.

Alginates have specific characteristics that differ from other natural polymers [15]. They are mainly composed of two uronic acid residues: β-D-mannuronic acid (M block) and α-L-guluronic acid (G block). These two acid units are linked by β-1,4 glycosidic bonds [16] according to a sequence of three types of blocks. The latter can be homo-polymer blocks (-M-M-M- or -G-G-G-) or co-polymer blocks (-M-G-M-) (Figure 1).

It is important to note that the distribution (poly-dispersity) and the length of the blocks depend on the species and the part of the algae from which the alginic acid is extracted.

Alginate is characterized by several properties such as low cost, non-toxicity, bio-compatibility, bio-degradability, liquid absorption capacity, non-immunogenicity, abundance of availability in nature, and ease of chemical derivatization. These properties make alginate one of the most commonly used bio-polymers over a wide application area, such as a thickener in many food products, pharmaceutical formulations, tissue engineering, cell culture, and textile printing agents [17].

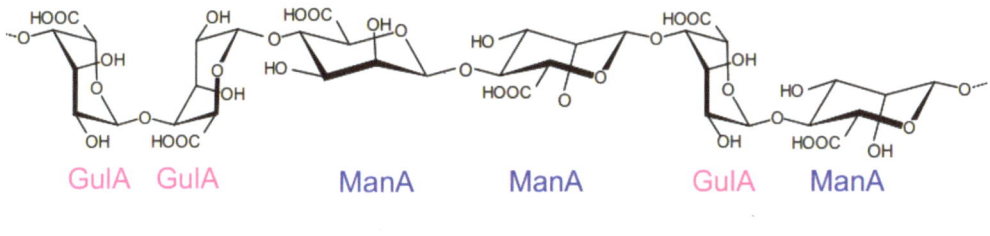

Figure 1. Alginate structure.

3. Properties of Alginate

3.1. Physico-Chemical Behaviors of Alginates

Pure alginic acid is insoluble in water. Its solubility or not in water depends on the type of salts associated with it. The sodium, ammonium, and potassium salts dissolve perfectly in aqueous solution, giving highly viscous solutions. In contrast, the magnesium and calcium salts are insoluble.

Sodium alginate is soluble in water at low ionic strength values. The solubilization of the polyanion becomes slower with an increase in the salinity of the medium.

The pH of the solution and the ionic strength of the solvent play an important role in the solubility of the alginate. Indeed, phase separation or even the formation of a hydrogel may occur when the pH of the solution is below the pKa of guluronic acid (pKa = 3.65) and mannuronic acid (pKa = 3.38) [18]. The rate of dissolution of the alginate decreases with an increase in the ionic strength [19].

Sodium alginate can have a stability of several months, and can be stored in a dry and cool place away from sunlight. At low temperatures, sodium alginate can be stored for several years without a significant reduction in its molecular weight. On the other hand, dry alginic acid has a very limited stability at ordinary temperatures due to its intramolecular degradation [20].

In the face of their multiple uses, it is important to be aware of the factors that determine and limit the stability of aqueous alginate solutions, and the chemical reactions responsible for their degradation.

The relative viscosity of an alginate solution can be severely reduced over a short period of time under conditions favoring degradation.

Alginates are increasingly being used because of their ability to form gels, and more precisely, hydrogels. There are two types of hydrogels: chemical hydrogels and physical hydrogels.

The nodes of the network are covalent bonds, in the case of chemical hydrogels. This type of hydrogel is irreversible, due to the irreversible nature of the covalent bond [21]. Physical hydrogels are formed by chains entanglements, hydrogen bonds, ionic bonds, and van der Waals interactions. These physical interactions have low energy, allowing the reversibility of the hydrogels [22].

Alginates have the ability to form a physical hydrogel through inter-chain interactions [23]. This interaction is described by the egg-box model, in which each divalent ion can interact with two adjacent units, G, or those belonging to two opposite chains [24].

The salt of the divalent cation that is generally used for alginate gelling is calcium chloride, because of its good solubility in aqueous medium, and the high availability of the calcium ions. The latter are retained in a kind of cage cavity, and interact with the carboxylate functions and oxygen atoms of the hydroxyl functions (Figure 2) [25,26].

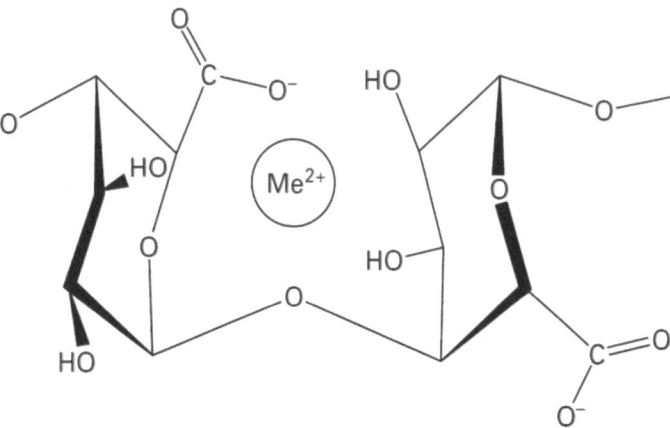

Figure 2. The stereo-chemical structure of the GG block [25].

The properties of alginate gel formation are influenced by the molecular weight, divalent ions, and chemical structure.

$$Me^{2+} = Ca^{2+}, Zn^{2+}, Cu^{2+}$$

3.2. Biological Properties of Alginate

A biocompatible material is one that is capable of not degrading the biological environment in which it is used [27]. The alginate composition has a very important role on its biocompatibility properties. In fact, alginates with a high M content have been reported to be immunogenic and more potent in inducing cytokine production compared to alginates with a high G content.

The immunogenic response during an alginate injection or at the implantation site could be attributed to the various impurities remaining in the alginate during its extraction. Nevertheless, the studies of Orive et al. [28], and Jangwook and Kuen Yong Lee [29] have shown that the purification of alginate did not induce an immunogenic response in animals.

Moreover, this natural substance has the advantage of having a hemostasis property. The exchange between sodium Na^+ ions present in the wound exudate and calcium Ca^{2+} ions will form a non-adherent hydrophilic gel that fills the wound and creates a microclimate favorable for healing [30].

Additionally, calcium alginate has healing properties. It has been described that alginate fibers are able to form a gel, which allows for the creation of a moist microenvironment favorable to the scarring process. Therefore, it has a high absorption capacity, superior to those of hydro-colloids and hydro-cellulars [31,32]. It has been shown that alginates rich in mannuronic units have positive impacts on cicatrization, while other authors believe that this activity also depends on the level of purification of the alginate [33,34].

The high water absorption capacity of alginate confirms its use at all wound-healing stages for highly exudative wounds [35].

The intra-fiber absorption of alginate allows for the textile support used for wound-healing, to limit the risk of wound infection. Indeed, the swelling of the fibers via the absorption of fluids facilitates the solubilization of the alginate, which makes it possible to immobilize the bacteria present on the textile.

The absorption of wound fluids by the calcium alginate compress allows for the formation of a highly viscous solution via ion exchange between the calcium of the alginate and the sodium present in the blood [36]. This moist microenvironment promotes healing.

4. Production of Alginate Fibers

4.1. Process

Fibers of alginate have attracted extensive interest because of their ease of handling, high surface area, and its ability to retain mechanical integrity [37].

The melt spinning process is not adopted for alginates. In fact, the latter are characterized by several hydrogen bonds, responsible for a glass transition and a melting temperature above their thermal decomposition temperature. Alginate fibers and alginate composite fibers are elaborated using several techniques: the wet-spinning method, electro-spinning, and the micro-fluidic system.

4.1.1. Fibers Formation via Wet-Spinning

The most commonly used spinning technique for alginates is wet-spinning (Figure 3) [38].

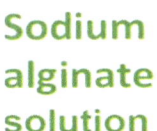
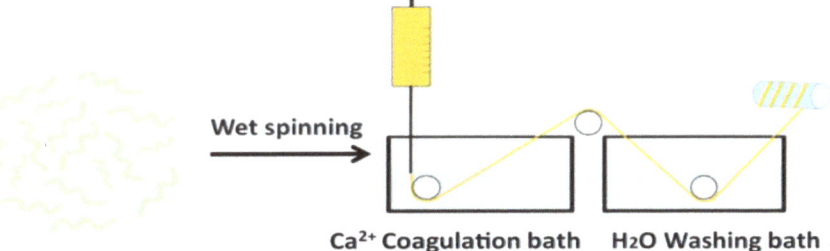

Figure 3. Schematization of the wet-spinning process for preparing alginate fibers.

The alginate powder can be dissolved in distilled water. After degassing, the sodium alginate solution can be extruded through a die immersed into a coagulation bath containing calcium chloride ($CaCl_2$) solution, to cross-link the alginate.

An interaction is performed between the Ca^{2+} cations and the negatively charged molecule of alginate, promoting the solidification of alginate networks (Figure 4) [39].

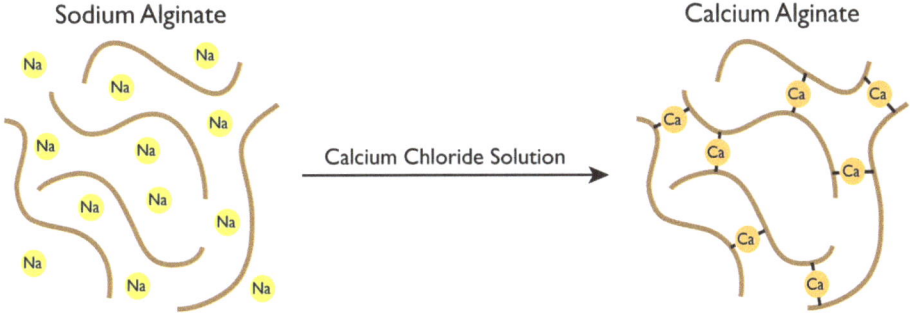

Figure 4. Gellification of sodium alginate solution.

After an appropriate immersion time, the calcium alginate filaments can be washed in a bath of demineralized water. These obtained filaments can then be stretched and dried at a temperature of 80 °C [40].

4.1.2. Fibers Formation via Electro-Spinning

Electro-spinning is a highly versatile spinning process from a technical point of view. Figure 5 represents the principle diagram of the electrostatic alginate spinning device. This device usually consists of a high-voltage supply system, a syringe with a metal needle, a syringe pump, and a target or collector system [41].

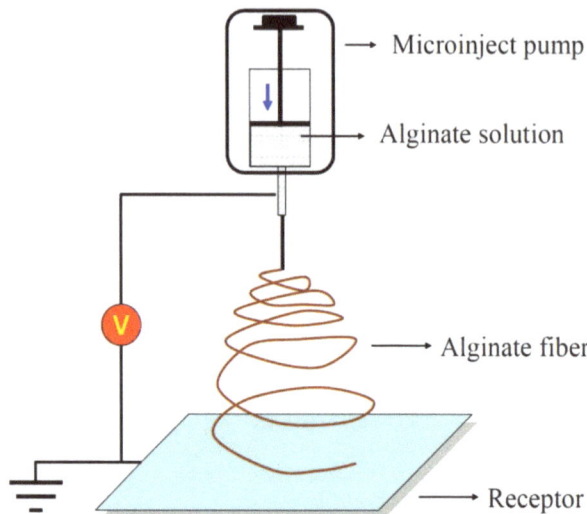

Figure 5. Illustration of the electro-spinning process for preparing alginate fibers.

Compared to the wet-spinning method, electro-spinning allows for the elaboration of ultrathin alginate fibers by placing a bio-polymer in between the charged electrode. When a strong electric field is applied, generally in the order of kilovolts (kV), the molten polymer is ejected towards the collector electrode. This allows for the formation of a fibrous mat consisting of individual fibers with nanometer- to micrometer-size diameters [42].

The electro-spinning technique of alginate has a large application in the biomedical domain because it can provide a larger specific surface area than the traditional wet-spinning technique [43].

However, it is difficult to manufacture continuous and uniform electro-spun fibers from neat alginate solutions using the method of electro-spinning, due to the rigid structure of the alginate and the lack of molecular entanglement.

4.1.3. Fibers Formation via the Microfluidic System

The microfluidic system gives precise control of the morphology and the dimensional characteristics of the elaborated fiber. This is attributed to the formation of a stable laminar flow during injection into the microchannel [44].

The pioneering work has produced alginate micro-fibers via the microfluidic system. They have based their work on the coaxial flow, whereby the sheath flow ($CaCl_2$) and the specimen flow (sodium alginate) meet at the intersection and leave the outlet in the form of continuous alginate micro-fibers [45]. This idea has been refined to make hollow alginate fibers by introducing a core fluid into the inner layer, resulting in the formation of a coaxial flow with three-layers: the sheath fluid, the specimen, and the core [46].

Compared to the wet-spinning and the electro-spinning methods, the microfluidic systems offer smoother and controllable micro-fiber manufacturing, making it possible to encapsulate cells within these fibers.

Figure 6 illustrates the concept of coaxial flow in the generation of continuous alginate micro-fibers [47,48].

4.2. Preparation of the Solutions

Alginate fibers are produced from alginate solution, extruded through a die immersed in a coagulation bath. Coagulation produces a physical hydrogel of alginate whose shape is determined by the extrusion die. The gel thus formed then undergoes various steps (drawing, drying, etc.) to form fibers.

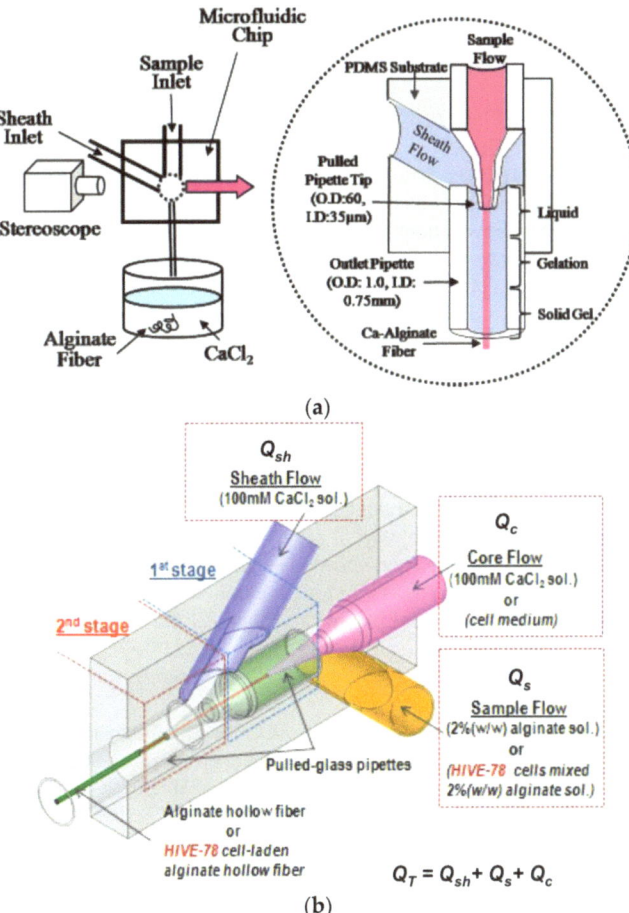

Figure 6. Microfluidic concept for (**a**) continuous alginate micro-fibers and (**b**) continuous alginate hollow micro-fibers.

Generally, alginate fibers are spun from an aqueous solution with a concentration between 4 and 10 wt% into a spinning bath. The content of the bath depends on the nature of these fibers to be spun. The coagulation bath usually contains calcium chloride, in the case of the preparation of fibers from calcium alginate. It is possible to use an acidic aqueous medium with calcium chloride and additives as a coagulation bath to improve the spinning process [49].

When spinning alginate acid fibers, the coagulation bath contains sodium sulfate and sulfuric acid. A 1–6 wt% sodium alginate solution can be extruded into a coagulation bath containing 0.2 M hydrochloric acid, depending on the desired viscosity [50]. After an appropriate immersion time, these alginate fibers are removed from the bath and washed in a water bath for approximately 20 s.

Calcium chloride ($CaCl_2$) is the most commonly used reagent that allows for the ionic cross-linking of sodium alginate. The ionic gelation induced by calcium chloride is poorly controlled, irreversible, and rapid [51].

The coagulation bath may also contain a combination of salts: (copper, magnesium, and zinc salts) [52]. In the work of Qianqian Wang et al. [53], zinc (Zn^{2+}), barium (Ba^{2+}), copper (Cu^{2+}), and aluminum (Al^{3+}) ions were mixed with calcium (Ca^{2+}) ions in the coagulation bath to improve the mechanical behaviors of these alginate fibers. A spinning solution was prepared by mixing 4 g of sodium alginate in 94 mL of distilled water. This solution was immersed in the coagulation bath, which contains the complex metal ions, calcium–zinc, calcium–aluminum, calcium–copper, and calcium–barium.

4.3. Effects of Process Parameters on Thermo-Mechanical and Physico-Chemical Properties

The structure of the fiber is highly dependent on the spinning conditions. In the case of wet-spinning, the low polymer concentration results in fibers that contain a large amount of solvent. The cross-sections and the structures of these fibers are influenced by the dehydration mechanism. In the work of B Niekraszewicz et al. [54], alginate fibers were fabricated using the wet-spinning technique. Morphological characterizations show that these elaborated fibers have porous structures and irregular cross-sections.

In their works, Su-Jung Shin et al. [55] elaborate continuous calcium alginate fibers with microfluidic devices. These authors investigated the variation of calcium alginate fiber diameters as a function of the sheath and sample flow rate. They showed that the fiber diameters increased almost linearly with increasing sample flow rates. The diameters of these elaborated fibers was regulated by changes in the sheath and sample flow rate.

Teresa Cuadros et al. [56] studied the impact of the residence time on the mechanical properties of alginate fibers. The concentrations of alginate and $CaCl_2$ were fixed to 2 wt% and 0.5 wt%, respectively. They indicated that these elaborated fibers rapidly reach gelation (or equilibrium) with the surrounding solution.

The extrusion speed also has an impact on the properties of alginate fibers [57]. The cross-head speeds inferior to 10 mm.m^{-1} allows for an erratic alginate stream. In addition, there are several bumps and kinks along the gel of the alginate fibers. These resulting fibers always break at these bumps and kinks. For speeds greater than 20 mm.m^{-1}, rapidly entangled coils are produced. Consequently, these authors used 15 mm.m^{-1} as the cross-head speed.

Magdalena Brzezińska et al. [58] studied the effect of the draw on the tenacity of alginate fibers. These authors showed that from a concentration of 12 wt%, alginate fibers resulted in a slightly higher tenacity (Table 1). It has also been shown that a drawing ratio of 100% gives a higher tenacity of around 15.54 cN·tex^{-1}. These results can be explained by the greater orientation of the alginate macromolecules in the coagulation bath.

Table 1. Conditions of fiber formation and alginate fiber properties.

Samples	Polymer Content (wt%)	Drawing Ration (%)	Total Drawing Ratio (%)	Tenacity (cN·tex^{-1})	Elongation at Break (%)
S1	12	50	70	14.66	5.37
S2	12	100	70	15.54	4.89
S3	13	50	74	14.39	4.2
S4	13	100	70	14.59	5.36

In another work, Lin H.Y. et al. [59] studied the impact of needle diameter and air pressure on the properties of the alginate fiber scaffold. They reported that from a pressure of 6 bar, the alginate solution was rapidly injected into the solution of the calcium. When the pressure has been decreased to 3 bar, the alginate solution flowed slowly into the calcium solution, allowing the formation of larger fibers. A slow injection speed allowed more time for the calcium ions to cross-link these alginate fibers.

In addition, it was shown that a larger alginate fiber was discovered when the diameter of the needle was augmented from 150 μm to 200 μm. A needle diameter of larger than 200 μm allowed the solution of alginate to seep out under the effect of gravity.

A higher water vapor transmission rate by alginate fibers was observed when increasing the air pressure and decreasing the needle size. These observations can be attributed to the enhanced adhesion between the contacting fibers, which generate a compact structure.

4.4. Effects of Solution Parameters on Thermo-Mechanical and Physico-Chemical Properties

Teresa Cuadros et al. [60] studied the effect of $CaCl_2$ and sodium alginate concentrations on calcium alginate fiber properties. It was shown that when the concentration of $CaCl_2$ was varied, the fiber diameters showed an oscillatory behavior. The amplitude of the diameter fluctuation was approximately 150 µm.

They showed that increasing the concentration of $CaCl_2$ increased the tensile stress. In fact, a $CaCl_2$ concentration of approximately 1.4 wt% allows for maximum fiber tensile stress. This effect can be justified by the fact that when the calcium ion concentration increased to a saturation point, the egg-box type sites were filled. A higher concentration of $CaCl_2$ allows for a reduction in tensile stress, which is attributed to a partial collapse of the network. These results are in contradiction with the results of J. Zhang et al. [61], who showed that for higher concentrations of $CaCl_2$, the tensile stress increased.

The pH of the solution also plays a role in the solubilization of alginates. A hydrogel can be formed when the pH of the alginate-containing solution is lower than the pKa of mannuronic acid (pKa = 3.38) and guluronic acid (pKa = 3.65). Each alginate has its own apparent pKa that depends on the distribution of G and M blocks on the chain, the alginate concentration, and the ionic strength of the aqueous solution [62].

A change in ionic strength in the solution can have a significant effect on the conformation of the polymer chain, and thus, on the viscosity of the solution. Furthermore, when the ionic strength increases, the solubilization rate of the alginate decreases. Thus, it is preferable to first solubilize the alginate in pure water before adding an ionic species under agitation [63].

It has been shown that the formation of hydrogels is due to the cross-linking of the carboxylate groups of the G residues with divalent cations. For this reason, a stiffer hydrogel can be formed by alginate fibers with high G, while a softer elastic hydrogel can be formed by fibers with high M [64].

In the presence of fluids, the alginate fibers with a high G content swell only slightly. Additionally, the structures of these fibers are not radically altered during processing. Calcium ions, Ca^{2+}, are easily exchanged for sodium ions, Na^+, when alginate fibers have a high M content. This allows these fibers to swell and become an elastic gel.

Yimin Qin et al. [65,66] have studied the properties of swelling of alginate fibers. The introduction of sodium ions into alginate fibers with high G content can regulate the swelling properties. The water absorption of calcium alginate fibers is lower than that of calcium–sodium alginate fibers. This is attributed to the ability of the sodium ions in these fibers to bind water.

Alginate fibers with a high G content and sodium–calcium alginate fibers are characterized by a lower salt solution absorption compared to fibers formed by calcium alginate with a high M content. These results can be attributed to the excellent gelling ability of fibers with high M loading [65].

Struszczyk H. [67] used a pilot-scale spinning apparatus to elaborate the alginate fibers. These fibers were obtained by immersing a sodium alginate solution in an acid coagulation bath. The impact of calcium ion concentration on the properties of the elaborated fibers was investigated (Table 2). Fibers with a moisture content of 16–26%, an elongation of 13–21%, and a tenacity of 16–23 cN·tex^{-1} were obtained when the substitution of the calcium ion was superior to 30%.

Table 2. Impact of calcium ion concentration on alginate fiber properties.

Calcium Ion Concentration (wt%)	Substitution Degree (%)	Water Retention Value (%)
3.2	31	561
10.2	99	97

Rhoda Au Yeung et al. [68] elaborated calcium alginate fibers using two types of sodium alginate: alginate with a high G content and alginate with a high M content.

Their results showed that the calcium content of alginate fibers with a high G content was 2.79 µmoles·mg^{-1}, whereas alginate fibers with a high M content had only 2.58 µmoles·mg^{-1}. This can be explained by the lower binding capacity of M residues compared to G residues.

According to mechanical tests, these authors indicated that there was no significant difference of the Young's modulus between alginate with a high G content and alginate with a high M content. Indeed, the Young's modulus value for alginate fibers with a high M content was approximately 5.63 GPa, and it was only 4.99 GPa for alginate fibers with a high G content.

It appears that the yield strength of alginate fibers with a high M content is approximately 15% higher, compared to alginate fibers with a high G content. This indicates that these fibers are resistant to irreversible deformation.

5. Alginate and Bio-Composites

Substances were mixed with alginate to improve the properties of alginate fibers. Alginate bio-composites are made by adding inorganic compounds such as hydroxyapatite (HA) and tetraethylorthosilicate (TEOS), synthetic polymers such as polypyrrole and polylactide, and natural polymers such as gelatin, chitosan, and collagen [69,70]. The mixing of other types of materials, such as bio-glass, ceramics, and inorganic materials based on carbon, and inorganic nano-particles, has also been studied [71,72].

5.1. Alginate–Polymer Blends

5.1.1. Synthetic or Artificial Polymers

Weidong Zhou et al. [73] used polyethylene glycol diacrylate (PEGDA) as a filler in order to improve the fluidity of sodium alginate (SA). Their rheological results showed that the loss modulus G" and storage modulus G' decreased with increasing PEGDA concentration. The decrease in G' can be explained by the decrease in the degree of interaction between the SA molecular chains caused by the increase in molecular spacing. The decrease in G" can be attributed to the reduction in internal friction between the intermolecular force and SA molecules caused by PEGDA.

In another work, Wei-Wen Hu et al. [74] used electric field treatment to enhance gene transfection into alginate fibers reinforced by poly (caprolactone). It was found that the treatment enhanced the fluorescence intensity and the number of transfected cells, compared with the untreated group. These improvements were greater when the voltage values were smaller, to 1.5 V.

In order to facilitate the spinnability of an alginate bio-polymer, Xu, W. et al. [75] were used polylactic acid. These authors dissolved polylactic acid in chloroform and alginate in distilled water. Then, these two solutions were blended together to obtain emulsions. It was found that the tensile strength of the resulting fibers increased from 0.25 to 3.13 MPa.

Jie Liu et al. [76] developed sodium alginate fibers reinforced by cellulose nano-crystals, in order to enhance the mechanical behaviors of sodium alginate fibers. From the mechanical strength tests, these authors concluded that the mechanical properties of sodium alginate fibers were improved by the incorporation of cellulose nano-crystals. Indeed, the incorporation of cellulose nano-crystals increased the elongation at break and the tensile strength from 8.29% to 15.05% and from 1.54 to 2.05 cN·dtex^{-1}, respectively. These observations can be attributed to the uniform distribution of the cellulose nano-crystals in the polymer matrix.

To ameliorate the flame-retardancy of alginate fibers, Xian-Sheng Zhang et al. [77] incorporated flame retardant viscose (FRV) into alginate fibers. The evolution of the heat release rate (HRR) and the total heat release (THR) are reported in Figure 7. These authors reported that alginate/FRV possessed a higher time to ignition (TTI). In contrast, bio-composite fibers showed lower HRR and THR values compared to neat alginate fibers. This effect may be related to the metal ions in alginate fibers that are considered to be flame retardant.

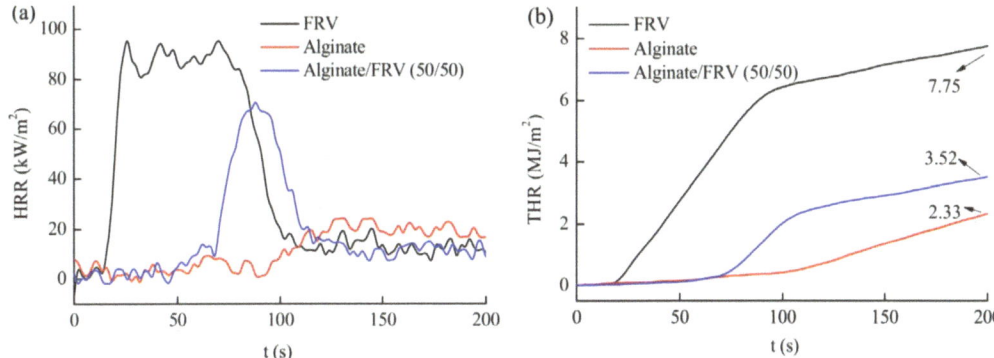

Figure 7. Evolution of (**a**) HRR and (**b**) THR as function of time of prepared fibers [77].

5.1.2. Natural Polymers

Wawro, D. et al. [78] studied the structure of alginate fibers. Their results showed that alginate fibers have a porous structure and irregular cross-sections. To resolve this problem, these authors were used a chitosan bio-polymer in order to ameliorate the morphologies of alginate fibers. Indeed, significant differences in the cross-sections and fiber surface views are clearly visible when comparing alginate fibers and alginate/chitosan fibers. The flat indentations in the case of alginate/chitosan fibers are significantly smaller than in alginate fibers, and their cross-sections are more rounded.

Recently, Wang Bin et al. [79] have been successful in the elaboration of alginate/cotton blended fibers. Through morphological analyses, these authors have confirmed that the surfaces of alginate and cotton fibers are not very smooth. Additionally, the surface morphologies of alginate and cotton fibers are not regular cylindrical surfaces. In contrast, scanning electron microscope photographs of cotton/alginate blended fibers showed that the surface micro-morphologies between alginate fibers and cotton fibers are similar. In order to study the effect of chitin on the thermal properties of calcium alginate fibers, J. L. Shamshina et al. [80] examined the thermal properties of alginate/chitin fibers. Their thermo-gravimetric analysis results showed that the thermal stability of calcium alginate fibers was enhanced with the addition of chitin.

A few years later, Ma Xiaomei et al. [81] were used cellulose nano-crystals and its oxidized derivative as nano-fillers in order to enhance the flame retardant properties of alginate fibers. These author indicated that an enhancement in flame-retardancy was greater when using cellulose nano-crystals than its oxidized derivative. Indeed, the addition of a small amount of cellulose based nano-crystals decreased the limiting oxygen index of alginate composite fibers.

Furthermore, Wang Bin et al. [79] also studied the impact of alginate fibers on the flame retardant and combustion properties of alginate/cotton blended fibers (Figure 8). They concluded that alginate fibers improved the fire behavior and flame retardant properties of the elaborated alginate/cotton blended fibers.

In recent years, injectable hydrogels have attracted interest because of their capability to blend homogeneously with therapeutic agents and cells. Despite this advantage, the utilization of injectable hydrogels is nowadays severely restricted by the difficulty in improving bone regeneration, mimicking the natural environment of modified cells, and facilitating cell proliferation.

To remedy these problems, Bin Liu et al. [82] developed an injectable nanocomposite hydrogel composed of alginate reinforced by gelatin. These authors showed that the encapsulated rat bone marrow mesenchymal stem cells survived in the elaborated nanocomposite hydrogel. This study proves that the developed material can be used as a candidate for orthopedic applications.

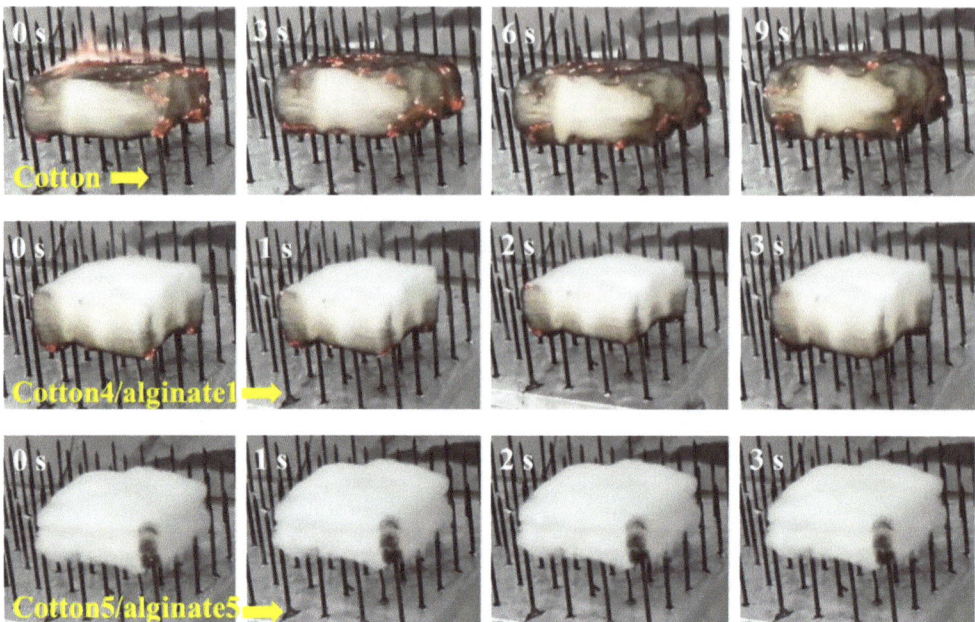

Figure 8. Photos of a flammability test for specimens recorded at several time points [79].

5.2. Alginate/Nano-Particle Composites

5.2.1. Zinc

Alginate fibers reinforced by nano-particles have recently received much attention. In studies by Andrea Dodero et al. [83], alginate fibers were loaded with zinc oxide nano-particles (ZnO-NPs). These nano-particles were produced using the sol–gel technique. The incorporation of ZnO-NP improved the rheological properties of the alginate. This is attributed to the electrostatic interactions and intermolecular hydrogen bonding between ZnO-NPs and the polysaccharide.

In addition, these authors indicated that it is preferable to use alginates with a high G content and a medium molecular weight, or with a high M content and a low molecular weight, when reinforced with ZnO-NPs. Indeed, a high G content allows for cavities along the chains of the polymer, which prevent the formation of interactions between these chains and the nano-particle molecules. Consequently, the impact of ZnO-NPs is nearly negligible. On the contrary, for M-rich alginates, the availability of establishing strong interactions with ZnO-NPs is important, due to the exposure of a high number of carboxyl and hydroxyl groups.

In another work, Guangyu Zhang et al. [84] coated calcium alginate nonwoven fabric with ZnO nano-particles, using the method of ion exchange.

Calcium alginate nonwoven fabrics were first immersed in $Zn(NO_3)_2$ solution, in order to obtain zinc calcium alginate fabrics. Indeed, the high-Zn^{2+} concentration solution allows part of the Ca^{2+} on the calcium alginate fibers to undergo an ion exchange reaction with Zn^{2+} (Figure 9a). Then, the zinc calcium alginate fabric was immersed in amino hyperbranched HBP solutions. Zn^{2+} was obtained in the solution after an ion exchange of Zn^{2+} with NH^{3+}. A high temperature of 80 °C can convert Zn^{2+} into $Zn(OH)_4^{2-}$, and then, the ZnO-NPs are obtained. Finally, ZnO-NPs were bonded to the surface of calcium alginate fabrics (Figure 9b). Indeed, the force of attraction between the positive groups of the ZnO-NPs and the negative groups of the alginate fabric, and the interactions of the hydrogen bonds between the amino groups on the ZnO-NPS, and the hydroxyl and carboxyl groups on the fabric allow for the attachment of the ZnO-NPs on the alginate fabric.

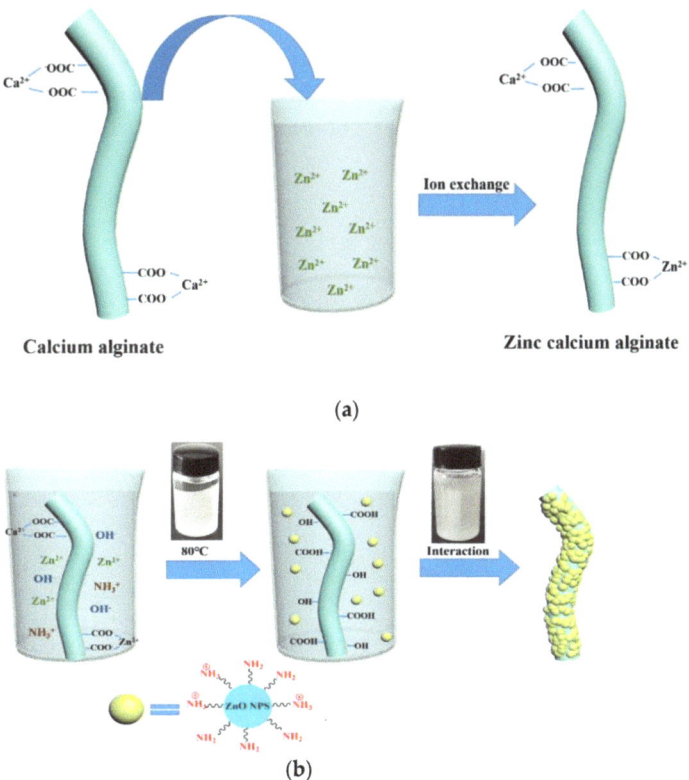

Figure 9. The coating of ZnO−NPs on an alginate fabric [84]. (**a**) Ion exchange reaction between Ca^{2+} and Zn^{2+}; (**b**) Synthesis of the ZnO NP−coated alginate fabric.

5.2.2. Silver

X. H. Zhao at al. [85] embedded alginate fibers with silver nano-particles (Ag-NPs) using the method of in situ reduction. First, alginate fibers and a silver nitrate ($AgNO_3$) aqueous solution are mixed together. Ion exchange between the silver ions (Ag^+) and the sodium or calcium ions in the alginate allows for the diffusion of Ag^+ ions into alginate fibers. The fixation of Ag^+ ions is performed thanks to an electrostatic attraction between the negative groups of the alginate and the positively charged Ag^+ ions. The silver ions were then reduced in situ, so that the metallic silver generated can adhere to these elaborated fibers.

These authors indicate that in an aqueous medium, these elaborated fibers allow for the reduction of 4-nitrophenol 4-NP to 4-aminophenol 4-AP. The catalytic reduction is performed by the Ag-NPs by relaying the electrons from the BH_4^- donor to the 4-NP acceptor (Figure 10).

In the first stage, metal hydride formation was achieved via the adsorption of BH_4^- and its reaction with the surface of the elaborated alginate/Ag-NPs fibers (1) [86,87]. Due to the strong adsorption of the alginate/Ag-NPs fibers, 4-nitrophenol 4-NP can transport to the surface of the Ag-NPs (2). The desorption/adsorption equilibrium of the reactants on the surface of alginate/Ag-NPs fibers is fast. Then, the interaction of the adsorbed 4-NP with the silver nano-particles reduces the 4-NP [87]. The reduction reaction allows the formation of the 4-aminophenol 4-AP (3) [88]. A new reduction cycle (4) will take place when the 4-AP reactant is desorbed from the surfaces of the Ag-NPs [89].

In the work of Maila Castellano et al. [90], the alginate polymer was reinforced with Ag-NPs to elaborate nano-textured mats. According to morphological and spectroscopic tests, these authors confirmed that Ag-NPs were formed within an alginate polymer. The resulting material was then mixed with polyethylene oxide to produce alginate fibers using the electro-spinning technique. It has been demonstrated that these elaborated nano-fibers are insensitive to physical treatments. This allows for the use of ultraviolet light or heat as a sterilization method. On the other hand, basic or oxidizing reagents affect the stability of the prepared material, which confirms its sensitivity to chemical products.

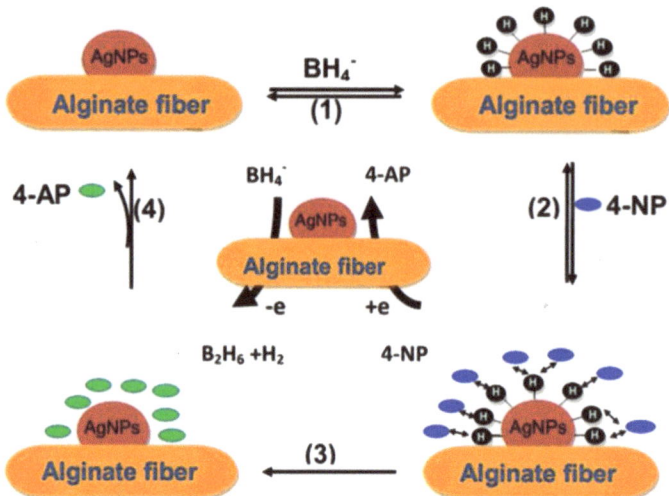

Figure 10. The mechanism of 4−NP reduction in the presence of the alginate/Ag−NP fibers [85].

Kaczmarek-Pawelska et al. [91] also studied the mechanical properties of alginate-based hydrogels reinforced by Ag nano-particles. These authors confirmed that the elaborated hydrogels are biomechanically compatible. Indeed, the obtained material showed mechanical properties very close to those of human skin. However, the increase in alginate concentration decreases the Young modulus. In fact, it is 8 MPa when the concentration of alginate is 0.1 mg/mL, whereas it is only 1.2 MPa when the concentration is 0.2 mg/mL.

5.2.3. Graphene

Linhai Pan et al. [92] reinforced alginate fibers with graphene oxide as a filler in order to remove Cu^{2+} ions and Pb^{2+} ions from waste-water. Alginate fibers reinforced with graphene oxide nano-particles present a very high affinity with Pb^{2+} ions. The high adsorption for Cu^{2+} and Pb^{2+} is 102.4 and 386.5 mg·g^{-1}. This high adsorption can be explained by the interaction of the oxygen atoms of alginate/graphene oxide fibers with the Cu^{2+} and Pb^{2+} ions. Based on the analysis of the adsorption mechanism, these authors confirmed that the chemical coordination and ion exchange effects (Figure 11) are responsible for the combination of heavy metals by alginate/graphene oxide fibers.

In other works, Xingzhu Fu et al. [93] used polymeric ionic liquids (PILs) to coat the surfaces of fibers based on calcium alginate using graphene, in order to develop conductive fibers.

To elaborate these fibers, these authors were used two coagulation baths. The first coagulation bath comprised calcium chloride. The second coagulation bath comprised a graphene aqueous dispersion. In fact, graphene aqueous dispersion was obtained by the dispersion of graphene by PIL.

In the step of secondary coagulation, the positively charged groups (imidazole ring) of LIP and the negatively charged groups (carboxylate ion) of alginate were linked together. The graphene layers attach to the surface of alginate fibers through the interactions of cation-π and π-π between graphene and PIL.

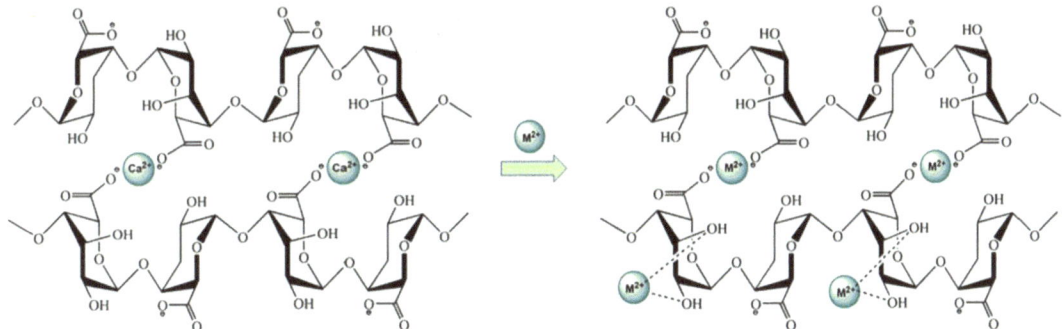

Figure 11. Mechanism interaction between alginate/graphene oxide fibers and heavy metals [92]. M^{2+}: Cu^{2+} or Pb^{2+} ions.

5.2.4. Magnesium Oxide

In the work of De Silva et al. [94], alginate is reinforced with magnesium oxide nanoparticles (MgO-NPs) with the aim of realizing nano-fibrous scaffolds. These authors have elaborated alginate fibers loaded by magnesium oxide whose diameters vary between 60 and 250 nm, using NPs of quasi-spherical shape. The tensile tests revealed that the mechanical behaviors of these obtained fibers were enhanced using the incorporation of MgO-NPs. Indeed, the reinforcement of alginate fibers by a 10 wt% of MgO-NPs resulted in a higher elastic modulus (E) and tensile strength (σ_U) among the studied samples.

In order to elaborate the sodium alginate scaffolds loaded by MgO-NPs, Bijan Nasri-Nasrabad et al. [95] used a two-step technique: poly (vinyl alcohol) leaching and film casting. Their results showed that after the leaching step, the incorporation of 4 wt% MgO-NPs resulted in better mechanical properties (Table 3). Indeed, the incorporation of 4 wt% MgO-NPs improved the Young's modulus of sodium alginate scaffolds by approximately 44%, compared to that of the neat sample. These improvements can be explained by the strong interaction between the alginate chains and the molecules of the nano-particles, and the decrease in the mobility of the polymer macromolecular chains.

Table 3. Mechanical and bacterial behaviors of the sodium alginate scaffolds loaded with different MgO-NP concentrations.

Nano-Particles Concentration (wt%)	Young Modulus (kPa)	Average Diameter of the Inhibitory Zone (mm²)
0	180.4 ± 15.2	13.6 ± 1.8
1	190.5 ± 25.2	13.1 ± 2.1
2	230.1 ± 27.8	11.7 ± 1.3
3	250.8 ± 30.4	9.8 ± 1.7
4	260.3 ± 19.6	9.6 ± 1.9

Moreover, with an increase in the nano-particle content, the antibacterial properties of the scaffolds have been enhanced, with an increase in the concentration of the MgO-NPs (Table 3). With an introduction of 1 and 2 wt% MgO-NPs, the average diameter of the bacterial zone of the scaffold samples is more than 10 mm², suggesting sensitive anti-microbial behavior. Whereas, with the incorporation of 3 wt% and 4 wt% MgO-NPs, the average diameter of the bacterial zone is reduced to less than 10 mm², compared with the pure sodium alginate, which exhibits anti-microbial insensitivity. The improved antibacterial behavior can be explained by the anti-microbial effects of MgO-NPs protecting the obtained materials, in contrast to *P. aeruginosa* and *S. aureus*.

5.2.5. Carbon Nanotubes

In order to have fibers with high electrical properties, Vijoya Sa et al. [96] prepared fibers based on alginate reinforced with carbon nano-tubes (CNTs) as a nano-filler, using a wet-spinning technique. This laboratory scale process could be used to produce industrial fibers via the addition of drawing/stretching steps. The use of calcium as a reticulation agent allows for electrostatic assembly between alginates and sodium dodecyl sulfate (SDS)-coated nano-tubes.

Both the alginate and the carbon nano-tubes are negatively charged. Thus, a repulsion will take place between the alginate solution and the aqueous solution of nano-tubes coated with SDS when they are added together. This prevents the agglomeration of the nano-tubes, and consequently allows for homogeneity in the spinning solution.

The interaction mechanism between CNT and the alginate is shown in Figure 12. The prepared solution is extruded into a coagulation bath containing an aqueous calcium chloride solution. Gel formation occurs when the spinning solution and the solution of calcium chloride are in contact. The coordination of Ca^{+2} ions in the cavities formed by the guluronate sequence pairs allows the formation of a calcium cage [97,98]. When the nano-tubes coated by SDS exist in the solution, Ca^{+2} ions link the nano-tubes and the alginate chains.

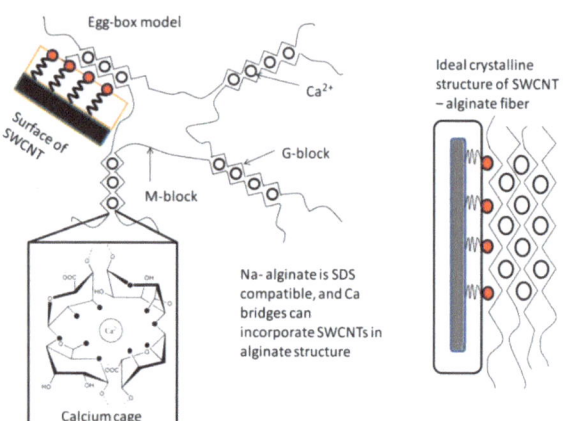

Figure 12. The interaction mechanism between CNT and alginate [96].

Recently, Aline Lima et al. [99] synthesized porous scaffolds based on CNT and hydroxamic alginate (HX). The HX was synthesized using the nucleophilic attack of hydroxylamine at the alginate carboxylic groups with dicyclohexylcarbodiimide. The partial modification of alginate with a derivative that is in acidic form facilitates its interaction with positively charged compounds.

From the study of mechanical behavior, these authors indicated that the HX/CNT scaffolds exhibited an improvement in their mechanical properties. According to FTIR and Raman spectroscopy, these authors confirmed an interaction between the alginate and the CNT cross-linked with the calcium.

5.2.6. Hydroxyapatite

Fuqiang Wan et al. [100] elaborated alginate fibers loaded by hydroxyapatite (HAP) as a nano-filler, using the technique of spin-coating. These fibers are characterized by an anisotropic structure. The alignment of nano-filler wires and the formation of the anisotropic structure are achieved via mechanical force.

Gel formation occurs upon contact between the alginate spinning solution and calcium ions, Ca^{2+}. Fiber formation takes place, after the covering of the glass substrate by the excess hydrogel (Figure 13).

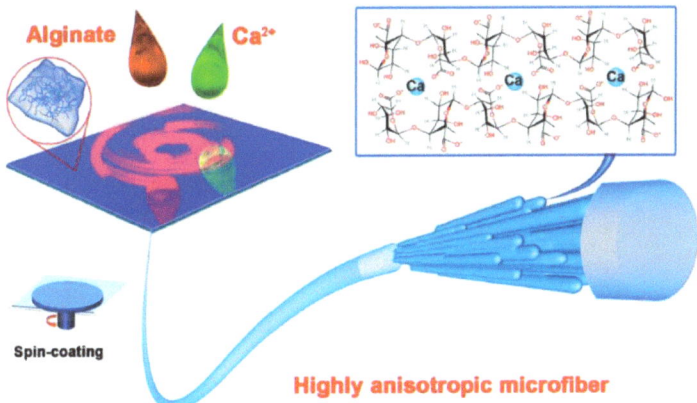

Figure 13. Elaboration of alginate fibers with anisotropic structure [100].

Excess water is removed during the spin-coating step, followed by an additional gelling step. The mechanical stresses exerted by the centrifugal force and the incorporation of HAP nanowires are responsible for the formation of a fiber with interesting mechanical and physical properties.

In the research studies of Peilong Ni et al. [101] a composite fiber membrane based on sodium alginate/polyvinyl alcohol/hydroxyapatite was elaborated. In order to avoid precipitation and agglomeration of the HAP in the spinning solution, the latter was ultrasonically suspended in a solution, using an alginate polymer as a stabilizer. Their results showed that the distribution of HAP nano-particles is uniform with the absence of agglomerates when the concentration of alginate–hydroxyapatite in these fibers is augmented from 1.64 wt% to 6.25 wt%. In contrast, poor particle distribution with the appearance of agglomerates is observed when the alginate—hydroxyapatite loading is between 7.70 wt% and 9.10 wt%.

5.2.7. Silica

Using the technique of microfluidic spinning, Zhang et al. [102] developed a novel fiber based on alginate reinforced by silica (SiO_2). The morphology characterizations of these elaborated fibers were studied. They showed that alginate/silica composite fibers had a certain additive on the surface, whereas the alginate fibers had a smooth surface.

These authors also studied the performance of SiO_2 nano-particles in improving the mechanical properties of alginate fibers. Indeed, they indicated that the addition of silica nano-particles improved the breaking stress of alginate fibers.

The application of alginate fibers is very limited in the biomedical field, because of their limited mechanical performance. To solve this problem, Lin Weng et al. [103] reinforced alginate fibers with SiO_2 nano-particles using a microfluidic spinning technique. Mechanical tests showed that neat alginate fibers exhibited behaviors similar to brittle materials, with low elongation and stress. The addition of SiO_2 nano-particles allows for the production of hybrid fibers with excellent mechanical performances, compared with the original alginate fibers. The elongation at break of the alginate fibers reinforced by SiO_2 is 52.08%, whereas it is only 7.32% for the original alginate fibers. The breaking strength of the alginate fibers is augmented from 0.76 MPa to 4.96 MPa for the alginate/SiO_2 fibers. These enhancements can be explained by the fact that the surface defects of alginate fibers are reduced by the SiO_2 nano-particles.

6. Applications of Alginate

Nowadays, alginate fibers are widely used in various applications. Among the most important areas are cosmetic, hygiene, and medical textile materials, etc.

6.1. Cosmeto Textiles

Cosmeto textiles are materials with cosmetic behaviors. However, these textile materials can also have other functions, such as UV protection agents, medical properties, and odor reducers.

Cosmetic textiles are an industry developed to ensure the well-being and health of consumers. Textile fibers are used to deliver a large variety of microencapsulated ingredients such as vitamin E, aloe vera, caffeine, retinol, etc. [104]. The new generation of cosmetic textile products uses innovative new techniques to provide medical, anti-aging, and stress relief benefits through clothing, textiles, and other products. In this regard, alginate fibers are highly biocompatible and hydrophilic, making them ideal products for the elaboration of face masks. In addition, alginate fibers can be used to carry several bioactive substances, allowing for sustained release on the skin.

6.2. Waste-Water Treatment

The elimination of dye molecules from waste-water is a complicated mechanism because of their inertness [105]. The active hydroxyl and carboxyl groups of alginate fibers have been investigated for the removal of dye molecules from effluents [75]. Electro-spun alginate fibers show interesting properties such as a high specific surface area and porosity. These properties make these materials suitable for waste-water treatment. A recent study by Zhao, X. et al. [106] demonstrated that fibrous sodium alginate/chitosan composite foam presents the potential to eliminate anionic and cationic dyes from waste-waters (Figure 14). Its high adsorption capacity can be attributed to its inter-connective pores and microscale fibers. According to the adsorption kinetics, these authors indicated that the adsorption rates of anionic and cationic dyes were initially rapid and then progressively slowed down to equilibrium.

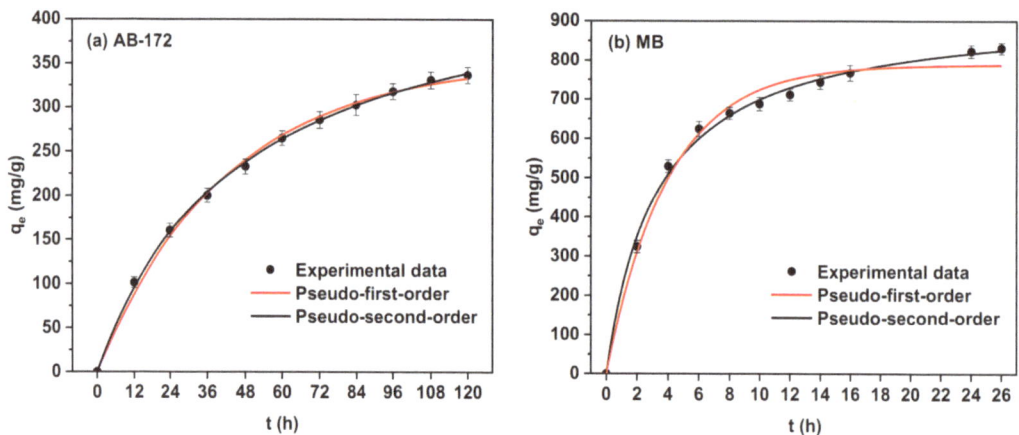

Figure 14. Adsorption kinetics, pseudo-first-order, and pseudo-second-order non-linear fitting curves of MB and AB-172 on the elaborated composite foams [106].

Additionally, alginate fibers were used to prepare nano-fibrous membranes for metal adsorption. Mokhena et al. [107] fabricated alginate fibers using polyethylene oxide as the carrier polymer for copper adsorption. They found that these obtained fibers were characterized by a high porosity and a large surface area. These properties allow these macro-porous fibrous membranes to exhibit an excellent capacity for removing copper ions from the aqueous medium. The prepared membrane can also recover nickel and copper ions, since it shows better selectivity upon these ions. In other research works, the authors coated alginate fibers with cellulose, in order to eliminate chromium from effluents [108]. It was shown that these prepared fibers for a greater than 80% rejection of chromium ions.

These observations can be attributed to the existence of the carboxyl and hydroxyl groups of alginate, as well as the hydroxyl and sulfate groups of cellulose.

In the works of F. Sun et al. [109], alginate fibers were used as filtration membranes in order to separate oil from water. These authors modified the surface properties of alginate fibers by incorporating acrylonitrile into alginate fibers. They obtained fibers with a high oil affinity. It was found that the angle of contact with the water was augmented from 56° to 70° with increasing acrylonitrile content. These effects can be explained by the strong interaction between molecular chains in the developed material, due to the existence of highly polar -CN substituents. The hydrophilic properties of calcium Ca^{2+} ion cross-linked alginate fibers have been examined for the retention of oil by Mokhena and co-workers [108].

6.3. Wound-Dressing

Wound-healing, which is one of the most complicated processes, involves a series of events, such as cell response, growth, and differentiation [110]. Consequently, the products used for the treatment of wounds must be characterized by durability, non-toxicity, and flexibility. Moisture-regulating and oxygen-transporting porosity is one of the supplementary properties conferred by the electro-spinning technique for wound-healing. Additionally, the production of multifunctional materials through the incorporation of bioactive compounds facilitates wound-healing [111].

Alginate is one of the most commonly used bio-polymers for dressing products, thanks to its interesting properties such as biodegradability, bio-compatibility, great absorption capacities, low toxicity, and low cost [112]. Alginate fibers have ion exchange behaviors when in contact with wound exudates. In fact, calcium ions are replaced by sodium ions from the body fluid, allowing for the development of a moist gel on the wound surface. The addition of other compounds to alginate fibers offers the possibility for producing advanced materials with several advantages such as a gel-forming ability and hemostatic capability. Alginate fibers loaded with chitosan for cancer stem-like cell enrichment were developed by Kievit's group [113,114]. They mixed chitosan and alginate and freeze-dried them in order to elaborate a porous sponge. The obtained materials allowed the enrichment of cancer stem cells by hepatocellular carcinoma and glioma-stoma. Nevertheless, the ratio of the composition is very limited, due to the restricted compatibility between the compounds. Tumor niches produce external factors that control the fate determination and the numbers of the stem cells. For this reason, it is necessary to prepare new materials allowing for the customization of the scaffold behaviors for cancer stem cells.

In this context, Wei-Wen Hu et al. [115] elaborated alginate fibers loaded with poly (caprolactone) using the method of co-electrospinning. They concluded that a low ratio of poly(caprolactone) considerably enhanced the cancer stem cell behaviors of these elaborated fibers. This can be justified by the spatial separation of cell populations by the sparse poly(caprolactone), which allows for the concentration of the cancer stem cells. In addition, it was found that neat alginate fibers and alginate fibers reinforced by poly(caprolactone) significantly decreased the wound area compared to tissue culture polystyrene and poly(caprolactone).

According to wound-healing studies, J. L. Shamshina et al. [80] indicated that the addition of chitin to calcium alginate fibers accelerated wound closure. It has been found that the wound sites covered by calcium alginate/chitin fibers have undergone normal wound-healing.

Yimin Qin [116] incorporated silver (Ag) into alginate fibers in order to ameliorate their anti-microbial properties. It has been demonstrated that alginate fibers reinforced with Ag nano-particles can be used to make highly absorbent and anti-microbial dressings. This effect can be explained by the liberation of Ag ions used as nano-fillers, which have anti-microbial properties. Bacteria trapped in alginate dressings will be killed by the Ag ions (Figure 15).

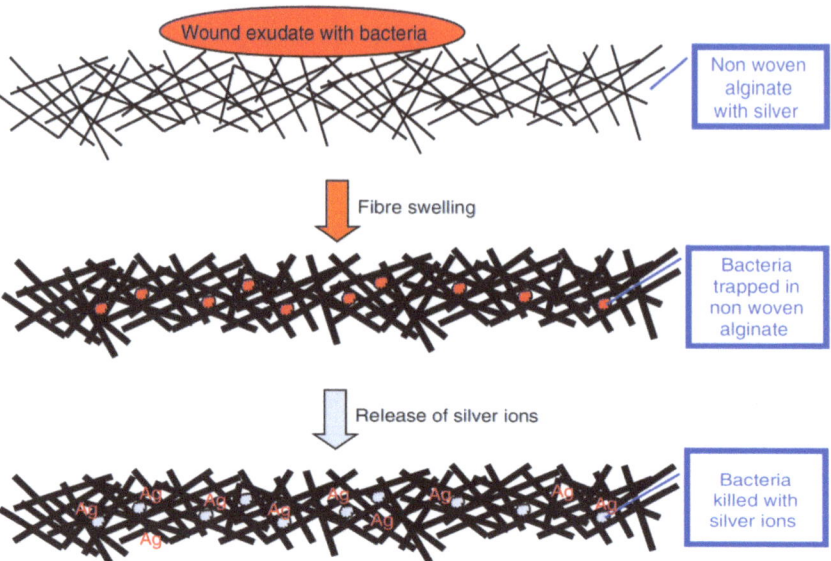

Figure 15. The anti-microbial mechanism of the elaborated wound-dressings [116].

Additionally, alginate-based hydrogels reinforced with Ag nano-particles have been used for wound-healing in various animal models, and they prevent contamination. Alginate nanocomposite hydrogels showed antibacterial activity in the long term, and sustained Ag release [117]. In the same context, Diniz, F.R. et al. elaborated alginate/gelatin hydrogels reinforced with Ag nano-particles for wound-healing. The obtained results showed that the elaborated product is characterized by antibacterial activity against *P. aeruginosa* and *S. aureus*, and is non-toxic against fibroblasts [118].

6.4. Tissue Engineering

Tissue engineering aims to replace, maintain, or improve the function of human tissues, thanks to tissue substitutes. It is therefore a matter of elaborating artificial tissues, using cell cultures, biomaterials, and growth factors, in order to obtain a hybrid biomaterial.

Alginate gels have many applications in tissue engineering. Indeed, alginate is a non-toxic, inert, and non-immunogenic substance which thus presents the required characteristics to constitute a good scaffold for tissue engineering [119].

The applications of alginate fibers in tissue engineering are very limited according to cell adhesion and viability. The addition of bioactive substances onto alginate fibers in order to ameliorate cell adhesion and proliferation makes these fibers ideal materials for the elaboration of scaffolds.

Jeong S. I. et al. [120] developed alginate fibers covalently bonded with a cellular adhesive in order to improve cell growth and viability. The results of these authors show that the addition of an adhesive peptide improved the propagation and adhesion of cells without changing the morphology of these fibers. In another study, Jeong S. I. et al. [121] elaborated a polyionic complex based on alginate and chitosan to achieve cell attachment and proliferation. The swelling rate in the deionized medium decreased with an increase in the concentration of chitosan content. Mouse pre-osteoblastic cells adhered to the alginate/chitosan nano-fibrous membrane and showed substantial proliferation.

To combine the properties of alginate and chitosan, Xinxin and Christopher [122] developed a process to treat the surface of alginate fibers with an aqueous solution of chitosan. After the absorption of chitosan on the surface of the calcium alginate fibers, these alginate/chitosan fibers are then freeze-dried. These fibrous materials are subsequently used as scaffolding materials in tissue engineering. Figure 16 gives us an idea regarding the applications of alginate/chitosan fibers in the area of tissue engineering.

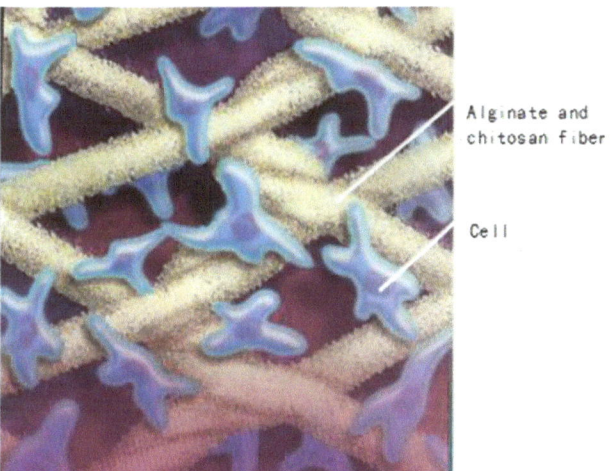

Figure 16. Chitosan and alginate fibers and their applications in tissue engineering.

In addition, the reinforcement of alginate fibers resulted in the improvement of mechanical behaviors, and therefore, in the usefulness of these materials for tissue engineering. In their studies, Tao, F. et al. [123] prepared fibers based on sodium alginate, carboxy-methyl chitosan, and biodegradable polymer poly(caprolactone) via the electro-spinning technique. According to tensile test analyses, the tensile strength of the elaborated micron-fibers was significantly higher than those of the poly(caprolactone)/carboxymethyl chitosan and poly(caprolactone)/sodium alginate micron-fibers. Further, these authors concluded that these obtained fibers can be used for periosteal tissue engineering.

In another work, alginate/chitosan–polylactide fibers were elaborated by Wu Hua et al. [124] for applications of neural tissue engineering. The mechanical properties of dry chitosan fibers showed that the modulus, tenacity, and elongation were around 25 cN·dtex^{-1}, 1.5 cN·dtex^{-1}, and 10%, respectively. The chitosan–polylactide fibers showed a higher modulus, tenacity, and elongation in the dry and wet states compared to the chitosan fibers. These results can be attributed to the polylactide component in chitosan–polylactide fibers, which is hydrophobic and mechanically strong. In addition, it found that the modulus and the tenacity of the alginate/chitosan–polylactide fibers are higher than that of chitosan–polylactide fibers in the dry state, whereas the modulus, tenacity, and elongation are still similar in the wet state.

Alginates are also reported as natural polymers utilized in hydrogel-based nanocomposites. A technique widely used in the elaboration of alginate-based nanocomposites is the chemical modification of the polymer in order to improve the interaction between the polymer matrix and the nano-particles, and thus to form a hydrogel characterized by stable mechanical properties. Nevertheless, chemical modification of the natural polymers can alter their biocompatibilities, which is a reason for using unmodified bio-polymers.

In this context, Rebeca Leu Alexa et al. have studied the different manufacturing methods for obtaining alginate–natural clay hydrogel-based nanocomposites adapted to 3D printing processes. The 3D multilayered scaffolds were obtained by printing the nanocomposite inks using the extrusion technique. The properties of the obtained materials confirm their use in tissue engineering. According to the biological analysis, these authors showed that the addition of unmodified clay into the alginate polymer allows for the development of cells [125].

6.5. Anti-Microbial Activity

In their work, Dumont et al. [126] studied the antibacterial activity of alginate-reinforced chitosan fibers, which were prepared using the technique of wet-spinning.

It was found that the inclusion of chitosan offers antibacterial properties, and that alginate gives healing properties and good hemostatic properties to these elaborated fibers (Figure 17). They concluded that the addition of chitosan on alginate fibers provides antibacterial activities contrary to Escherichia coli, Staphylococcus epidermidis, and various strains of *Staphylococcus aureus*, namely Healthcare Associated Methicillin Resistant *Staphylococcus aureus* (HA-MRSA), Methicillin Sensitive *Staphylococcus aureus* (MSSA), and Community Associated Methicillin Resistant *Staphylococcus aureus* (CA-MRSA).

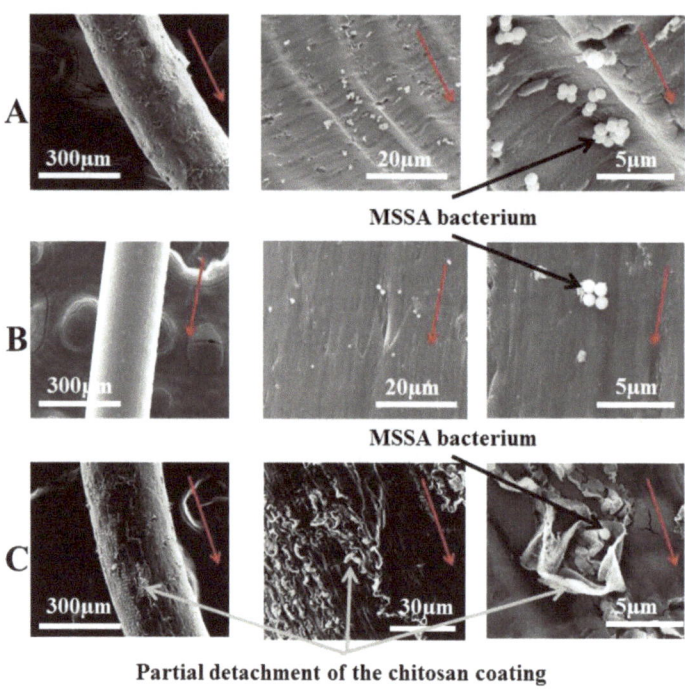

Figure 17. Scanning electron microscope observation of alginate fibers (**A**), chitosan fibers (**B**), and alginate-reinforced chitosan fibers (**C**) [126].

Other authors such as Sibaja Bernal et al. [127] indicated that alginate/chitosan fibers showed an excellent degree of inhibition of *Escherichia coli* growth, according to the test of bacterial inhibition. In fact, a great bacterial growth inhibition area was observed around these fibers loaded with the sulfathiazole drug after 24 h of incubation at 37 °C.

In the studies of Batista, M. P. et al. [128], a new route towards hybrid alginate/chitosan fibers via the emulsion gelation technique was developed. According to the standard tests, these authors indicated a clear antibacterial activity of these alginate/chitosan fibers against *Klebsiella pneumonia* and *Staphylococcus aureus*. To achieve this study, these authors were used two different methodologies with various contact times between the selected bacterial inocula and the specimens.

Recently, there have been several studies that investigate the antibacterial properties of alginate-based composite hydrogels. In general, changes in the structure of chitosan decrease its anti-microbial properties [129]. The incorporation of alginate hydrogels into chitosan solution showed a greater than 99% anti-microbial activity, as compared to the neat alginate hydrogel. Additionally, the addition of the chitosan into alginate hydrogels enhanced their anti-microbial properties [130].

It has been shown that the introduction of a anti-microbial peptide into sodium composite hydrogels based on alginate/polyethylene glycol provides good biocompatibility and an improvement in antibacterial activity [131]. A hydrogel loaded with an ultrashort peptide has been formulated for the treatment of the eyes and skin infections, and for the prevention of biomaterial infections [132]. An amphiphilic anti-bacterial hydrogel has been developed for skin wound treatments, and shows antimicrobial properties against bacteria, such as *S. aureus* and *E. coli* [133].

6.6. Sensors and Energy

Fibers elaborated using the electro-spinning method have interesting characteristics such as a large surface area, ease of modification of surface functionality, and malleable mechanical properties. These unique characteristics allow these fibers to provide a novel platform for the design of new sensors with high portability, sensitivity, and selectivity [134,135]. In order to improve the sensitivity, the response time, and the detection level, many sensing agents have been incorporated into electro-spun fibers. Alginate has been functionalized with heavy metal-sensitive compounds. Its carboxyl and hydroxyl groups can bind to multivalent ions, allowing them to be easily detected [135,136]. For this purpose, alginate fibers can be labeled with fluorescent sensors that are greatly selective in detecting metal ions found in aqueous solutions [136].

Wei-Peng Hu et al. [137] reinforced alginate fibers with silver nano-particles to obtain moisture-sensitive materials for respiratory sensors to monitor breathing during exercise and changes in emotion. It was demonstrated that the obtained material was able to detect the respiratory rhythm during a race, by fixing it on the seal of the mask exhalation valve. Additionally, these authors showed that the masks are characterized by stability and reusability, because they yielded the same results after 3 months. In the case of changes in emotion, it was shown that the mask was able to distinguish between sadness and pleasure by monitoring breathing frequencies.

Chen, H. et al. [138] developed alginate fibers reinforced with gels for the preparation of skin intelligence, such as ionic sensors. It was found that the gels maintained good electrical conductivity and mechanical deformability when exposed to long-term storage under ambient conditions or extreme conditions. The gels could be stretched and knotted without any damage to the structure. After storage for 6 h at 40 °C or −18 °C, they were capable of illuminating an LED bulb (Figure 18a). Then, the change in the conductivity of the gels was discovered in the range of the temperature between −20 °C and 40 °C, or at 25 °C for 6 days (Figure 18b,c).

In research by Ying He et al. [134], the authors elaborated fluorescent fibers based on alginate reinforced using gold nanoclusters and chicken egg white using the wet-spinning technique. These prepared fibers present a fluorescent sensor of great selectivity for detecting Hg^{2+} and Cu^{2+} ions among several metal ions in an aqueous solution. Among 11 kinds of common cations, Hg^{2+} cations could completely extinguish the fluorescence, while Cu^{2+} cations caused an obvious decrease in the intensity of the fluorescence.

In the same focus, these authors used these fluorescent fibers as an anti-counterfeiting label into cotton textiles, using the knitting technique. These smart fibers can be used in the design of novel flexible optical sensors and wearable optical sensors.

Another pressure sensor characterized by interesting sensing and mechanical properties has been developed from a composite hydrogel based on sodium alginate/polyacrylamide nanofibrils [139]. In fact, the compression, tension strength, stretchability, toughness, and elasticity of the obtained material are 4 MPa, 0.750 MPa, 3120%, 4.77 MJ m^{-3}, and 100%, respectively. The ionic conductors have been presented as sensors with deformation variations of between 0.3 and 1800%, a low applied voltage of up to 0.04 V, and a high sensitivity to pressure, equal to 1.45 kPa^{-1}. These ionic sensors could be used in sports tracking applications, soft robotics, and machine/human interfaces.

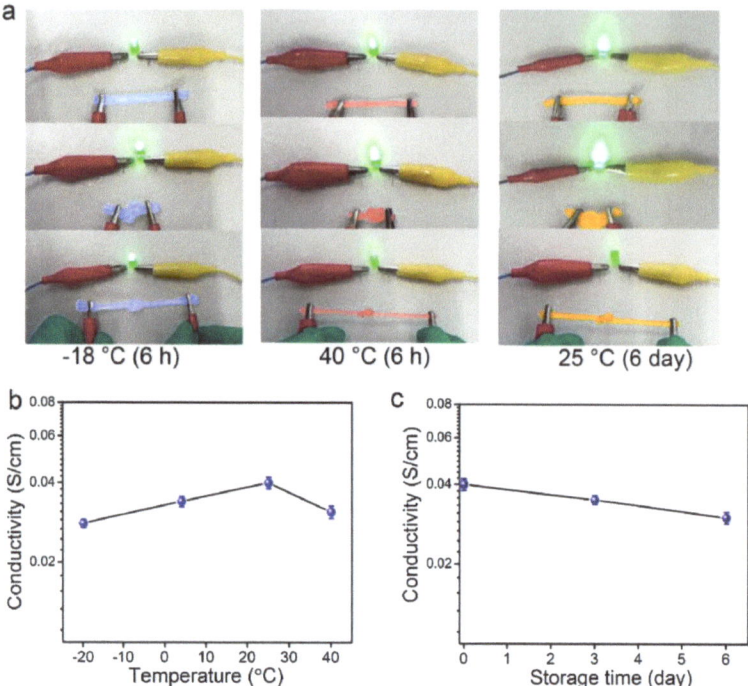

Figure 18. Stretched and knotted gels illuminated an LED bulb (**a**); The conductivity changed with temperature (**b**); The conductivity changed with storage time (**c**) [138].

6.7. Antiviral Activity

According to the antiviral performance of ionic polysaccharides, like alginate, they will play significant roles in the anti-COVID-19 field [140]. These bio-polymers are able to cause a slow release, prevent antigen degradation, and improve their stability, thereby enhancing immunogenicity. They can interact directly with the surface of viruses, and inhibit their infectivity or murder them [141].

Dental impressions, dentures, occlusal records, and trays can be contaminated with viruses and bacteria. The studies carried out do not prove the contamination or survival of the virus and the dental impression; however, salivary contamination suggests the possibility of the presence of a viral biological load in addition to those of yeasts and bacteria [142].

Generally, principal recommendations are linked to the danger of infections when a dental impression is carried out using conventional procedures. Nevertheless, there are no contamination or specific notes on disinfectants for tips, impression, scanner, cast, and hardware for computer aided manufacturing/computer aided design [143]. It can be noted that when taking dental impressions using both digital and traditional procedures, dental personnel are at risk of danger, due to close contact with the droplets and the aerosols of patients; however, they experience a different degree of exposure to aerosol-generating procedures and oral fluids [144].

The main advantages of digital techniques are the close-contact minimization of dental personnel with patients, and limited transmission through aerosol-generating procedures and respiratory droplets. These features are very essential for COVID-19 prevention, and particularly for dental care [145].

It is recommended that fewer objects are left on surfaces, in order to decrease the possibility of contamination of the surfaces, equipment, and environment. Computer keyboards must be covered with films based on polyethylene. Additionally, surfaces contaminated with biological particles should be disinfected using appropriate detergents [146]. According to in vitro studies, the chemical disinfection of alginate by sodium hypochlorite, glutaraldehyde, alcohol, and chlorhexidine reduced microbial counts on the surface without altering the dimensional stability of alginate impressions. Therefore, these disinfectant agents could be exploited to decrease the cross-contamination of alginate impressions [147,148].

7. Conclusions and Future Perspectives

In this work, we gave the state of knowledge regarding the definition, source, structure, and specific properties of alginate bio-polymers. Additionally, we have detailed the different strategies of alginate spinning and the influences of processes and solution parameters on the properties of alginate fibers. The article also discusses the influence of a wide range of materials on the properties of the obtained fibers from these bio-polymers. Finally, the potential applications of these bio-composite fibers of alginate are discussed.

Alginates are ecologically and environmentally friendly. The particular properties of these bio-polymers allow fibers produced from them to find more and more uses in special applications, especially for medical uses. Alginate fibers are commonly made by extruding a sodium alginate solution into a calcium chloride bath, producing calcium alginate fibers. Calcium alginate fibers should have a high potential for some specific applications. The incorporation of fillers into calcium alginate fibers seems to be one of the most successful solutions. These fillers allow for the improvement of the physical, thermal, mechanical, and wound-healing properties of calcium alginate fibers.

Although the spinning of alginate fibers and the reinforcement of these fibers by substances of different natures in order to enlarge their fields of application have been widely studied over the last few decades, there are still other important topics that deserve to be further investigated. For example, the different techniques of the mixed spinning of alginate and chitosan fibers. In addition, the properties of these mixed fibers and their applications in various potential applications also deserve further analysis. All of this remains within the framework of encouraging researchers and industries to develop innovative and sustainable materials.

Author Contributions: Conceptualization, K.Z., A.C. and F.S.; methodology, K.Z., A.C. and F.S.; validation, K.Z., A.C., A.E. (Adel Elamri) and F.S.; investigation, K.Z., A.C. and F.S.; writing—original draft preparation, K.Z., A.C. and F.S.; writing—review and editing K.Z., A.C. and F.S.; visualization, K.Z., A.C., F.S., A.E. (Adel Elamri) and A.E. (Annaëlle Erard); supervision, K.Z., A.C. and F.S.; project administration, A.C. All authors have read and agreed to the published version of the manuscript.

Funding: This research received no external funding.

Institutional Review Board Statement: Not applicable.

Informed Consent Statement: Not applicable.

Data Availability Statement: Not applicable.

Conflicts of Interest: The authors declare no conflict of interest.

References

1. Ma, C.; Liu, L.; Hua, W.; Cai, Y.; Yao, J. Fabrication and characterization of absorbent and antibacterial alginate fibers loaded with sulfanilamide. *Fibers Polym.* **2015**, *16*, 1255–1261. [CrossRef]
2. Tian, G.; Ji, Q.; Xu, D.; Tan, L.; Quan, F.; Xia, Y. The effect of zinc ion content on flame retardance and thermal degradation of alginate fibers. *Fibers Polym.* **2013**, *14*, 767–771. [CrossRef]
3. Boguń, M.; Mikołajczyk, T.; Szparaga, G.; Kurzak, A. Water-soluble nanocomposite sodium alginate fibres. *Fibers Polym.* **2010**, *11*, 398–405. [CrossRef]
4. Agarwal, A.; McAnulty, J.F.; Schurr, M.J.; Murphy, C.J.; Abbott, N.L. Polymeric materials for chronic wound and burn dressings. In *Advanced Wound Repair Therapies*; Elsevier: Amsterdam, The Netherlands, 2011; pp. 186–208. ISBN 978-1-84569-700-6.
5. Miraftab, M.; Iwu, C.; Okoro, C.; Smart, G. Inherently Antimicrobial Alchite Fibres Developed for Wound Care Applications. In *Medical and Healthcare Textiles*; Elsevier: Amsterdam, The Netherlands, 2010; pp. 76–83. ISBN 978-1-84569-224-7.

6. Shang, L.; Yu, Y.; Liu, Y.; Chen, Z.; Kong, T.; Zhao, Y. Spinning and Applications of Bioinspired Fiber Systems. *ACS Nano* **2019**, *13*, 2749–2772. [CrossRef]
7. Xu, G.K.; Liu, L.; Yao, J.M. Fabrication and Characterization of Alginate Fibers by Wet-Spinning. *AMR* **2013**, *796*, 87–91. [CrossRef]
8. Gady, O.; Poirson, M.; Vincent, T.; Sonnier, R.; Guibal, E. Elaboration of light composite materials based on alginate and algal biomass for flame retardancy: Preliminary tests. *J. Mater. Sci.* **2016**, *51*, 10035–10047. [CrossRef]
9. Phillips, G.O.; Williams, P.A. *Handbook of Hydrocolloids*, 2nd ed.; Woodhead publishing in food science, technology and nutrition; CRC Press: Boca Raton, FL, USA; Woodhead Publishing: Oxford, UK, 2009; ISBN 978-1-4398-0820-7.
10. Riaz, S.; Rehman, A.; Ashraf, M.; Hussain, T.; Hussain, M.T. Development of functional alginate fibers for medical applications. *J. Text. Inst.* **2017**, *108*, 2197–2204. [CrossRef]
11. Chowdhury, S.; Chakraborty, S.; Maity, M.; Hasnain, M.S.; Nayak, A.K. Biocomposites of Alginates in Drug Delivery. In *Alginates in Drug Delivery*; Elsevier: Amsterdam, The Netherlands, 2020; pp. 153–185. ISBN 978-0-12-817640-5.
12. Goff, H.D.; Guo, Q. Chapter 1: The Role of Hydrocolloids in the Development of Food Structure. In *Food Chemistry, Function and Analysis*; Spyropoulos, F., Lazidis, A., Norton, I., Eds.; Royal Society of Chemistry: Cambridge, UK, 2019; pp. 1–28. ISBN 978-1-78801-216-4.
13. Liang, B.; Zhao, H.; Zhang, Q.; Fan, Y.; Yue, Y.; Yin, P.; Guo, L. Ca^{2+} Enhanced Nacre-Inspired Montmorillonite–Alginate Film with Superior Mechanical, Transparent, Fire Retardancy, and Shape Memory Properties. *ACS Appl. Mater. Interfaces* **2016**, *8*, 28816–28823. [CrossRef]
14. Sultan, M.T.; Rahman, A.; Islam, J.M.M.; Khan, M.A.; Rahman, N.; Alam, N.A.; Hakim, A.K.M.A.; Alam, M.M. Preparation and Characterization of an Alginate/Clay Nanocomposite for Optoelectronic Application. *Adv. Mater. Res.* **2010**, *123*, 751–754. [CrossRef]
15. Gao, X.; Guo, C.; Hao, J.; Zhao, Z.; Long, H.; Li, M. Adsorption of heavy metal ions by sodium alginate based adsorbent-a review and new perspectives. *Int. J. Biol. Macromol.* **2020**, *164*, 4423–4434. [CrossRef]
16. Paredes JuÃ¡rez, G.A.; Spasojevic, M.; Faas, M.M.; de Vos, P. Immunological and Technical Considerations in Application of Alginate-Based Microencapsulation Systems. *Front. Bioeng. Biotechnol.* **2014**, *2*, 26. [CrossRef]
17. Szekalska, M.; Puciłowska, A.; Szymańska, E.; Ciosek, P.; Winnicka, K. Alginate: Current Use and Future Perspectives in Pharmaceutical and Biomedical Applications. *Int. J. Polym. Sci.* **2016**, *2016*, 7697031. [CrossRef]
18. Niemelä, K.; Sjöström, E. Alkaline degradation of alginates to carboxylic acids. *Carbohydr. Res.* **1985**, *144*, 241–249. [CrossRef]
19. Varaprasad, K.; Jayaramudu, T.; Kanikireddy, V.; Toro, C.; Sadiku, E.R. Alginate-based composite materials for wound dressing application: A mini review. *Carbohydr. Polym.* **2020**, *236*, 116025. [CrossRef]
20. Porrelli, D.; Berton, F.; Camurri Piloni, A.; Kobau, I.; Stacchi, C.; Di Lenarda, R.; Rizzo, R. Evaluating the stability of extended-pour alginate impression materials by using an optical scanning and digital method. *J. Prosthet. Dent.* **2021**, *125*, 189.e1–189.e7. [CrossRef]
21. Joanny, J.F. Flow birefringence at the sol-gel transition. *J. De Phys.* **1982**, *43*, 467–473. [CrossRef]
22. Pathak, T.S.; Kim, J.S.; Lee, S.-J.; Baek, D.-J.; Paeng, K.-J. Preparation of Alginic Acid and Metal Alginate from Algae and their Comparative Study. *J. Polym. Environ.* **2008**, *16*, 198–204. [CrossRef]
23. Topuz, F.; Henke, A.; Richtering, W.; Groll, J. Magnesium ions and alginate do form hydrogels: A rheological study. *Soft Matter* **2012**, *8*, 4877. [CrossRef]
24. Cao, L.; Lu, W.; Mata, A.; Nishinari, K.; Fang, Y. Egg-box model-based gelation of alginate and pectin: A review. *Carbohydr. Polym.* **2020**, *242*, 116389. [CrossRef]
25. Zhang, H.; Cheng, J.; Ao, Q. Preparation of Alginate-Based Biomaterials and Their Applications in Biomedicine. *Mar. Drugs* **2021**, *19*, 264. [CrossRef]
26. Bissantz, C.; Kuhn, B.; Stahl, M. A Medicinal Chemist's Guide to Molecular Interactions. *J. Med. Chem.* **2010**, *53*, 5061–5084. [CrossRef]
27. Williams, D.F. On the mechanisms of biocompatibility. *Biomaterials* **2008**, *29*, 2941–2953. [CrossRef]
28. Orive, G.; Ponce, S.; Hernández, R.M.; Gascón, A.R.; Igartua, M.; Pedraz, J.L. Biocompatibility of microcapsules for cell immobilization elaborated with different type of alginates. *Biomaterials* **2002**, *23*, 3825–3831. [CrossRef]
29. Lee, J.; Lee, K.Y. Local and Sustained Vascular Endothelial Growth Factor Delivery for Angiogenesis Using an Injectable System. *Pharm. Res.* **2009**, *26*, 1739–1744. [CrossRef]
30. Taskin, A.K.; Yasar, M.; Ozaydin, I.; Kaya, B.; Bat, O.; Ankarali, S.; Yildirim, U.; Aydin, M. The hemostatic effect of calcium alginate in experimental splenic injury model. *Turk. J. Trauma Emerg. Surg.* **2013**, *19*, 195–199. [CrossRef]
31. Hampton, S. The role of alginate dressings in wound healing. *Diabet Foot* **2004**, *7*, 162–167.
32. Thomas, S. Alginate dressings in surgery and wound management—Part 1. *J. Wound Care* **2000**, *9*, 56–60. [CrossRef]
33. Rashedy, S.H.; Abd El Hafez, M.S.M.; Dar, M.A.; Cotas, J.; Pereira, L. Evaluation and Characterization of Alginate Extracted from Brown Seaweed Collected in the Red Sea. *Appl. Sci.* **2021**, *11*, 6290. [CrossRef]
34. Zimmermann, U.; Klöck, G.; Federlin, K.; Hannig, K.; Kowalski, M.; Bretzel, R.G.; Horcher, A.; Entenmann, H.; Sieber, U.; Zekorn, T. Production of mitogen-contamination free alginates with variable ratios of mannuronic acid to guluronic acid by free flow electrophoresis. *Electrophoresis* **1992**, *13*, 269–274. [CrossRef]
35. Qin, Y. Absorption characteristics of alginate wound dressings. *J. Appl. Polym. Sci.* **2004**, *91*, 953–957. [CrossRef]

36. Venkatesan, J.; Lowe, B.; Anil, S.; Manivasagan, P.; Kheraif, A.A.A.; Kang, K.-H.; Kim, S.-K. Seaweed polysaccharides and their potential biomedical applications: Seaweed polysaccharides and their potential biomedical applications. *Starch Stärke* **2015**, *67*, 381–390. [CrossRef]
37. Notin, L.; Viton, C.; Lucas, J.; Domard, A. Pseudo-dry-spinning of chitosan. *Acta Biomater.* **2006**, *2*, 297–311. [CrossRef] [PubMed]
38. Zhao, Z.; Geng, C.; Zhao, X.; Xue, Z.; Quan, F.; Xia, Y. Preparation of CdTe/Alginate Textile Fibres with Controllable Fluorescence Emission through a Wet-Spinning Process and Application in the Trace Detection of Hg^{2+} Ions. *Nanomaterials* **2019**, *9*, 570. [CrossRef] [PubMed]
39. Hwang, T.-Y.; Choi, Y.; Song, Y.; Eom, N.S.A.; Kim, S.; Cho, H.-B.; Myung, N.V.; Choa, Y.-H. A noble gas sensor platform: Linear dense assemblies of single-walled carbon nanotubes (LACNTs) in a multi-layered ceramic/metal electrode system (MLES). *J. Mater. Chem. C* **2018**, *6*, 972–979. [CrossRef]
40. Nguyen, K.D. Temperature Effect of Water Coagulation Bath on Chitin Fiber Prepared through Wet-Spinning Process. *Polymers* **2021**, *13*, 1909. [CrossRef]
41. Sun, J.; Tan, H. Alginate-Based Biomaterials for Regenerative Medicine Applications. *Materials* **2013**, *6*, 1285–1309. [CrossRef]
42. Greiner, A.; Wendorff, J.H. Electrospinning: A Fascinating Method for the Preparation of Ultrathin Fibers. *Angew. Chem. Int. Ed.* **2007**, *46*, 5670–5703. [CrossRef]
43. Lee, K.Y.; Jeong, L.; Kang, Y.O.; Lee, S.J.; Park, W.H. Electrospinning of polysaccharides for regenerative medicine. *Adv. Drug Deliv. Rev.* **2009**, *61*, 1020–1032. [CrossRef]
44. Yamada, M.; Seki, M. Multiphase Microfluidic Processes to Produce Alginate-Based Microparticles and Fibers. *J. Chem. Eng. Jpn.* **2018**, *51*, 318–330. [CrossRef]
45. Nishimura, K.; Morimoto, Y.; Mori, N.; Takeuchi, S. Formation of Branched and Chained Alginate Microfibers Using Theta-Glass Capillaries. *Micromachines* **2018**, *9*, 303. [CrossRef]
46. Lee, K.H.; Shin, S.J.; Park, Y.; Lee, S.-H. Synthesis of Cell-Laden Alginate Hollow Fibers Using Microfluidic Chips and Microvascularized Tissue-Engineering Applications. *Small* **2009**, *5*, 1264–1268. [CrossRef]
47. Meng, Z.-J.; Wang, W.; Xie, R.; Ju, X.-J.; Liu, Z.; Chu, L.-Y. Microfluidic generation of hollow Ca-alginate microfibers. *Lab Chip* **2016**, *16*, 2673–2681. [CrossRef]
48. Kang, E.; Wong, S.F.; Lee, S.-H. Microfluidic "On-the-Fly" Fabrication of Microstructures for Biomedical Applications. In *Microfluidic Technologies for Human Health*; World Scientific: Hackensack, NJ, USA, 2013; pp. 293–309. ISBN 978-981-4405-51-5.
49. *Natural Fibers, Plastics and Composites*; Wallenberger, F.T.; Weston, N.E. (Eds.) Springer: Boston, MA, USA, 2004; ISBN 978-1-4613-4774-3.
50. Knill, C.J.; Kennedy, J.F.; Mistry, J.; Miraftab, M.; Smart, G.; Groocock, M.R.; Williams, H.J. Alginate fibres modified with unhydrolysed and hydrolysed chitosans for wound dressings. *Carbohydr. Polym.* **2004**, *55*, 65–76. [CrossRef]
51. Larsen, B.E.; Bjørnstad, J.; Pettersen, E.O.; Tønnesen, H.H.; Melvik, J.E. Rheological characterization of an injectable alginate gel system. *BMC Biotechnol* **2015**, *15*, 29. [CrossRef]
52. Qin, Y.; Hu, H.; Luo, A. The conversion of calcium alginate fibers into alginic acid fibers and sodium alginate fibers. *J. Appl. Polym. Sci.* **2006**, *101*, 4216–4221. [CrossRef]
53. Wang, Q.; Zhang, L.; Liu, Y.; Zhang, G.; Zhu, P. Characterization and functional assessment of alginate fibers prepared by metal-calcium ion complex coagulation bath. *Carbohydr. Polym.* **2020**, *232*, 115693. [CrossRef]
54. Niekraszewicz, B.; Niekraszewicz, A. The structure of alginate, chitin and chitosan fibres. In *Handbook of Textile Fibre Structure*; Elsevier: Amsterdam, The Netherlands, 2009; pp. 266–304. ISBN 978-1-84569-730-3.
55. Shin, S.-J.; Park, J.-Y.; Lee, J.-Y.; Park, H.; Lee, K.-B.; Whang, C.-M.; Lee, S.-H. "On the Fly" Continuous Generation of Alginate Fibers Using a Microfluidic Device. *Langmuir* **2007**, *23*, 9104–9108. [CrossRef]
56. Cuadros, T.R.; Skurtys, O.; Aguilera, J.M. Mechanical properties of calcium alginate fibers produced with a microfluidic device. *Carbohydr. Polym.* **2012**, *89*, 1198–1206. [CrossRef]
57. Marcos, B.; Gou, P.; Arnau, J.; Comaposada, J. Influence of processing conditions on the properties of alginate solutions and wet edible calcium alginate coatings. *LWT* **2016**, *74*, 271–279. [CrossRef]
58. Brzezińska, M.; Szparaga, G. The Effect Of Sodium Alginate Concentration On The Rheological Parameters Of Spinning Solutions. *Autex Res. J.* **2015**, *15*, 123–126. [CrossRef]
59. Lin, H.-Y.; Wang, H.-W. The influence of operating parameters on the drug release and antibacterial performances of alginate fibrous dressings prepared by wet spinning. *Biomatter* **2012**, *2*, 321–328. [CrossRef] [PubMed]
60. Cuadros, T.R.; Skurtys, O.; Aguilera, J. Fibers of Calcium Alginate Produced by a Microfluidic Device and Its Mechanical Properties. 2011. Available online: https://www.semanticscholar.org/paper/Fibers-of-calcium-alginate-produced-by-a-device-and-Cuadros-Skurtys/fde7027ef622827538983247a530f52dc76163a3 (accessed on 6 July 2022).
61. Zhang, J.; Daubert, C.R.; Foegeding, E.A. Fracture Analysis of Alginate Gels. *J. Food Sci.* **2005**, *70*, e425–e431. [CrossRef]
62. Haug, A.; Larsen, B.; Smidsrød, O.; Haug, A.; Hagen, G. Alkaline Degradation of Alginate. *Acta Chem. Scand.* **1967**, *21*, 2859. [CrossRef]
63. Haug, A.; Claeson, K.; Hansen, S.E.; Sömme, R.; Stenhagen, E.; Palmstierna, H. Fractionation of Alginic Acid. *Acta Chem. Scand.* **1959**, *13*, 601–603. [CrossRef]
64. Ahmad Raus, R.; Wan Nawawi, W.M.F.; Nasaruddin, R.R. Alginate and alginate composites for biomedical applications. *Asian J. Pharm. Sci.* **2021**, *16*, 280–306. [CrossRef]
65. Qin, Y. Gel swelling properties of alginate fibers. *J. Appl. Polym. Sci.* **2004**, *91*, 1641–1645. [CrossRef]

66. Qin, Y. The gel swelling properties of alginate fibers and their applications in wound management. *Polym. Adv. Technol.* **2008**, *19*, 6–14. [CrossRef]
67. Struszczyk, H. Some Aspects on Preparation and Properties of Alginate and Chitosan Fibres. *MRS Online Proc. Libr.* **2001**, *702*, U2.3.1–U2.3.8. [CrossRef]
68. Yeung, R.A.; Kennedy, R.A. A comparison of selected physico-chemical properties of calcium alginate fibers produced using two different types of sodium alginate. *J. Mech. Behav. Biomed. Mater.* **2019**, *90*, 155–164. [CrossRef]
69. Venkatesan, J.; Bhatnagar, I.; Manivasagan, P.; Kang, K.-H.; Kim, S.-K. Alginate composites for bone tissue engineering: A review. *Int. J. Biol. Macromol.* **2015**, *72*, 269–281. [CrossRef]
70. Lee, K.Y.; Mooney, D.J. Alginate: Properties and biomedical applications. *Prog. Polym. Sci.* **2012**, *37*, 106–126. [CrossRef]
71. Neibert, K.; Gopishetty, V.; Grigoryev, A.; Tokarev, I.; Al-Hajaj, N.; Vorstenbosch, J.; Philip, A.; Minko, S.; Maysinger, D. Wound-Healing with Mechanically Robust and Biodegradable Hydrogel Fibers Loaded with Silver Nanoparticles. *Adv. Healthc. Mater.* **2012**, *1*, 621–630. [CrossRef]
72. Khan, I.; Saeed, K.; Khan, I. Nanoparticles: Properties, applications and toxicities. *Arab. J. Chem.* **2019**, *12*, 908–931. [CrossRef]
73. Zhou, W.; Zhang, H.; Liu, Y.; Zou, X.; Shi, J.; Zhao, Y.; Ye, Y.; Yu, Y.; Guo, J. Preparation of calcium alginate/polyethylene glycol acrylate double network fiber with excellent properties by dynamic molding method. *Carbohydr. Polym.* **2019**, *226*, 115277. [CrossRef]
74. Hu, W.-W.; Ting, J.-C. Gene immobilization on alginate/polycaprolactone fibers through electrophoretic deposition to promote in situ transfection efficiency and biocompatibility. *Int. J. Biol. Macromol.* **2019**, *121*, 1337–1345. [CrossRef]
75. Xu, W.; Shen, R.; Yan, Y.; Gao, J. Preparation and characterization of electrospun alginate/PLA nanofibers as tissue engineering material by emulsion eletrospinning. *J. Mech. Behav. Biomed. Mater.* **2017**, *65*, 428–438. [CrossRef]
76. Liu, J.; Zhang, R.; Ci, M.; Sui, S.; Zhu, P. Sodium alginate/cellulose nanocrystal fibers with enhanced mechanical strength prepared by wet spinning. *J. Eng. Fibers Fabr.* **2019**, *14*, 1558925019847553. [CrossRef]
77. Zhang, X.-S.; Xia, Y.-Z.; Shi, M.-W.; Yan, X. The flame retardancy of alginate/flame retardant viscose fibers investigated by vertical burning test and cone calorimeter. *Chin. Chem. Lett.* **2018**, *29*, 489–492. [CrossRef]
78. Wawro, D.; Niekraszewicz, A. Research into the Process of Manufacturing Alginate-Chitosan Fibres. *Fibres Text. East. Eur.* **2006**, *14*, 25–31.
79. Wang, B.; Li, P.; Xu, Y.-J.; Jiang, Z.-M.; Dong, C.-H.; Liu, Y.; Zhu, P. Bio-based, nontoxic and flame-retardant cotton/alginate blended fibres as filling materials: Thermal degradation properties, flammability and flame-retardant mechanism. *Compos. Part B Eng.* **2020**, *194*, 108038. [CrossRef]
80. Shamshina, J.L.; Gurau, G.; Block, L.E.; Hansen, L.K.; Dingee, C.; Walters, A.; Rogers, R.D. Chitin–calcium alginate composite fibers for wound care dressings spun from ionic liquid solution. *J. Mater. Chem. B* **2014**, *2*, 3924–3936. [CrossRef]
81. Ma, X.; Li, R.; Zhao, X.; Ji, Q.; Xing, Y.; Sunarso, J.; Xia, Y. Biopolymer composite fibres composed of calcium alginate reinforced with nanocrystalline cellulose. *Compos. Part A Appl. Sci. Manuf.* **2017**, *96*, 155–163. [CrossRef]
82. Liu, B.; Li, J.; Lei, X.; Miao, S.; Zhang, S.; Cheng, P.; Song, Y.; Wu, H.; Gao, Y.; Bi, L.; et al. Cell-loaded injectable gelatin/alginate/LAPONITE® nanocomposite hydrogel promotes bone healing in a critical-size rat calvarial defect model. *RSC Adv.* **2020**, *10*, 25652–25661. [CrossRef]
83. Dodero, A.; Alloisio, M.; Vicini, S.; Castellano, M. Preparation of composite alginate-based electrospun membranes loaded with ZnO nanoparticles. *Carbohydr. Polym.* **2020**, *227*, 115371. [CrossRef]
84. Zhang, G.; Xiao, Y.; Yan, J.; Zhang, H. Fabrication of ZnO nanoparticle-coated calcium alginate nonwoven fabric by ion exchange method based on amino hyperbranched polymer. *Mater. Lett.* **2020**, *270*, 127624. [CrossRef]
85. Zhao, X.H.; Li, Q.; Ma, X.M.; Xiong, Z.; Quan, F.Y.; Xia, Y.Z. Alginate fibers embedded with silver nanoparticles as efficient catalysts for reduction of 4-nitrophenol. *RSC Adv.* **2015**, *5*, 49534–49540. [CrossRef]
86. Wu, X.-Q.; Wu, X.-W.; Huang, Q.; Shen, J.-S.; Zhang, H.-W. In situ synthesized gold nanoparticles in hydrogels for catalytic reduction of nitroaromatic compounds. *Appl. Surf. Sci.* **2015**, *331*, 210–218. [CrossRef]
87. Lin, H.-L.; Sou, N.-L.; Huang, G.G. Single-step preparation of recyclable silver nanoparticle immobilized porous glass filters for the catalytic reduction of nitroarenes. *RSC Adv.* **2015**, *5*, 19248–19254. [CrossRef]
88. Wunder, S.; Polzer, F.; Lu, Y.; Mei, Y.; Ballauff, M. Kinetic Analysis of Catalytic Reduction of 4-Nitrophenol by Metallic Nanoparticles Immobilized in Spherical Polyelectrolyte Brushes. *J. Phys. Chem. C* **2010**, *114*, 8814–8820. [CrossRef]
89. Liu, B.H.; Li, Z.P. A review: Hydrogen generation from borohydride hydrolysis reaction. *J. Power Sources* **2009**, *187*, 527–534. [CrossRef]
90. Castellano, M.; Alloisio, M.; Darawish, R.; Dodero, A.; Vicini, S. Electrospun composite mats of alginate with embedded silver nanoparticles: Synthesis and characterization. *J. Therm. Anal. Calorim.* **2019**, *137*, 767–778. [CrossRef]
91. Kaczmarek-Pawelska, A. Alginate-Based Hydrogels in Regenerative Medicine. In *Alginates—Recent Uses of This Natural Polymer*; Pereira, L., Ed.; IntechOpen: London, UK, 2020; ISBN 978-1-78985-641-5.
92. Pan, L.; Wang, Z.; Zhao, X.; He, H. Efficient removal of lead and copper ions from water by enhanced strength-toughness alginate composite fibers. *Int. J. Biol. Macromol.* **2019**, *134*, 223–229. [CrossRef]
93. Fu, X.; Liang, Y.; Wu, R.; Shen, J.; Chen, Z.; Chen, Y.; Wang, Y.; Xia, Y. Conductive core-sheath calcium alginate/graphene composite fibers with polymeric ionic liquids as an intermediate. *Carbohydr. Polym.* **2019**, *206*, 328–335. [CrossRef]

94. De Silva, R.T.; Mantilaka, M.M.M.G.P.G.; Goh, K.L.; Ratnayake, S.P.; Amaratunga, G.A.J.; de Silva, K.M.N. Magnesium Oxide Nanoparticles Reinforced Electrospun Alginate-Based Nanofibrous Scaffolds with Improved Physical Properties. *Int. J. Biomater.* **2017**, *2017*, 1391298. [CrossRef]
95. Nasri-Nasrabadi, B.; Kaynak, A.; Heidarian, P.; Komeily-Nia, Z.; Mehrasa, M.; Salehi, H.; Kouzani, A.Z. Sodium alginate/magnesium oxide nanocomposite scaffolds for bone tissue engineering. *Polym. Adv. Technol.* **2018**, *29*, 2553–2559. [CrossRef]
96. Sa, V.; Kornev, K.G. A method for wet spinning of alginate fibers with a high concentration of single-walled carbon nanotubes. *Carbon* **2011**, *49*, 1859–1868. [CrossRef]
97. Qin, Y. Alginate fibres: An overview of the production processes and applications in wound management. *Polym. Int.* **2008**, *57*, 171–180. [CrossRef]
98. Fang, Y.; Al-Assaf, S.; Phillips, G.O.; Nishinari, K.; Funami, T.; Williams, P.A.; Li, L. Multiple Steps and Critical Behaviors of the Binding of Calcium to Alginate. *J. Phys. Chem. B* **2007**, *111*, 2456–2462. [CrossRef]
99. Lima, A.M.F.; de Lima, M.F.; Assis, O.B.G.; Raabe, A.; Amoroso, H.C.; de Oliveira Tiera, V.A.; de Andrade, M.B.; Tiera, M.J. Synthesis and Physicochemical Characterization of Multiwalled Carbon Nanotubes/Hydroxamic Alginate Nanocomposite Scaffolds. *J. Nanomater.* **2018**, *2018*, 4218270. [CrossRef]
100. Wan, F.; Ping, H.; Wang, W.; Zou, Z.; Xie, H.; Su, B.-L.; Liu, D.; Fu, Z. Hydroxyapatite-reinforced alginate fibers with bioinspired dually aligned architectures. *Carbohydr. Polym.* **2021**, *267*, 118167. [CrossRef]
101. Ni, P.; Bi, H.; Zhao, G.; Han, Y.; Wickramaratne, M.N.; Dai, H.; Wang, X. Electrospun preparation and biological properties in vitro of polyvinyl alcohol/sodium alginate/nano-hydroxyapatite composite fiber membrane. *Colloids Surf. B Biointerfaces* **2019**, *173*, 171–177. [CrossRef]
102. Zhang, X.; Huang, C.; Zhao, Y.; Jin, X. Preparation and characterization of nanoparticle reinforced alginate fibers with high porosity for potential wound dressing application. *RSC Adv.* **2017**, *7*, 39349–39358. [CrossRef]
103. Weng, L.; Zhang, X.; Fan, W.; Lu, Y. Development of the inorganic nanoparticles reinforced alginate-based hybrid fiber for wound care and healing. *J. Appl. Polym. Sci.* **2021**, *138*, 51228. [CrossRef]
104. Wijesinghe, W.A.J.P.; Jeon, Y.-J. Biological activities and potential cosmeceutical applications of bioactive components from brown seaweeds: A review. *Phytochem. Rev.* **2011**, *10*, 431–443. [CrossRef]
105. Zhijiang, C.; Cong, Z.; Ping, X.; Jie, G.; Kongyin, Z. Calcium alginate-coated electrospun polyhydroxybutyrate/carbon nanotubes composite nanofibers as nanofiltration membrane for dye removal. *J. Mater. Sci.* **2018**, *53*, 14801–14820. [CrossRef]
106. Zhao, X.; Wang, X.; Lou, T. Preparation of fibrous chitosan/sodium alginate composite foams for the adsorption of cationic and anionic dyes. *J. Hazard. Mater.* **2021**, *403*, 124054. [CrossRef]
107. Mokhena, T.C.; Jacobs, N.V.; Luyt, A.S. Electrospun alginate nanofibres as potential bio-sorption agent of heavy metals in water treatment. *Express Polym. Lett.* **2017**, *11*, 652–663. [CrossRef]
108. Mokhena, T.C.; Jacobs, N.V.; Luyt, A.S. Nanofibrous alginate membrane coated with cellulose nanowhiskers for water purification. *Cellulose* **2018**, *25*, 417–427. [CrossRef]
109. Sun, F.; Guo, J.; Liu, Y.; Yu, Y. Preparation, characterizations and properties of sodium alginate grafted acrylonitrile/polyethylene glycol electrospun nanofibers. *Int. J. Biol. Macromol.* **2019**, *137*, 420–425. [CrossRef] [PubMed]
110. Jeong, S.; Jeon, O.; Krebs, M.; Hill, M.; Alsberg, E. Biodegradable photo-crosslinked alginate nanofibre scaffolds with tuneable physical properties, cell adhesivity and growth factor release. *Eur. Cells Mater.* **2012**, *24*, 331–343. [CrossRef] [PubMed]
111. Dodero, A.; Scarfi, S.; Pozzolini, M.; Vicini, S.; Alloisio, M.; Castellano, M. Alginate-Based Electrospun Membranes Containing ZnO Nanoparticles as Potential Wound Healing Patches: Biological, Mechanical, and Physicochemical Characterization. *ACS Appl. Mater. Interfaces* **2020**, *12*, 3371–3381. [CrossRef] [PubMed]
112. Hu, W.-W.; Wu, Y.-C.; Hu, Z.-C. The development of an alginate/polycaprolactone composite scaffold for in situ transfection application. *Carbohydr. Polym.* **2018**, *183*, 29–36. [CrossRef] [PubMed]
113. Leung, M.; Kievit, F.M.; Florczyk, S.J.; Veiseh, O.; Wu, J.; Park, J.O.; Zhang, M. Chitosan-Alginate Scaffold Culture System for Hepatocellular Carcinoma Increases Malignancy and Drug Resistance. *Pharm. Res.* **2010**, *27*, 1939–1948. [CrossRef] [PubMed]
114. Kievit, F.M.; Florczyk, S.J.; Leung, M.C.; Veiseh, O.; Park, J.O.; Disis, M.L.; Zhang, M. Chitosan–alginate 3D scaffolds as a mimic of the glioma tumor microenvironment. *Biomaterials* **2010**, *31*, 5903–5910. [CrossRef]
115. Hu, W.-W.; Lin, C.-H.; Hong, Z.-J. The enrichment of cancer stem cells using composite alginate/polycaprolactone nanofibers. *Carbohydr. Polym.* **2019**, *206*, 70–79. [CrossRef]
116. Qin, Y. Silver-containing alginate fibres and dressings. *Int. Wound J.* **2005**, *2*, 172–176. [CrossRef]
117. Carpa, R.; Remizovschi, A.; Culda, C.A.; Butiuc-Keul, A.L. Inherent and Composite Hydrogels as Promising Materials to Limit Antimicrobial Resistance. *Gels* **2022**, *8*, 70. [CrossRef]
118. Diniz, F.R.; Maia, R.C.A.P.; Rannier Andrade, L.; Andrade, L.N.; Vinicius Chaud, M.; da Silva, C.F.; Corrêa, C.B.; de Albuquerque, R.L.C., Jr.; Pereira da Costa, L.; Shin, S.R.; et al. Silver Nanoparticles-Composing Alginate/Gelatine Hydrogel Improves Wound Healing In Vivo. *Nanomaterials* **2020**, *10*, 390. [CrossRef]
119. Guo, J.; Zhang, Q.; Cai, Z.; Zhao, K. Preparation and dye filtration property of electrospun polyhydroxybutyrate–calcium alginate/carbon nanotubes composite nanofibrous filtration membrane. *Sep. Purif. Technol.* **2016**, *161*, 69–79. [CrossRef]
120. Jeong, S.I.; Krebs, M.D.; Bonino, C.A.; Khan, S.A.; Alsberg, E. Electrospun Alginate Nanofibers with Controlled Cell Adhesion for Tissue Engineeringa: Electrospun Alginate Nanofibers. *Macromol. Biosci.* **2010**, *10*, 934–943. [CrossRef]

121. Jeong, S.I.; Krebs, M.D.; Bonino, C.A.; Samorezov, J.E.; Khan, S.A.; Alsberg, E. Electrospun Chitosan–Alginate Nanofibers with In Situ Polyelectrolyte Complexation for Use as Tissue Engineering Scaffolds. *Tissue Eng. Part A* **2011**, *17*, 59–70. [CrossRef]
122. Shao, X.; Hunter, C.J. Developing an alginate/chitosan hybrid fiber scaffold for annulus fibrosus cells. *J. Biomed. Mater. Res.* **2007**, *82*, 701–710. [CrossRef]
123. Tao, F.; Cheng, Y.; Tao, H.; Jin, L.; Wan, Z.; Dai, F.; Xiang, W.; Deng, H. Carboxymethyl chitosan/sodium alginate-based micron-fibers fabricated by emulsion electrospinning for periosteal tissue engineering. *Mater. Des.* **2020**, *194*, 108849. [CrossRef]
124. Wu, H.; Liu, J.; Fang, Q.; Xiao, B.; Wan, Y. Establishment of nerve growth factor gradients on aligned chitosan-polylactide/alginate fibers for neural tissue engineering applications. *Colloids Surf. B Biointerfaces* **2017**, *160*, 598–609. [CrossRef]
125. Leu Alexa, R.; Ianchis, R.; Savu, D.; Temelie, M.; Trica, B.; Serafim, A.; Vlasceanu, G.M.; Alexandrescu, E.; Preda, S.; Iovu, H. 3D Printing of Alginate-Natural Clay Hydrogel-Based Nanocomposites. *Gels* **2021**, *7*, 211. [CrossRef]
126. Dumont, M.; Villet, R.; Guirand, M.; Montembault, A.; Delair, T.; Lack, S.; Barikosky, M.; Crepet, A.; Alcouffe, P.; Laurent, F.; et al. Processing and antibacterial properties of chitosan-coated alginate fibers. *Carbohydr. Polym.* **2018**, *190*, 31–42. [CrossRef]
127. Sibaja, B.; Culbertson, E.; Marshall, P.; Boy, R.; Broughton, R.M.; Solano, A.A.; Esquivel, M.; Parker, J.; Fuente, L.D.L.; Auad, M.L. Preparation of alginate–chitosan fibers with potential biomedical applications. *Carbohydr. Polym.* **2015**, *134*, 598–608. [CrossRef]
128. Batista, M.P.; Gonçalves, V.S.S.; Gaspar, F.B.; Nogueira, I.D.; Matias, A.A.; Gurikov, P. Novel alginate-chitosan aerogel fibres for potential wound healing applications. *Int. J. Biol. Macromol.* **2020**, *156*, 773–782. [CrossRef]
129. Szymańska, E.; Winnicka, K.; Wieczorek, P.; Sacha, P.; Tryniszewska, E. Influence of Unmodified and β-Glycerophosphate Cross-Linked Chitosan on Anti-Candida Activity of Clotrimazole in Semi-Solid Delivery Systems. *Int. J. Mol. Sci.* **2014**, *15*, 17765–17777. [CrossRef]
130. Noppakundilograt, S.; Sonjaipanich, K.; Thongchul, N.; Kiatkamjornwong, S. Syntheses, characterization, and antibacterial activity of chitosan grafted hydrogels and associated mica-containing nanocomposite hydrogels: Characterization of Grafted CS/mica Nanocomposite Hydrogels. *J. Appl. Polym. Sci.* **2013**, *127*, 4927–4938. [CrossRef]
131. Wang, G.; Zhu, J.; Chen, X.; Dong, H.; Li, Q.; Zeng, L.; Cao, X. Alginate based antimicrobial hydrogels formed by integrating Diels–Alder "click chemistry" and the thiol–ene reaction. *RSC Adv.* **2018**, *8*, 11036–11042. [CrossRef]
132. Albadr, A.; Coulter, S.; Porter, S.; Thakur, R.; Laverty, G. Ultrashort Self-Assembling Peptide Hydrogel for the Treatment of Fungal Infections. *Gels* **2018**, *4*, 48. [CrossRef]
133. Atefyekta, S.; Blomstrand, E.; Rajasekharan, A.K.; Svensson, S.; Trobos, M.; Hong, J.; Webster, T.J.; Thomsen, P.; Andersson, M. Antimicrobial Peptide-Functionalized Mesoporous Hydrogels. *ACS Biomater. Sci. Eng.* **2021**, *7*, 1693–1702. [CrossRef]
134. He, Y.; Du, E.; Zhou, X.; Zhou, J.; He, Y.; Ye, Y.; Wang, J.; Tang, B.; Wang, X. Wet-spinning of fluorescent fibers based on gold nanoclusters-loaded alginate for sensing of heavy metal ions and anti-counterfeiting. *Spectrochim. Acta Part A Mol. Biomol. Spectrosc.* **2020**, *230*, 118031. [CrossRef]
135. Mokhena, T.C.; Jacobs, V.; Luyt, A.S. A review on electrospun bio-based polymers for water treatment. *Express Polym. Lett.* **2015**, *9*, 839–880. [CrossRef]
136. Zhang, J.; Wang, X.-X.; Zhang, B.; Ramakrishna, S.; Yu, M.; Ma, J.-W.; Long, Y.-Z. In Situ Assembly of Well-Dispersed Ag Nanoparticles throughout Electrospun Alginate Nanofibers for Monitoring Human Breath—Smart Fabrics. *ACS Appl. Mater. Interfaces* **2018**, *10*, 19863–19870. [CrossRef]
137. Hu, W.-P.; Zhang, B.; Zhang, J.; Luo, W.-L.; Guo, Y.; Chen, S.-J.; Yun, M.-J.; Ramakrishna, S.; Long, Y.-Z. Ag/alginate nanofiber membrane for flexible electronic skin. *Nanotechnology* **2017**, *28*, 445502. [CrossRef] [PubMed]
138. Chen, H.; Gao, Y.; Ren, X.; Gao, G. Alginate fiber toughened gels similar to skin intelligence as ionic sensors. *Carbohydr. Polym.* **2020**, *235*, 116018. [CrossRef] [PubMed]
139. Zhang, X.; Sheng, N.; Wang, L.; Tan, Y.; Liu, C.; Xia, Y.; Nie, Z.; Sui, K. Supramolecular nanofibrillar hydrogels as highly stretchable, elastic and sensitive ionic sensors. *Mater. Horiz.* **2019**, *6*, 326–333. [CrossRef]
140. Mallakpour, S.; Azadi, E.; Hussain, C.M. Chitosan, alginate, hyaluronic acid, gums, and β-glucan as potent adjuvants and vaccine delivery systems for viral threats including SARS-CoV-2: A review. *Int. J. Biol. Macromol.* **2021**, *182*, 1931–1940. [CrossRef] [PubMed]
141. Gherlone, E.; Polizzi, E.; Tetè, G.; Capparè, P. Dentistry and COVID-19 pandemic: Operative indications post-lockdown. *New Microbiol* **2021**, *44*, 1–11.
142. Nasiri, K.; Dimitrova, A. Comparing saliva and nasopharyngeal swab specimens in the detection of COVID-19: A systematic review and meta-analysis. *J. Dent. Sci.* **2021**, *16*, 799–805. [CrossRef]
143. Amato, A.; Caggiano, M.; Amato, M.; Moccia, G.; Capunzo, M.; De Caro, F. Infection Control in Dental Practice During the COVID-19 Pandemic. *Int. J. Environ. Res. Public Health* **2020**, *17*, 4769. [CrossRef]
144. Barenghi, L.; Barenghi, A.; Garagiola, U.; Di Blasio, A.; Giannì, A.B.; Spadari, F. Pros and Cons of CAD/CAM Technology for Infection Prevention in Dental Settings during COVID-19 Outbreak. *Sensors* **2021**, *22*, 49. [CrossRef]
145. Crook, H.; Raza, S.; Nowell, J.; Young, M.; Edison, P. Long COVID—Mechanisms, risk factors, and management. *BMJ* **2021**, *374*, n1648. [CrossRef]
146. Capparè, P.; D'Ambrosio, R.; De Cunto, R.; Darvizeh, A.; Nagni, M.; Gherlone, E. The Usage of an Air Purifier Device with HEPA 14 Filter during Dental Procedures in COVID-19 Pandemic: A Randomized Clinical Trial. *Int. J. Environ. Res. Public Health* **2022**, *19*, 5139. [CrossRef]

147. Hardan, L.; Bourgi, R.; Cuevas-Suárez, C.E.; Lukomska-Szymanska, M.; Cornejo-Ríos, E.; Tosco, V.; Monterubbianesi, R.; Mancino, S.; Eid, A.; Mancino, D.; et al. Disinfection Procedures and Their Effect on the Microorganism Colonization of Dental Impression Materials: A Systematic Review and Meta-Analysis of In Vitro Studies. *Bioengineering* **2022**, *9*, 123. [CrossRef]
148. Demajo, J.; Cassar, V.; Farrugia, C.; Millan-Sango, D.; Sammut, C.; Valdramidis, V.; Camilleri, J. Effectiveness of Disinfectants on Antimicrobial and Physical Properties of Dental Impression Materials. *Int. J. Prosthodont.* **2016**, *29*, 63–67. [CrossRef]

Journal of Functional Biomaterials

Article

The Mineralization of Various 3D-Printed PCL Composites

Artem Egorov [1,2], Bianca Riedel [1], Johannes Vinke [2], Hagen Schmal [1], Ralf Thomann [3], Yi Thomann [3] and Michael Seidenstuecker [1,*]

[1] G.E.R.N. Center of Tissue Replacement, Regeneration & Neogenesis, Department of Orthopedics and Trauma Surgery, Medical Center–Albert-Ludwigs-University of Freiburg, Faculty of Medicine, Albert-Ludwigs-University of Freiburg, Hugstetter Straße 55, 79106 Freiburg, Germany

[2] Institute for Applied Biomechanics, Offenburg University, Badstraße 24, 77652 Offenburg, Germany

[3] Freiburg Center for Interactive Materials and Bioinspired Technologies (FIT), Albert-Ludwigs-University Freiburg, Georges-Koehler-Allee 105, 79110 Freiburg, Germany

* Correspondence: michael.seidenstuecker@uniklinik-freiburg.de

Abstract: In this project, different calcification methods for collagen and collagen coatings were compared in terms of their applicability for 3D printing and production of collagen-coated scaffolds. For this purpose, scaffolds were printed from polycaprolactone PCL using the EnvisionTec 3D Bioplotter and then coated with collagen. Four coating methods were then applied: hydroxyapatite (HA) powder directly in the collagen coating, incubation in $10\times$ SBF, coating with alkaline phosphatase (ALP), and coating with poly-L-aspartic acid. The results were compared by ESEM, μCT, TEM, and EDX. HA directly in the collagen solution resulted in a pH change and thus an increase in viscosity, leading to clumping on the scaffolds. As a function of incubation time in $10\times$ SBF as well as in ALP, HA layer thickness increased, while no coating on the collagen layer was apparently observed with poly-L-aspartic acid. Only ultrathin sections and TEM with SuperEDX detected nano crystalline HA in the collagen layer. Exclusively the incubation in poly-L-aspartic acid led to HA crystals within the collagen coating compared to all other methods where the HA layers formed in different forms only at the collagen layer.

Keywords: PCL scaffolds; 3D printing; collagen coating; hydroxyapatite; alkaline phosphatase; poly-L aspartic acid

1. Introduction

Demographic change is currently a much-discussed phenomenon whose consequences appear more clearly every year [1,2]. In Germany, every second person is 45 years and one in five is older than 66 years of age [3]. In the EU, the average age of the population continues to rise and one in five is already older than the age of 65 [4]. A well-known and well-followed consequence is the increase of musculoskeletal diseases and the related increase of clinical interventions on the musculoskeletal system. For example, the use of an endoprosthesis in the hip is the sixth most frequent operation in Germany [5]. Thus, the need for clinically proven and effective bone graft substitutes is also increasing. However, there are still known and documented problems with the use of metal implants. Among other things, the unequal relationship of elastic moduli between the metals used and human bone can lead to undesirable or too weak bone growth (so-called stress shielding) [6]. In addition, a metallic implant is usually installed in the body for a long period of time. Depending on the age of the patient at initial implantation, re-implantation of an endoprosthesis may be required after 10–15 years (knee), 15–20 years (hip), or 10 years (shoulder), depending on the site of implantation [7,8]. Such reoperations pose enormous risks to the health of patients, especially at advanced ages. Bone substitutes made of biodegradable biomaterials represent an innovative alternative, which ideally are dissolved by the body after preservation of mechanical stability and healing of the bone [9]. One such biodegradable biomaterial is polycaprolactone (PCL) [10]. It is a biodegradable semi-crystalline polymer. Its good formability at relatively low temperatures makes it ideal for additive manufacturing processes in the context of medical use [9]. Additive manufacturing processes offer enormous potential for individual patient care. If, for example, bone is missing after a fracture, this individual lesion can be converted into a digital three-dimensional construct using clinical imaging techniques (CT, MRI). In a further step, this three-dimensional (3D) model of the damaged bone can be used to develop a bone substitute that is perfectly adapted to the individual case. This should ensure mechanical stability and thus take over the function of the damaged bone. This individual replacement piece made of PCL also has the advantage that, as a biodegradable polymer, it can be degraded by the body once it has fulfilled its task and as the bone heals. This eliminates the need for a second surgery to remove the structure, with all the associated risks. To enable even faster recovery and new bone formation, autologous bone-forming cells can be applied to bone substitutes made of PCL (further scaffolds). These cells accelerate new bone formation within the printed scaffolds. Thus, the printed PCL scaffold serves as a supporting and guiding structure during bone regeneration. Moreover, the surface of the printed PCL scaffolds plays an important role in regeneration. This surface has hydrophobic properties and thus inhibits cell colonization. Cell adhesion and further cell profiling does not occur with pure PCL [10]. To improve cell adhesion and cell profiling, the printed scaffolds should provide an "in vivo"-like surface. This should be based on the microstructure of bone, with main components consisting of collagen and HA. In theory, coating with collagen has improved cell adhesion [11] and cell profiling. The formation of HA on the collagen layer provides further improvement of cell profiling [12]. The integration of HA would provide a biomimetic layer [13]. In the present study, different ways of preparing this collagen HA surface layer were applied and investigated. Thus, various methods for generating HA layers on or within the collagen coatings of 3D-printed PCL scaffolds have been investigated as a possible use as bone substitutes. Ordered nano-crystalline HA layers within the collagen are already one step closer to the generation of bone tissue in the petri dish.

2. Materials and Methods

2.1. Materials

Ethanol, magnesium-chloride-hexahydrate, and PCL (Mn = 45,000; Art. No. 704105) were purchased by Merck (Merck KGaA, Darmstadt, Germany). Hydroxyapatite (Art. No 677418), 1-ethyl-3-(3-dimethylaminopopyl)-carbodiimid hydrochloride (Art. No. SLBZ0862), potassium chloride (Art. No P5405), Trizma base (Art. No. T6065), magnesium sulphate (Art. No. M2643-100GM) and Alkaline phosphatase (Art. No. P7923-2KU) were purchased by Sigma-Aldrich (Sigma-Aldrich, St. Louis, MO, USA). Collagen I (Art. No. 354236) was purchased by Corning (Corning, NY, USA). Calciumchloride (Art. No. CN93.2), Sodium-di-hydrogenphosphate (Art. No. K300.1), sodiumhydrogencarbonate (Art. No. 6885.2), Sodiumhydroxide (Art. No. 6771.1), di-sodium-hydrogenphosphate (Art. No. T876.1), Sodiumdihydrogenphosphate monohydrate (Art. No. K300.1), Sodiumchloride (Art. No. HN00.2), and di-ammoniumhydrogenphosphate (Art. No. P736.1) were purchased by Carl Roth (Carl Roth GmbH, Karlsruhe, Germany). Poly-L-aspartic acid sodium salt was purchased by Alamanda Polymers (Alamanda Polymers Inc., Huntsville, AL, USA). Beta-glycerphosphate di-sodium salt pentahydrate was purchased by EMD Millipore (EMD Millipore Corp., Billerica, MA, USA). Goat anti mouse Alexa 488 antibody (Art. No: A-11001) was purchased by ThermoFisher Scientific (Thermo Fisher Scientific Inc., Waltham, MA, USA).

2.2. Methods

2.2.1. 3D Printing of PCL Scaffolds

The PCL scaffolds were manufactured by using a 3D-BioPlotter (EnvisionTEC, Gladbeck, Germany) according to our previous work [11,14]. For this purpose, a 3D model of the scaffold was created in Autodesk Inventor (Autodesk Inventor 2019; Autodesk,

San Raphael, CA, USA). This was assigned to the printer in the subsequent step. The 3D printing parameters used are shown in Table 1. Before 3D printing, the PCL used was dried for one hour in a desiccator under vacuum in silica gel. It was then transferred inside a cartridge into the print head of the 3D-BioPlotter, which had been preheated to 80 °C. After one hour, the PCL became liquid and thus ready for 3D printing. A glass plate, previously cleaned with isopropanol, was placed underneath as a printing surface. The 3D print was made by plotting 12 layers. Each layer contained a frame and an inner structure, the so-called base pattern, with a layer height of 0.17 mm. Thus, with 12 layers, the total height was 2.04 mm. This resulted in a cuboid scaffold. The dimensions of a scaffold were 8.4 mm × 8.4 mm × 2.04 mm.

Table 1. Parameters for the 3D printing of PCL.

Used Needle	Pressure	Temperature	Speed	Needle-Offset	Pre/Post Flow	Temperature Underground
24G	4–5 Bar	80 °C	1.0 mm/s	0.19 mm	0.07 s pre 0.10 s post	17 °C

The finished scaffolds were visually inspected for print defects and contamination, with discarding of defective specimens and storage of acceptable specimens at room temperature.

2.2.2. Collagen Coating

The final printed pre-sorted scaffolds were dried and simultaneously sterilized by five immersions in an ascending alcohol series (30/50/70/80/96 and 100% ethanol) before coating with collagen. After drying the scaffolds, their surfaces were activated by plasma (Piezobrush PZ3, Relyon Plasma GmbH, Regensburg, Germany). For surface activation, the scaffolds were plasma treated with the Piezobrush 3 at 80% power according to the manufacturer instructions at room temperature in air. Each scaffold was always treated for 30 s, cooled for 10 s, and then treated again for 20 s. Immediately after activation, the activated scaffolds were transferred to a 24-well plate containing 600 µL of collagen solution per well. The well plate (with scaffolds and collagen) was then incubated on a moving plate at 4 °C for 72 h. The scaffolds were then removed from the collagen solution and dried at 37 °C for 24 h. To cross-link the collagen layer, the dried scaffolds were transferred to a cross-linking solution in the subsequent step. 10 mL of cross-linking solution consisted of 10 mL of 95% ethanol containing 95.85 mg of 1-ethyl-3-(3-dimethylaminopropyl)carbodiimide hydrochloride (EDC) powder. The solution was shaken using a shaker (IKA Shaker MS 3 Basic) until the EDC powder was completely dissolved. Then, the cross-linking solution was added to the dried scaffolds at 1 mL per well (thus also 1 mL per scaffold) and incubated for 16 h. After the incubation period, the final scaffolds were cleaned five times with deionized water and then five times with 70% ethanol and dried on a filter paper. For stable storage, the dry scaffolds were stored in a new corrugated plate at room temperature.

2.2.3. Methods for Inserting Hydroxyapatite

For the following experiments, at least 10 scaffolds were coated with the different methods. All experiments were repeated at least three times.

Collagen-Hydroxyapatite Coating

The procedure was similar to the collagen coating process presented in the previous chapter. First, 5% by weight of HA was added to the collagen as a nanopowder. The collagen-HA solution was further placed in an ultrasonic bath (Elmasonic P 60H, Elma Schmidbauer GmbH, Singen, Germany) at a frequency of 120 Hz for 1 h before addition to prevent rapid agglomeration. During the ultrasonic bath, the rising temperature of the water was steadily cooled down to 20–25 °C with double distilled water ice to prevent denaturation of the collagen. Since the addition of HA leads to gel formation [15] in the type of collagen used (Corning), caused by the basic effect of HA, 0.02 molar acetic acid was added. The acetic acid corresponded to the solvent of the Corning collagen solution. Furthermore, the amount of HA was reduced to 2% by weight so that the gel could be liquefied again. The liquid collagen-HA solution was again shaken with a Minivortex (Roth) to prevent agglomeration of the HA. In the subsequent step, 600 µL of this collagen-HA solution per well was added to the scaffolds. The following steps were identical to those for collagen coating. With 5% w/v HA, collagen gel formation occurred. Only when using 2% w/v HA nanopowder in the collagen and diluting with 0.02 M acetic acid could gel formation be prevented.

Immersion in SBF

Based on the findings of Poh et al. [16] and Vaquette et al. [17] immersion of objects in simulated body fluid (further referred to as SBF) leads to the formation of an HA layer on the surface of the inserted objects. According to Gomori et al. [18], this results from the well-known reaction of phosphates in calcium-rich environments. Both are sufficiently present in SBF, above the amounts required for the formation of HA. Furthermore, the formation time of the HA layer is shortened by increasing the ion concentration [16]. This was used in this variant for application for collagen coated scaffolds by increasing the ion concentration tenfold. First, $10\times$ SBF was prepared according to the formulation of Yang et al. [19]. For this purpose, the substances shown in Table 2 were added sequentially in the order and amount listed until they dissolved completely in 1 L double distilled water. To do this, the sequence must be followed and the substances must be completely dissolved before the next substance is added to avoid precipitation of the dissolved substances. The final $10\times$ SBF solution can be stored at 4 °C for several weeks without precipitation of the dissolved substances.

Table 2. Order and amount of substances required for $10\times$ SBF.

Order	Substance	Amount (for 1 L $10\times$ SBF)
1	NaCl	58.430 g
2	KCl	0.373 g
3	$CaCl_2$–2 H_2O	3.675 g
4	$MgCl_2$–6 H_2O	1.016 g
5	Na_2HPO_4–H_2O	1.633 g

Before storing the coated scaffolds in $10\times$ SBF, the pH of the $10\times$ SBF solution was adjusted to pH 6 with $NaHCO_3$. Then, 1 mL of the $10\times$ SBF was added per scaffold to a 24-well plate loaded with scaffolds (including collagen coating) in the subsequent step. The well plate was then transferred to a drying oven at 37 °C and the well was gently shaken after 15 min. After 30 min, the medium was renewed by pipetting off the old solution and pipetting in fresh $10\times$ SBF with pH 6. This step was continued throughout the duration of the experiment. To study different film thicknesses, different samples were stored in $10\times$ SBF for 1 h ($10\times$ SBF 1 h), 2 h ($10\times$ SBF 2 h), 4 h ($10\times$ SBF 4 h), and 8 h ($10\times$ SBF 8 h), respectively. Then, 30 min after the last change of medium, the samples were transferred to 0.5 M NaOH for 30 min. This step homogenizes the resulting HA layer. After the time elapsed, the final samples were washed with double distilled water until the pH value of the water being washed out reached approximately pH 7.

Surface Coating with Hydroxyapatite by Addition of ALP

According to Jaroscewicz et al. [20], it is possible to produce an HA layer with the help of ALP. ALP present in the body catalyzes the hydrolysis of phosphate monoesters. Starting materials of these reactions are inorganic phosphate and an alcohol [18]. In an environment enriched with the necessary amount of calcium, a reaction of the free calcium and the inorganic phosphate now occurs. This leads to the formation and precipitation of HA. This reaction was already observed by Gomori [18] in 1953. For its preparation, a solution enriched with the necessary concentrations of substances had to be prepared. The composition of the solution called "phosphatase incubation medium" (further referred to as PIM) was prepared as listed in Table 3 according to the work of Jaroszewicz et al. [20]. A total of 45 mL of PIM was obtained. The indicated substances were added one by one to double distilled water in the indicated order, using a magnetic stirrer. The next substance in the sequence was added only when the previous substance had completely dissolved. A measure of 45 mL of double distilled water was used as solvent (see Table 3).

Table 3. Sequence and amount of substances necessary for PIM.

Sequence	Substance	Amount (for 45 mL PIM)
1	TRIS buffer	545.13 mg
2	$C_3H_7Na_2O_6P\ 5\ H_2O$	300 mg
3	$CaCl_2$	200 mg
4	$MgSO_4$	50 mg
5	NaN_3	9 mg

Freshly collagen-coated scaffolds were prepared for addition by storage in 1 mL 1× PBS/scaffold. The prepared 45 mL PIM were mixed with ALP. For this purpose, 10 nL of ALP per scaffold should be added to the PIM just before incubation. To achieve this small amount of ALP, the ALP was diluted with PIM to the extent that 10 nL of ALP is effectively dissolved in one milliliter. After adding the ALP to the PIM, 1 mL per scaffold of the prepared solution was immediately added to a 24 well plate loaded with scaffolds. The well plate was then transferred to a slow-moving moving plate at 37 °C in a drying oven. Here, the scaffolds remained for 1, 3, and 6 days, respectively, without changing the medium. After the scaffolds were removed, they were washed three times for 15 min in double-distilled water on a moving plate in a drying oven at 37 °C and then air-dried on filter paper.

Mineralization of Collagen with Poly-L-Aspartic Acid

Based on the work of Deshpande et al. [21], it is shown that the presence of polyaspartic acid (further polyASP) leads to biomineralization of the collagen itself. Using this approach, four initial solutions were prepared for the generation of mineralization of collagen. For the first solution, 10× phosphate buffered saline (further referred to as 10× PBS) was prepared following the sequence listed in Table 2. This was then diluted to 3.4× PBS. For the second and third starting solution solutions, a 6.8 mM $CaCl_2$ and 4 mM $(NH_4)_2HPO_4$ solutions were prepared. Poly-L-aspartic acid as the fourth starting solution was brought to a 500 µg/mL solution by dissolving and diluting from a 5 mg/mL solution. These prepared starting solutions were mixed in equal parts (25 µL each) to make a total of 100 µL of mineralization solution. Then, 20 µL of the mineralization solution per scaffold was pipetted as a drop on a Petri dish with a diameter of 5 cm and 5 scaffolds each. Here, the drop was added to the center of the scaffold. The Petri dish was placed inside a home-made humidity chamber. The humidity chambers were sealed and placed in the drying oven (Memmert UM200, Memmert GmbH & Co KG; Schwabach, Germany) for incubation at 37 °C for 6 h. At the end of the 6 h, the finished scaffolds were briefly washed with double distilled water warmed to 37 °C and transferred to filter paper for drying.

2.2.4. Characterization of the Scaffolds and Coatings

Characterization by 3D Laser Scanning Microscopy

The 3D printed scaffolds were characterized before and after plasma treatment using a 3D laser scanning microscope (KEYENCE, VK-X210, KEYENCE, Osaka, Japan). This captures both images of the surface and laser scans. Surface-specific values such as surface roughness (center roughness Sa) can then be measured, analogous to our previous work [11,14]. To investigate the surface roughness before and after plasma treatment of the scaffolds, images were acquired using the 3D laser microscope and the surface roughness was measured at four random locations. A scaffold was examined before plasma treatment and after plasma treatment.

Characterization by Immunoassay for Collagen I

To prove the successful coating of the scaffolds with collagen, immunofluorescence staining with an antibody against collagen I was performed. For this purpose, collagen-coated scaffolds and uncoated scaffolds were used as control group. The scaffolds were first washed with phosphate buffered saline (PBS). After washing, they were incubated for 45 min at room temperature in a so-called blocking solution. The blocking solution consisted of PBS, 1% BSA, and 0.1% Triton-x 100. Subsequently, the scaffolds were washed again 3 times with PBS. The primary antibody was then added, which was previously dissolved 1:1000 in a buffer (lowcross buffer). For the detection of collagen, the mouse anti-human collagen 1 (company info, ab6308) mouse antibody at a concentration of 1.5 mg/mL was used as the primary antibody. After addition of the primary antibody, the scaffolds were incubated at 4 °C overnight. Subsequently, the scaffolds were washed again 3 times with PBS. The scaffolds were then incubated in the secondary antibody. A goat anti-mouse Alexa 488 antibody (ThermoFisher Scientific) dissolved 1:500 in a lowcross buffer for 60 min at room temperature. These antibodies were labeled with a fluoreschrome, allowing them to be examined later with a fluorescence microscope (Olympus BX-53, 500 ms illumination time) (excitation at 495 nm (blue), emission 519 nm (green)).

Characterization by ESEM

To study the mineralization of the collagen layer, this sample had to be prepared. A cryo-fracture had to be prepared. To do this, finished samples were held in liquid nitrogen at −196 °C for 30 s, following the procedures presented. The frozen samples were broken into two pieces using a 10 mm chisel. Breaking in the frozen state resulted in an optimal breaking edge. The resulting fracture edge also allowed a view into the coating. For this purpose, the edge areas of the strands oriented perpendicular to the fracture were viewed. Furthermore, otherwise only optically significant and deviating areas were examined more closely in order to record further information as well as anomalies. The measurement parameters in the FEI Quanta 250 FEG were: 20 kV accelerating voltage, the use of a Large Field Gaseous, detector (LFD) for secondary electrons, low vacuum of 130 Pa.

Characterization by MicroCT

Samples examined with the μCT must undergo several preparation steps before examination. First, they were cryo-fractured, comparable to the ESEM samples (see chapter before). Unlike the preparation of the ESEM images, however, this was not used to expose individual strands to allow insight into the coating, but merely to reduce the size of the sample. Sections of approximate size 1 mm × 1 mm × 2.04 mm were taken. The now cryo-broken sample pieces were thoroughly rinsed three times with PBS and then transferred to an ethanol series for dehydration. Here, all samples were transferred sequentially to 30%, 50%, 70%, 80%, 90%, 96%, and 100% ethanol for 30 min each. To fix the samples, the samples were transferred to a 24 well plate containing 1 mL/well ($\geq$99.0%) hexamethyldisalazane for 4 h [22]. Finally, samples were removed from the hexamethyldisalazane and dried under a fume hood for 8 h. Dry samples were attached to the sample holder with superglue and then examined using μCT. The parameters used for the examinations with the μCT Skyscan 1272 system were as follows:

- Tube voltage: 40 kV
- Tube current: 250 μA
- Exposure time: 1815 ms

- Additional filtering: No additional filtering
- Binning: 1 × 1 (projection size: 4032 × 2688)
- Voxel size: 2.0 μm
- Rotation step: 0.15 degrees
- Frame averaging: 5
- 360° scan
- Random movement off

Characterization by TEM/EDX

For the TEM and EDX investigations, the PCL moldings were coated with collagen and treated with poly-L-aspartic acid. Subsequently, the specimens were cold-embedded in PELCO® 10505 silicone embedding molds in Rencast FC52 (Goessl-Pfaff GmbH, Karlskron, Germany). Some samples were incubated in osmium (OsO4) for better detection of the collagen layer. Thin sections of the embedded samples were then prepared on the PowerTome XL ultramicrotome (Boeckeler Instruments Inc., Tucson, AZ, USA).

2.3. Statistics

Data are expressed as mean ± standard deviation of the mean and were analyzed by one-way analysis of variance (ANOVA). The level of statistical significance was set at $p < 0.05$. For statistical calculations, the Origin 2020 Professional SR1 (OriginLab, Northampton, MA, USA) was used.

3. Results

3.1. Characterization of the Scaffolds and Coatings

3.1.1. Characterization by 3D Laser Scanning Microscopy

No significant difference in surface roughness was found between the two groups (without and with plasma treatment). The mid-surface roughness Sa for the non-plasma treated specimens was 4.19 + 0.22 μm and for the plasma treated specimens 4.70 + 0.79 μm. The length, height, pore size, and strand width were also measured by means of 3D Laser-scanning microscopy (3D LSM) (pls see Table 4). The pore size was measured on the 3D LSM images by using ImageJ. However, there was a significant difference in surface roughness between uncoated (4.1 ± 0.1 μm) and coated scaffolds (3.35 ± 0.3 μm) with $p < 0.05$.

Table 4. Dimensions of the 3D-printed scaffolds.

Parameter	PCL Scaffold
Length (mm)	8.42 ± 0.01
Height (mm)	2.04 ± 0.03
Pore size (μm)	295.4 ± 9.8
Strand width (μm)	300 ± 12.6
Porosity (%)	31.9

3.1.2. Characterization of Collagen Coating by Immunoassay

The immunoassay showed that a collagen layer had formed on the scaffolds. In the following Figure 1, the intrinsic fluorescence of the unloaded scaffold is shown in (a) and the fluorescence of the loaded scaffolds is shown in (b). It is evident that the intrinsic fluorescence is very weak, whereas the fluorescence of the samples is very pronounced depending on the thickness of the collagen layer.

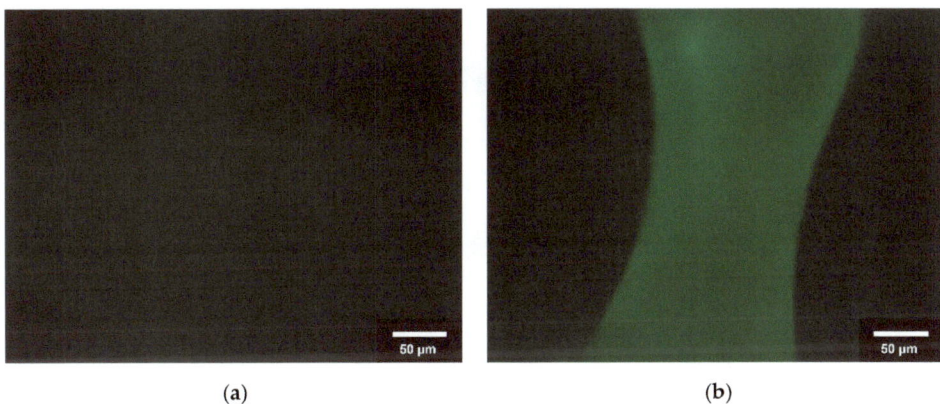

Figure 1. Overview of immunostaining; (**a**): uncoated scaffolds; (**b**): collagen-coated scaffolds. Images taken with Olympus BX-53 Fluorescence microscope @ 500 ms illumination time.

3.1.3. Characterization by Means of ESEM

Classical Collagen Coating

In the classical form of collagen coating, incubation in the collagen solution, collagen was deposited homogeneously on the surface of the scaffolds (see Figure 2). The thickness of the layer depended on the incubation time. In addition, it can be seen in Figure 2 that the uncoated scaffolds have gaps between the fused PCL particles, which were closed in a and b by the coating.

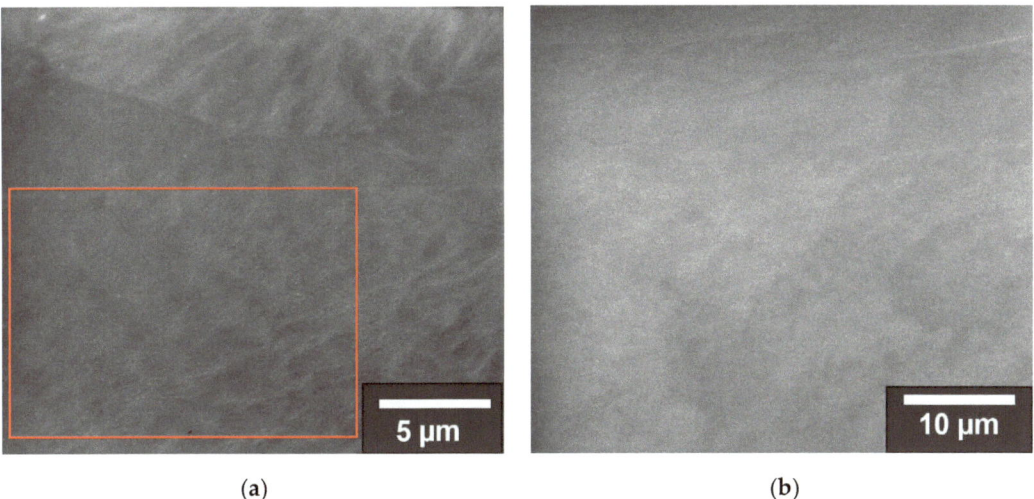

Figure 2. *Cont.*

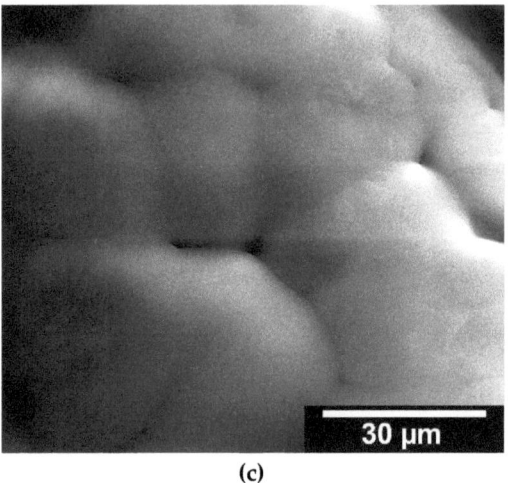

(c)

Figure 2. ESEM images of collagen coating by incubation; the red area in (**a**) is shown enlarged in (**b**); the uncoated scaffold is shown in (**c**).

Collagen-HA Coating

Due to the change in pH from slightly acidic to basic, the first mixing of collagen solution with nano HA resulted in the gelation of the suspension. To avoid this, all further tests were carried out with the addition of acetic acid. Figure 3 shows the inhomogeneous coating with gelled collagen HA coating. This is contrast with the collagen HA coating diluted with acetic acid. However, the dilution step changed the concentration of collagen + HA, which is reflected in a less effective coating.

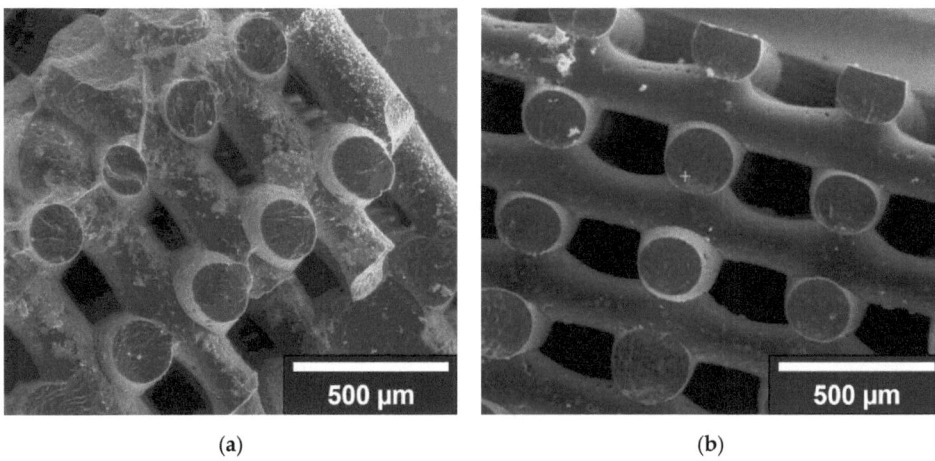

(a)   (b)

Figure 3. Comparison of collagen-HA coatings: (**a**): gelled collagen-HA coating; (**b**): collagen-HA coating, which has been diluted with acetic acid.

Surface Coating by Incubation in 10× SBF

The results of storage in 10× SBF for one hour showed no major agglomerations of crystalline structures, see overview image (Figure 4). However, a surface coating can already be recognized by the lighter shimmering. Looking at the individual results, crystalline structures are already clearly visible both in the analysis of the fracture edge and on the surface of a strand. In the top view of the fracture edge, the ridge line of protruding crystalline structures can also be seen. It is also clear that there is a lighter area between the ridge line and the PCL. This indicated that there is a layer between the PCL and the protruding crystalline structures. This is probably the collagen layer. When looking at the open area of a strand, crystalline structures were also visible, which are partially agglomerated. The results of the 2 h samples showed few changes compared to the samples immersed in 10× SBF for 1 h. In general, only an increased agglomeration of the crystalline structures already found in the samples of the 1 h storage is evident, this can also be seen in the overview images (Figure 4; left image). When looking at the fracture edge and the surface of a strand of the first sample, agglomerations of crystalline structures can be seen. On closer inspection of the second sample of the 2 h storage, clearly larger crystalline structures can be seen in the observation of the fracture edge; however, these could be related to the generally larger agglomerations that can be observed in the 2 h samples on the open strand. The results of the 4 h samples also showed comparable results with those of the 1 h and 2 h samples. The observed crystalline structures are larger and more defined compared to the 1 h and 2 h samples. Furthermore, the observed agglomerations of crystalline structures were also larger. Uncoated areas can be observed in both samples. A closer look at the fracture edge of the first sample again revealed crystalline structures on the surface. A top view of the open face of one strand showed an increased density and size of crystalline structures as well as some agglomerated structures. The samples that were incubated for 8 h in 10× SBF showed the largest agglomerations and thickest crystalline structures. Further magnification of both samples clearly showed, compared to the previous samples, even more defined and thicker crystalline structures as well as larger agglomerations in places. Thus, also when looking into the coating, much larger crystalline structures can be seen in both samples.

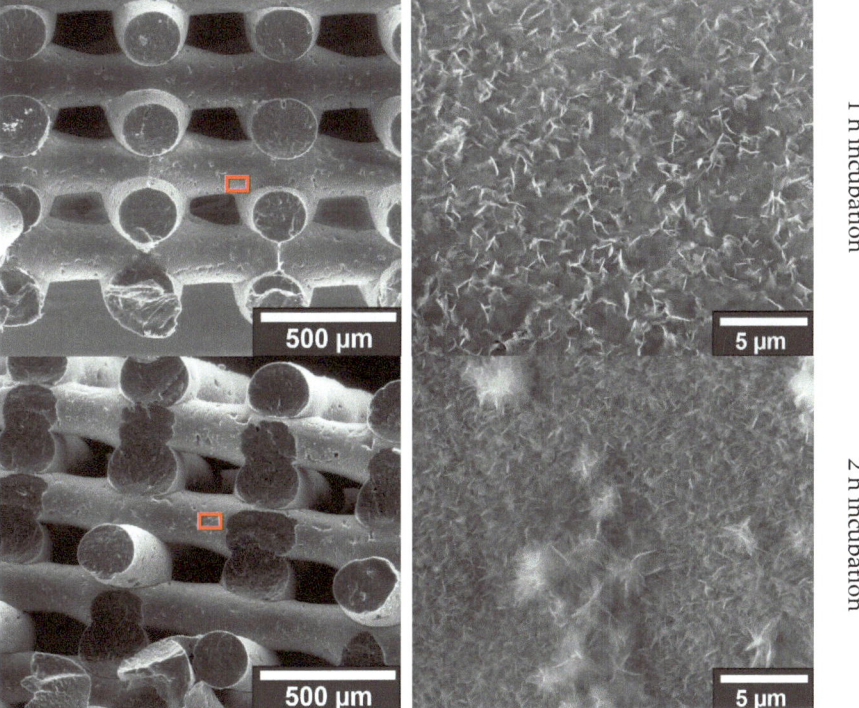

Figure 4. *Cont.*

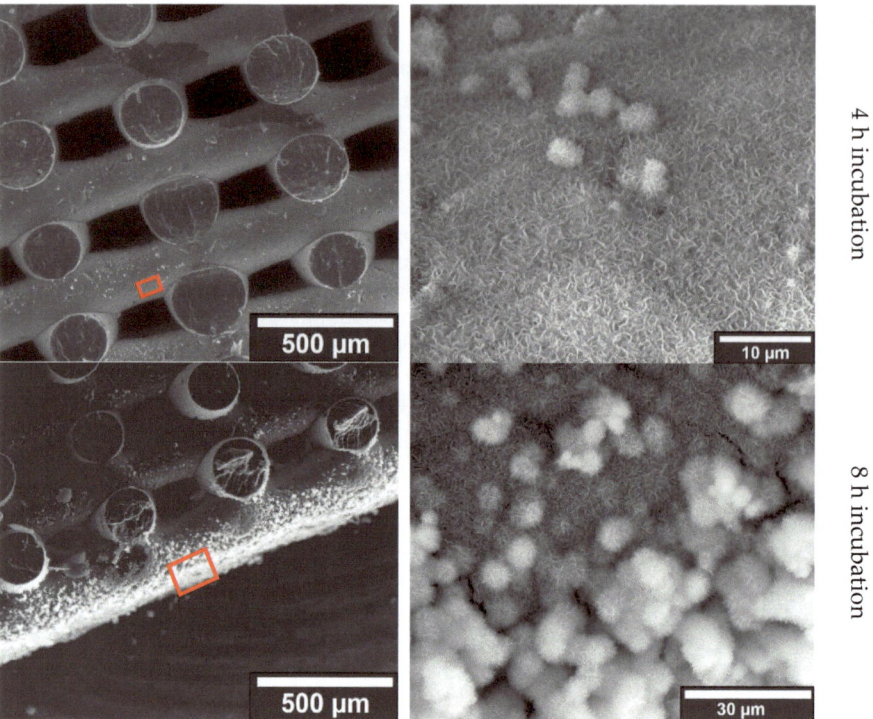

Figure 4. Comparative ESEM images of the coatings formed by incubation in 10× SBF. Incubation time varies from top to bottom: 1, 2, 4, and 8 h. The left image always shows the scaffold at 188× magnification, and the right image shows the crystalline structures and agglomerates on the surface. Because the crystalline structures and agglomerates increase over the incubation time, the magnification factor also varies. ESEM images were taken with a FEI Quanta FEG 250 @ 20 kV, 130 Pa, Large Field Detector.

Mineralization with ALP

The samples incubated for one day in PIM activated with ALP already showed first agglomerations of crystalline structures. Among and next to these agglomerated structures, crystals can also be found shining through as small dots on the ESEM images. The presence of small crystalline structures as well as agglomerations of these crystallites resembled, on a smaller scale, the results of the 10× SBF experiments. The samples incubated in PIM for 3 days continued to show crystalline structures under ESEM both in the observation of the fracture edge and on the surface of a strand. Compared to the 1-day sample, a finer distribution of small crystallites as well as more agglomeration were evident when looking at the surface. Looking at the fracture edge also showed a thicker surface layer compared to the 1-day sample. The images of the 6-day sample were comparable to the results observed in the 1-day and 3-day samples. The same small crystals as well as agglomerations were found when looking at the open area of a strand (blue). Again, the agglomeration increased with longer stored samples. However, the size of the observed crystals did not increase with longer storage. Observation of the fracture edge also revealed a larger layer of crystals on top of those protruding from the surface coating. When comparing the three different samples, the longer the storage in the ALP activated PIM, the stronger the agglomeration. This can be observed particularly well on the surface due to the occurrence of larger agglomerations (see Figure 5).

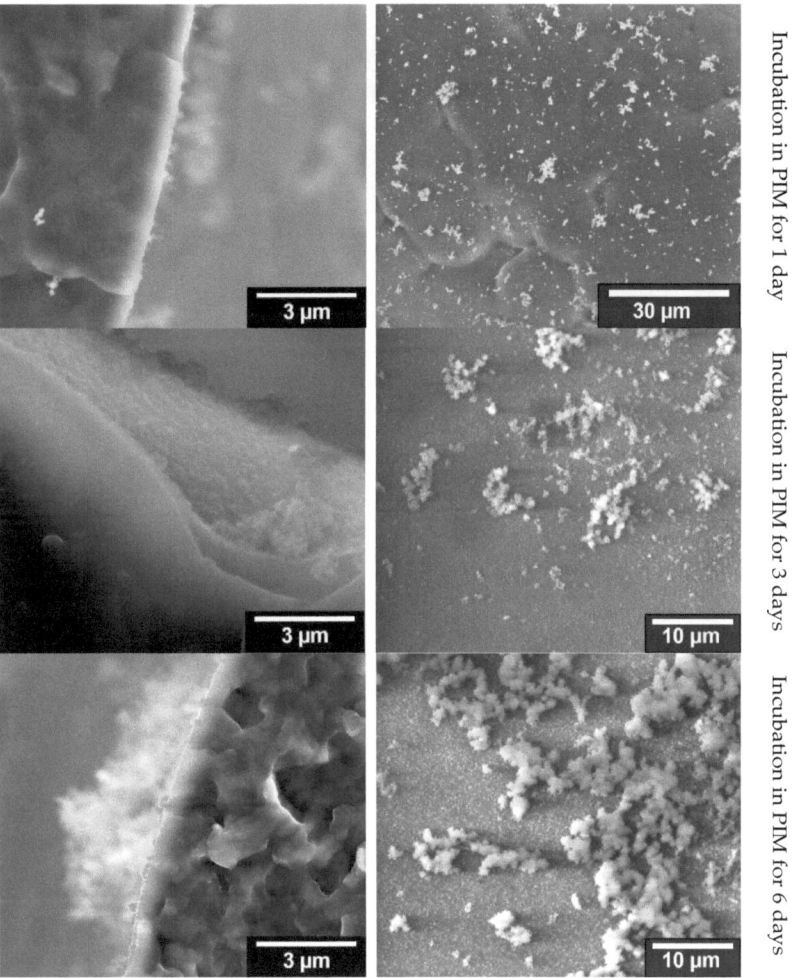

Figure 5. Comparison of the ESEM images for the different incubation times in PIM; (**left**): fracture edge of a cylindrical strand with coating on the outer surface; (**right**): surface of a strand with agglomerations of the crystalline structures.

Mineralization with Poly-L-Aspartic Acid

The samples treated with poly-L-aspartic acid showed no observable crystals under ESEM, and the surfaces appeared untreated except for a discernible collagen layer. The observable collagen layer through which the PCL shines through showed no signs of mineralization (cf. Figure 6). The uniformity of the coatings and the thickness of the coatings were summarized in Table 5.

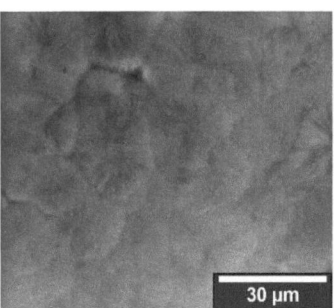 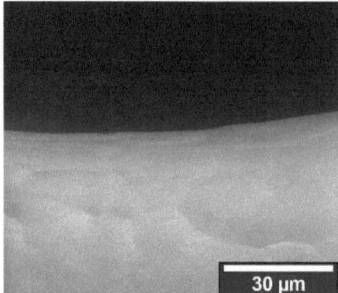 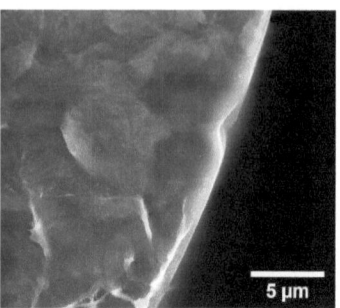

Figure 6. ESEM images of fracture edge (**left**) and surface of a strand (**center**), and magnification of the surface (**right**) of a sample treated with poly-L-aspartic acid.

Table 5. Comparison of the coating methods with regard to uniformity of the HA coating and thickness of the coating.

Coating Method	Uniformity of Coating	HA Coating Thickness (μm)
Collagen-HA	non-uniform coating, HA already clumps in the collagen solution, HA only on the collagen coating, not within	-
SBF (10×)	depending on the incubation time, short incubation (1 h) leads to uniform coating; moreover, formation of a uniform nanocrystalline layer with spots on the surface, whose expression increases with time, HA only on the collagen coating, not within	1 h: <1 μm 2 h: 1–3 μm 4 h: 3–6 μm 8 h: 10–30 μm
ALP	uniform coating, no nanocrystalline HA, as incubation time progresses, increased appearance of agglomerates on the surface, whose size and density increase with time, HA only on the collagen coating, not within	1 d: <1 μm 3 d: 1–2 μm 6 d: 2–4 μm
PolyASP	HA only detectable by high-res EDX within the collagen layer, no HA nanocrystals detectable at the outer collagen layer	-

3.1.4. Characterization by MicroCT

The investigations using μCT were only carried out for the ALP and poly-L-aspartic acid coatings.

Coatings with ALP

The results of the μCT examinations showed the presence of a surface coating on the ALP samples, but no crystals can be observed within the collagen layer. The surface coating is most evident when comparing the top and bottom surfaces of the constructs. Since the bottom side rested on the bottom of the corrugated sheet due to gravity during the coating with collagen and during storage in the ALP activated PIM, no surface coating could occur on the bottom side (cf Figure 7a).

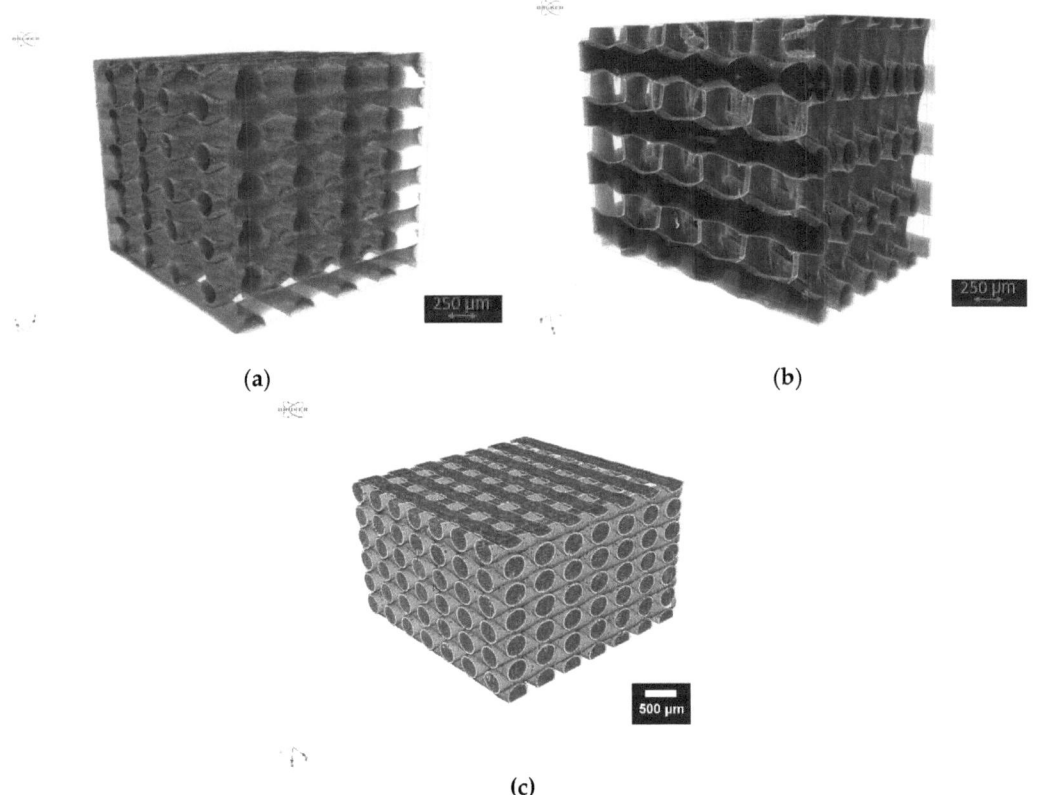

Figure 7. MicroCT images of samples incubated in: (**a**): ALP-activated PIM; (**b**): poly-ASP; (**c**): uncoated sample.

Coatings with Poly-L-Aspartic Acid

The results of the μCT examinations of the poly-ASP samples showed, as already the examinations of the ALP samples, a surface coating. However, no crystalline structures were visible here either (cf. Figure 7b) compared to the uncoated sample (cf. Figure 7c).

3.1.5. Characterization by TEM/EDX

Within the thin sections of the coated samples, HA nanocrystals have been detected in the collagen layer by TEM and subsequent EDX. The detection was done via EDX of Ca; this is neither contained in PCL nor in collagen (see Figure 8).

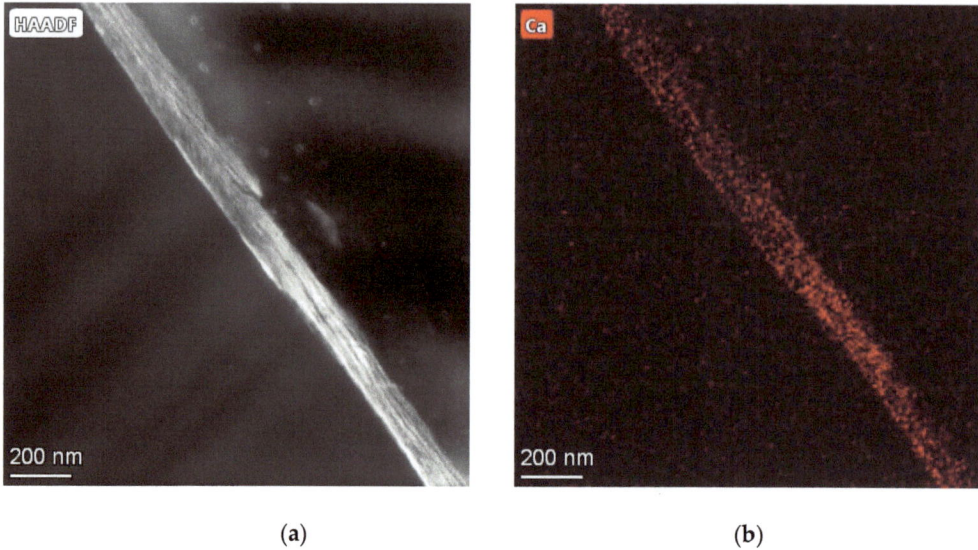

(a) (b)

Figure 8. TEM (**a**) and EDX (**b**) images of thin sections of collagen-coated PCL, post-treated with poly-L-aspartic acid; TEM image taken with TALOS 200X, 185,000× magnification, HFW 546 nm, STEM HAADF, EDX measuring with FEI SuperX EDX System.

4. Discussion

Collagen coating in the present project was performed analogously to work we have already published [11,14].

4.1. Collagen-HA Coatings

The addition of 5 wt% HA as a nanopowder did not result in a homogeneous surface coating consisting of collagen and HA. The HA did not distribute homogeneously and dried out and agglomerated on the surface of the scaffold. However, non-agglomerated homogeneous distribution of HA has been successful in other work [23]. This raises the question of why this was not successful in this work. The failure of homogeneous surface coating was due to the phase separation of the resulting collagen-HA gel. A possible gel formation had not been considered. Subsequently, a reference to this gel formation was found in the work of Oechsle et al. [24]. In this case, after gelation, the solid phase, the HA, separated and dried out in the subsequent step inhomogeneously on the surface. This gel formation was related to the pH, which could be confirmed by dissolving the gel when acetic acid was added. When the pH was changed, initiated by the addition of the HA nanopowder, electrostatic binding of the collagen molecule occurred, a phenomenon known for example with chitsoan molecules [25]. Since collagen behaves similarly to chitosan, this could be the reason for the unexpected gelation.

In the second set of experiments, in which only 2 wt% HA was added, similar gel formation was seen. Accordingly, the amount of HA added does not seem to be relevant to the formation of the gel. Consequently, the gel was dissolved by adding 0.2 M acetic acid. This 0.2 M acetic acid with a pH of 3.5 corresponds to the solvent of the commercially purchased collagen solution from Corning. The addition of the acetic acid lowered the pH, which dissolved the gel. This confirms that gelation is strongly linked to pH [25]. Twice the amount of acetic acid was needed to dissolve the gelation than collagen solution originally used. This, of course, also reduced the concentration of dissolved collagen and HA, and since 600 µL per scaffold continued to be added, insufficient collagen could be deposited on the surface, which is especially evident when comparing the ESEM results of the two sets of experiments. Working with pure collagen and other solvents may well lead to the desired HA/collagen coatings as demonstrated by the AG of Yeo et al. [23].

4.2. Incubation in 10× SBF

The experiments on the storage of collagen-coated scaffolds in 10× SBF showed that it was possible to obtain a homogeneous coating of HA on the collagen. The crystals and agglomerations produced in this regard are very similar to the HA crystals observed in the studies of Vaquette et al. [17], Poh et al. [16], and Yang et al. [19]. Furthermore, the expression and size of the crystals as well as the size of the agglomerations of the crystals could also be modulated by prolonged storage, a result that was also observed in the work of Yang et al. [19]. Furthermore, similar to the observations of Yang et al. [19] and Tas et al. [26], using 10× SBF for the preparation of a HA layer shortened the time to form this layer enormously, compared to normal SBF (i.e., 1× SBF without tenfold increased ion concentration). Thus, a very rapid formation of crystalline structures (within a few hours after storage) was also observable in the context of this work, which is a further an indication that the crystals formed are HA.

4.3. Coatings with ALP

The samples from the ALP experiments also showed successful surface coating with HA. Crystalline structures can be seen on the surface. These behaved similarly to the previously observed crystals of the 10× SBF samples. Thus, isolated crystals and agglomerations of different sizes are shown (Figures 4 and 5). These observed crystals are also optically similar to the HA crystals observed in the work of Jaroszewicz et al. [20].

4.4. Coatings with Poly ASP

The samples of the PolyASP experiments also showed a well recognizable collagen layer on the surface. This is further confirmed by the µCT images as well as the TEM + EDX images. Here, a surface coating can be seen on all images, which is different from the PCL. However, there were no crystalline structures on any of the images that indicate the presence of HA. In the work of Deshpande et al. [21], however, collagen was shown to mineralize successfully. The investigation with the TEM and EDX could prove the mineralization of the collagen. As mentioned earlier, the HA crystals observed by Deshpande et al. [21] were only a few 50–100 nm in size and were present in a layer above the collagen layer. This requires not only a view into the collagen layer (which was attempted with TEM + EDX and the ultra-thin sections required for this), but also the magnification required at the nanometer scale. This could be confirmed by our measurements. Calcium and oxygen were detected within the collagen layer by EDX, where collagen contains no calcium atoms. This would also correspond to the in vivo occurrence of HA in the form of crystals with a size of a few nanometers, which Deshpande et al. [21] tried to emulate with their method.

Implications with Respect to Application

Coating with HA crystals in collagen solution have proven to be impractical as the HA clumped in the solution and could not be applied homogeneously. For simple surface mineralization of collagen coatings but also simply for deposition of nanocrystalline HA on smooth surfaces, both incubation in SBF and ALP are suitable, with 10× SBF being the cheaper option with which thick HA layers can be generated in a short time. However, if the HA nanocrystals are to be present in the coating itself (similar to bone), the coating method with incubation in poly-ASP should be used. There is, of course, still a need for optimization with regard to the quantity and orientation (with regard to bones from the petri dish) of the HA nanocrystals within the collagen layer.

5. Conclusions

In the present work, we examined a wide variety of approaches to determine whether and where HA layers form on collagen-coated PCL scaffolds. The HA layers formed in various forms only at (not in) the collagen layer for all methods except incubation in poly ASP. However, through poly ASP treatment, nanocrystalline HA could also be detected within the collagen layer.

Author Contributions: Conceptualization, M.S., H.S. and J.V.; methodology, M.S. and J.V.; software, M.S.; validation, M.S., A.E. and B.R.; formal analysis, A.E.; investigation, A.E., Y.T. and R.T.; resources, M.S.; data curation, M.S., Y.T. and R.T.; writing—original draft preparation, M.S.; writing—review and editing, M.S. and A.E.; visualization, M.S.; supervision, M.S. and H.S.; project administration, M.S. All authors have read and agreed to the published version of the manuscript.

Funding: This research received no external funding.

Data Availability Statement: The data presented in this study are available on request from the corresponding author.

Acknowledgments: The work was still inspired by Anke Bernstein, who unfortunately passed away much too early due to an accident in June 2021. The authors would like to thank Melanie Lynn Hart for proofreading the manuscript.

Conflicts of Interest: The authors declare no conflict of interest.

References

1. Behrendt, H.; Runggaldier, K. A Problem Outline on Demographic Change in the Federal Republic of Germany. *Notf. + Rett.* **2009**, *12*, 45–50. [CrossRef]
2. Peters, E.; Pritzkuleit, R.; Beske, F.; Katalinic, A. Demografischer Wandel und Krankheitshäufigkeiten. *Bundesgesundheitsblatt-Gesundh.-Gesundh.* **2010**, *53*, 417–426. [CrossRef] [PubMed]
3. Destatis. Mitten im Demografischen Wandel. Available online: https://www.destatis.de/DE/Themen/Querschnitt/Demografischer-Wandel/demografie-mitten-im-wandel.html (accessed on 2 September 2020).
4. Eurostat. European Union: Age Structure in the Member States in 2019. Available online: https://de.statista.com/statistik/daten/studie/248981/umfrage/altersstruktur-in-den-eu-laendern/ (accessed on 14 March 2020).
5. Destatis. *Gesundheit-Fallpauschalenbezogene Krankenhausstatistik (DRG-Statistik) Operationen und Prozeduren der Vollstationären Patientinnen und Patienten in Krankenhäusern (4-Steller)*; Statistisches Bundesamt (Destatis): Wiesbaden, Germany, 2020.
6. Engh, C.A., Jr.; Young, A.M.; Engh, C.A., Sr.; Hopper, R.H., Jr. Clinical Consequences of Stress Shielding After Porous-Coated Total Hip Arthroplasty. *Clin. Orthop. Relat. Res.* **2003**, *417*, 157–163. [CrossRef] [PubMed]
7. Bublak, R. How long do artificial hips and knees last? *Orthopädie Und Rheuma* **2019**, *22*, 16–17. [CrossRef]
8. Fowler, T.J.; Blom, A.W.; Sayers, A.; Whitehouse, M.R.; Evans, J.T. How might the longer-than-expected lifetimes of hip and knee replacements affect clinical practice? *Expert Rev. Med. Devices* **2019**, *16*, 753–755. [CrossRef]
9. Epple, M. *Biomaterialien und Biomineralisation, Eine Einführung für Naturwissenschaftler, Mediziner und Ingenieure*; Teubner: Stuttgart, Germany, 2003. [CrossRef]
10. Patrício, T.; Domingos, M.; Gloria, A.; Bártolo, P. Characterisation of PCL and PCL/PLA Scaffolds for Tissue Engineering. *Procedia CIRP* **2013**, *5*, 110–114. [CrossRef]
11. Weingärtner, L.; Latorre, S.H.; Velten, D.; Bernstein, A.; Schmal, H.; Seidenstuecker, M. The Effect of Collagen-I Coatings of 3D Printed PCL Scaffolds for Bone Replacement on Three Different Cell Types. *Appl. Sci.* **2021**, *11*, 11063. [CrossRef]
12. Sousa, I.; Mendes, A.; Bártolo, P.J. PCL Scaffolds with Collagen Bioactivator for Applications in Tissue Engineering. *Procedia Eng.* **2013**, *59*, 279–284. [CrossRef]
13. Lüllmann-Rauch, R.; Asan, E. Zellenlehre. In *Taschenlehrbuch Histologie*, 6th ed.; Vollständig Überarbeitete Auflage ed.; Lüllmann-Rauch, R., Asan, E., Eds.; Georg Thieme Verlag: Stuttgart, Germany, 2019. [CrossRef]
14. Huber, F.; Vollmer, D.; Vinke, J.; Riedel, B.; Zankovic, S.; Schmal, H.; Seidenstuecker, M. Influence of 3D Printing Parameters on the Mechanical Stability of PCL Scaffolds and the Proliferation Behavior of Bone Cells. *Materials* **2022**, *15*, 2091. [CrossRef]
15. Ficai, A.; Andronescu, E.; Voicu, G.; Ghitulica, C.; Vasile, B.S.; Ficai, D.; Trandafir, V. Self-assembled collagen/hydroxyapatite composite materials. *Chem. Eng. J.* **2010**, *160*, 794–800. [CrossRef]
16. Poh, P.S.P.; Hutmacher, D.W.; Holzapfel, B.M.; Solanki, A.K.; Stevens, M.M.; Woodruff, M.A. In vitro and in vivo bone formation potential of surface calcium phosphate-coated polycaprolactone and polycaprolactone/bioactive glass composite scaffolds. *Acta Biomater.* **2016**, *30*, 319–333. [CrossRef] [PubMed]
17. Vaquette, C.; Ivanovski, S.; Hamlet, S.M.; Hutmacher, D.W. Effect of culture conditions and calcium phosphate coating on ectopic bone formation. *Biomaterials* **2013**, *34*, 5538–5551. [CrossRef] [PubMed]
18. Gomori, G.; Benditt, E.P. Precipitation of Calcium Phosphate in the histochemical Method for Phosphatase. *J. Histochem. Cytochem.* **1953**, *1*, 114–122. [CrossRef] [PubMed]
19. Yang, F.; Wolke, J.G.C.; Jansen, J.A. Biomimetic calcium phosphate coating on electrospun poly(ε-caprolactone) scaffolds for bone tissue engineering. *Chem. Eng. J.* **2008**, *137*, 154–161. [CrossRef]

20. Jaroszewicz, J.; Idaszek, J.; Choinska, E.; Szlazak, K.; Hyc, A.; Osiecka-Iwan, A.; Swieszkowski, W.; Moskalewski, S. Formation of calcium phosphate coatings within polycaprolactone scaffolds by simple, alkaline phosphatase based method. *Mater. Sci. Eng. C* **2019**, *96*, 319–328. [CrossRef] [PubMed]
21. Deshpande, A.S.; Beniash, E. Bioinspired Synthesis of Mineralized Collagen Fibrils. *Cryst. Growth Des.* **2008**, *8*, 3084–3090. [CrossRef] [PubMed]
22. Kestilä, I.; Folkesson, E.; Finnilä, M.A.; Turkiewicz, A.; Önnerfjord, P.; Hughes, V.; Tjörnstrand, J.; Englund, M.; Saarakkala, S. Three-dimensional microstructure of human meniscus posterior horn in health and osteoarthritis. *Osteoarthr. Cartil.* **2019**, *27*, 1790–1799. [CrossRef]
23. Yeo, M.G.; Kim, G.H. Preparation and Characterization of 3D Composite Scaffolds Based on Rapid-Prototyped PCL/β-TCP Struts and Electrospun PCL Coated with Collagen and HA for Bone Regeneration. *Chem. Mater.* **2012**, *24*, 903–913. [CrossRef]
24. Oechsle, A.M.; Wittmann, X.; Gibis, M.; Kohlus, R.; Weiss, J. Collagen entanglement influenced by the addition of acids. *Eur. Polym. J.* **2014**, *58*, 144–156. [CrossRef]
25. Avcu, E.; Baştan, F.E.; Abdullah, H.Z.; Rehman, M.A.U.; Avcu, Y.Y.; Boccaccini, A.R. Electrophoretic deposition of chitosan-based composite coatings for biomedical applications: A review. *Prog. Mater. Sci.* **2019**, *103*, 69–108. [CrossRef]
26. Tas, A.C.; Bhaduri, S.B. Rapid coating of Ti6Al4V at room temperature with a calcium phosphate solution similar to 10× simulated body fluid. *J. Mater. Res.* **2004**, *19*, 2742–2749. [CrossRef]

Review

Performance-Enhancing Materials in Medical Gloves

María José Lovato [1], Luis J. del Valle [1,2], Jordi Puiggalí [1,2] and Lourdes Franco [1,2,*]

1. Departament d'Enginyeria Química, Escola d'Enginyeria de Barcelona Est-EEBE, Universitat Politècnica de Catalunya, c/Eduard Maristany 10-14, 08019 Barcelona, Spain; maria.jose.lovato@upc.edu (M.J.L.); luis.javier.del.valle@upc.edu (L.J.d.V.); jordi.puiggali@upc.edu (J.P.)
2. Center for Research in Nano-Engineering, Universitat Politècnica de Catalunya, Campus Sud, Edifici C', c/Pasqual i Vila s/n, 08028 Barcelona, Spain
* Correspondence: lourdes.franco@upc.edu; Tel.: +34-93-401-1870

Abstract: Medical gloves, along with masks and gowns, serve as the initial line of defense against potentially infectious microorganisms and hazardous substances in the health sector. During the COVID-19 pandemic, medical gloves played a significant role, as they were widely utilized throughout society in daily activities as a preventive measure. These products demonstrated their value as important personal protection equipment (PPE) and reaffirmed their relevance as infection prevention tools. This review describes the evolution of medical gloves since the discovery of vulcanization by Charles Goodyear in 1839, which fostered the development of this industry. Regarding the current market, a comparison of the main properties, benefits, and drawbacks of the most widespread types of sanitary gloves is presented. The most common gloves are produced from natural rubber (NR), polyisoprene (IR), acrylonitrile butadiene rubber (NBR), polychloroprene (CR), polyethylene (PE), and poly(vinyl chloride) (PVC). Furthermore, the environmental impacts of the conventional natural rubber glove manufacturing process and mitigation strategies, such as bioremediation and rubber recycling, are addressed. In order to create new medical gloves with improved properties, several biopolymers (e.g., poly(vinyl alcohol) and starch) and additives such as biodegradable fillers (e.g., cellulose and chitin), reinforcing fillers (e.g., silica and cellulose nanocrystals), and antimicrobial agents (e.g., biguanides and quaternary ammonium salts) have been evaluated. This paper covers these performance-enhancing materials and describes different innovative prototypes of gloves and coatings designed with them.

Keywords: medical gloves; natural rubber; synthetic rubber; bio-filler; reinforcing filler; antimicrobial properties; performance-enhancing materials

1. Introduction

To minimize the risk of exposure to cross-infection between patients and healthcare workers, it is necessary to use personal protective equipment (PPE) such as disposable medical gloves, masks, or gowns [1]. Among these items, medical gloves were widely used by the population during the COVID-19 pandemic and played a key role as an infection prevention tool for medical staff and society in general. Microorganisms, infectious agents, and pathogens, such as bacteria, viruses, fungi, protozoa, and prions, live in the human body and the surrounding environment [2]. Most of these organisms do not pose a threat to the general population, but during an epidemic or in medical facilities, pathogenic microorganisms can be present at serious levels and cause illness. Hands are a major source of infection spread. Although hand washing is effective in eliminating most microorganisms, there are circumstances in which this practice is not sufficient, and exposure justifies the use of an additional layer of protection. For these reasons, medical gloves are mandatory when performing invasive procedures or coming into contact with sterile sites [3].

According to World Health Organization (WHO) recommendations, protective gloves should always be used in cases of contact with blood, mucous membranes, injured skin, or other potentially infectious material, as well as hazardous chemicals and drugs [1]. The aim of this work is to review the materials used in medical gloves due to their importance as an element of personal protection. The purpose is to compare the natural and synthetic rubbers used in their manufacture as well as identify performance-enhancing materials that can be added to medical glove formulations to improve their properties. These materials include biopolymers, eco-friendly additives, bio-based fillers, and antimicrobial agents [4]. Similarly, we intend to address several prototypes of medical gloves, blends, composites, and coatings made from these new materials.

1.1. History of Medical Gloves

Many healthcare workers were aware that accidental open lesions experienced while performing their duties could result in an infected wound, illness, and even death before the microbial nature of infection was established in the middle of the 19th century [5]. The exact time when protective gloves were first employed in the healthcare business is unknown. There are suggestions that an obstetrician called Walbaum covered his hands with sheep intestine as early as 1758 [6]. Other physicians used to cover their hands with cotton, silk, or leather gloves [5].

An important milestone in this field was the discovery of vulcanization by Charles Goodyear in 1839, when he was working at a rubber factory in Massachusetts and mixed a piece of rubber with sulfur on a hot stove [7,8]. He had discovered the vulcanization process, which turned natural rubber (NR) from a thermoplastic that could be softened by heat into a harder, more stable, and more durable product. Vulcanization consists of the development of a crosslinked rubber that is the product of the creation of bonds at several points of the individual NR chainlike molecules using sulfur as the crosslinking agent [9,10].

Vulcanized rubber quickly became the choice for coarse protective medical gloves. William Halsted of Johns Hopkins Hospital in Baltimore was likely one of the early promoters of sterile NR gloves in the operating room, but it is uncertain who initially encouraged their use. Halsted asked the Goodrich Rubber Company to make finer and less rudimentary NR gloves, although they were still quite stiff and difficult to handle. Over time, the NR gloves became even thinner and shorter. In 1897, the first article about sterile NR gloves in medical settings was published. This paper, entitled "Rubber gloves in the practice of surgery", was written by Werner von Manteuffel and appeared in a German surgical journal [11]. By the beginning of the 20th century, the use of sterile NR gloves had become widespread in surgical practice [5].

1.2. Market of Medical Gloves

The rising incidence of epidemic diseases such as swine flu (H1N1) and the more recent and widespread COVID-19 (SARS-CoV-2) has driven the growth of the global medical glove market. As reported by the Financial Times, during the latter pandemic, glove industry sales and profits increased by over 100% [12,13]. According to data provided by Global Market Insights, the worldwide market of medical gloves grew dramatically as a result of the first phase of the COVID-19 pandemic expansion, reaching over USD 4 billion in 2020 [14]. In 2021, when the infection was best understood and the supply of these products increased in line with demand, this market experienced a slight decline in profits and reached USD 12.31 billion in value. Nevertheless, it is expected to increase at a compound annual growth rate (CAGR) of 5.8% from 2022 to 2030 [15].

Figure 1 shows EU-27 imports of surgical gloves between January 2019 and December 2021. The graph was compiled from the Eurostat dataset "DS-1180622" for product code: "B3-40151100 Surgical gloves, of vulcanized rubber other than hard rubber (excluding fingerstalls)". In the graph, the business as usual (BAU) trend line was plotted using import data from January 2019 to March 2020, when the WHO proclaimed the global pandemic of COVID-19. To estimate the rise in medical glove imports during the COVID-19 pandemic, the over-BAU value was estimated using data from April 2020 to August 2021. The value of net imports in excess of BAU was approximately 62,000 Tons [16].

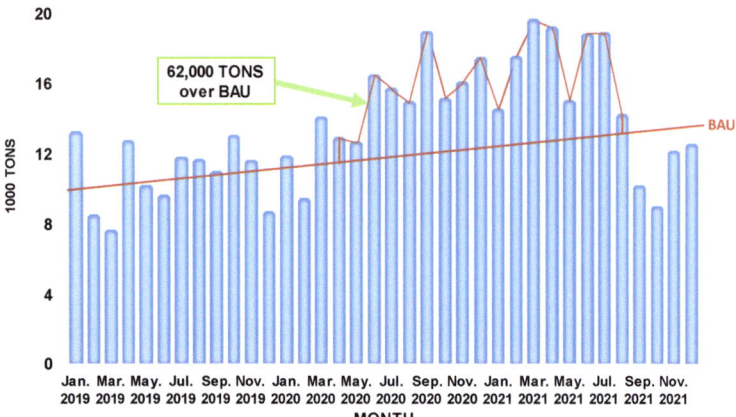

Figure 1. Imports of surgical gloves in the EU-27 from January 2019 to December 2021. Chart prepared by the authors based on Eurostat data [16].

MARGMA (Malaysian Rubber Glove Manufacturers Association) estimates that the global demand for gloves grew by almost 200 billion units in the first months of 2020 due to the COVID-19 pandemic [17]. In 2021, at the peak of this pandemic, the global demand for rubber gloves reached 492 billion units. The exports of rubber gloves from Malaysia in monetary value terms from 2014 to 2021 are illustrated in Figure 2. This graph clearly reflects the significant growth that has occurred. Prior to the pandemic, the value of exports in Malaysian ringgit (MYR) did not exceed MYR 20 billion; however, by 2020, exports had reached MYR 35.26 billion, and in 2021, they peaked at around MYR 54.81 billion [18].

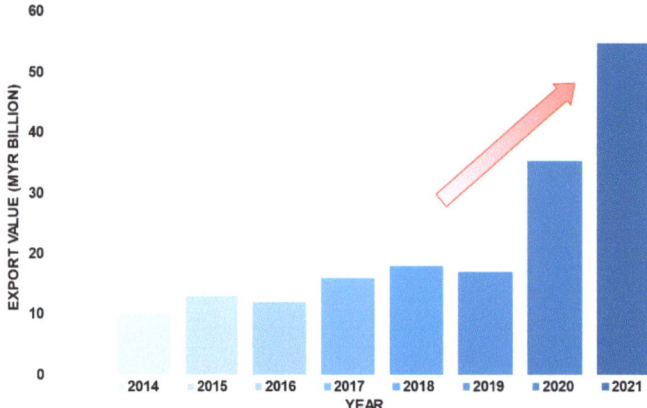

Figure 2. Exports of rubber gloves from Malaysia. The arrow indicates the sharp rise. Chart prepared by the authors based on MARGMA data shown in reference [18].

Major players in the glove market include Top Glove and Comfort Gloves [19]. Figure 3 shows the quarterly financial report of Top Glove Corporation Berhad with its earnings during the past pandemic period. In first quarter of 2021 (1Q-2021), this company achieved its maximum quarterly net profit of MYR 2.38 billion, and a high revenue of MYR 4.76 billion. The group's quarterly net profit, compared to the previous quarter (4Q-2020), increased 84% from MYR 1.292 billion, while revenue increased 53% from MYR 3.11 billion [20,21].

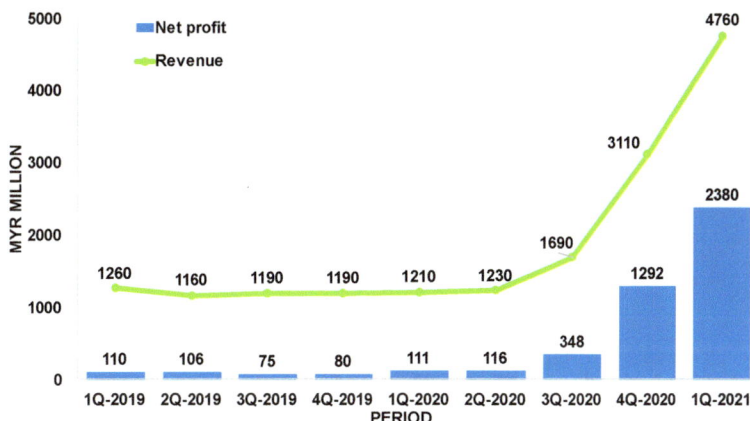

Figure 3. Quarterly financial report of Top Glove Corporation Berhad. Chart prepared by the authors based on Bursa Malaysia data shown in reference [21].

The quarterly financial report of Comfort Gloves Berhad is presented in Figure 4. This chart shows that the revenue increased from MYR 138.65 million in 1Q-2020 (before COVID-19) to MYR 541.24 million in 2Q-2021, which represents a rise of 290%. In the same quarters, the group net profit amounts were MYR 10.24 million and MYR 219.13 million, respectively, which means an increase of 2040% [17,22,23].

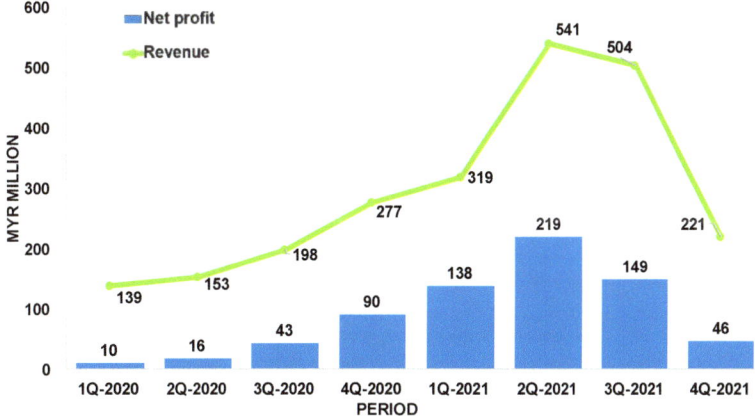

Figure 4. Quarterly financial report of Comfort Gloves Berhad. Chart prepared by the authors based on Bursa Malaysia data shown in reference [22].

In terms of the medical glove material market, natural rubber (NR) and acrylonitrile butadiene rubber (NBR) gloves are the most important sectors. NR gloves are the type that generates the highest revenues, due to their variety of applications in fields such as examinations and surgeries in the medical environment and as protection against chemicals and pathogens in the general industrial sector [24]. In the 2020 market share, the NR examination glove segment accounted for USD 5.1 billion, while the surgical glove segment reached USD 4 billion [14]. In 2021, the global NBR glove market was valued at USD 8.54 billion, and its size is expected to expand at a CAGR of 10.54% from 2022 to 2029. The NBR glove market attracted substantial new investments due to price incentives and increased demand resulting from the COVID-19 outbreak [25].

1.3. Production Process of Medical Gloves

The most common natural and synthetic rubber medical gloves are produced through the dipping process (Figure 5). Slowly, hand-shaped porcelain or metal molds are immersed in various tanks and subjected to different treatments. The main one is dipping in compounded latex, which consists of a mixture of natural or synthetic latex and compounding chemicals [26]. The compounding chemicals are the additives that must be included in medical glove formulations to achieve the required characteristics, such as mechanical strength, barrier integrity, color, aging protection, etc. [27]. These additives include vulcanizing agents, plasticizers, softeners, fillers, antioxidants, stabilizers, and different chemical compounds intended to improve processability [28,29].

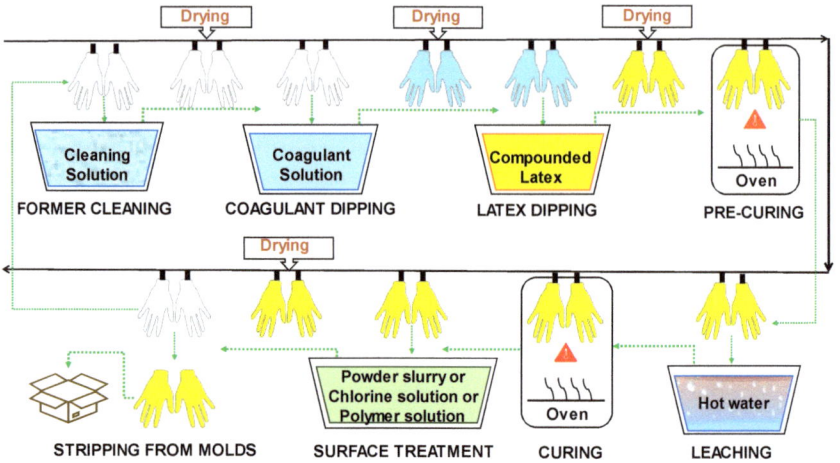

Figure 5. Production process of medical gloves by dipping.

The steps of the dipping process are briefly described below:

Former Cleaning: The procedure begins with washing and drying the hand-shaped molds. Alkaline solutions, acidic solutions, oxidizing agents, surfactants, and combinations of these can be employed as cleaning agents [28].

Coagulant Dipping: After cleaning the formers, they are coated with a coagulant, which is usually a polyvalent metal salt, an organic acid, or an organic acid salt [28]. The formers are dipped into the coagulant bath to promote adhesion and distribution of the compounded latex. The coagulant solution may also contain a separating agent, often calcium carbonate, which prevents the rubber from adhering to the molds. Subsequently, the molds are subjected to a drying process [26].

Latex Dipping: Next, the glove formers are dipped in a tank containing the compounded latex. The latter is a mixture of rubber suspension with several substances needed to form a glove, known as compounding chemicals. Formerly, the term "latex" referred to the white, milky sap gathered from the rubber tree; however, the terminology has also come to refer to dispersions of fine rubber particles in a liquid composed predominantly of water. Natural rubber (NR), polyisoprene rubber (IR), acrylonitrile butadiene rubber (NBR), and chloroprene rubber (CR) are mainly used in the dipping process [26].

Before adding other chemicals to commercial latexes, they must be stabilized to avoid alterations and variations in their ionic strength during the manufacturing process. The formulation ingredients must be integrated directly into an aqueous dispersion. For proper stabilization of the latex, the introduction of chemicals such as surfactants and rosin resins are required. Usually, two stabilization processes are needed; the provider performs the first stabilization step, but commercial latexes must be further stabilized before compounding chemicals are added. This second stabilization is mostly an electrostatic stabilization accomplished by altering the ionic strength of the latex. Table 1 presents some chemicals used for latex stabilization and their function [27].

Table 1. Typical chemicals used for latex stabilization [27].

Function	Description
pH increasing	Generally, KOH is added to latex to raise its pH to 10–11.
Surfactants	Suspensions of chemicals in water can be made more stable with the help of ionic and non-ionic additives.
Rosin resins	Some synthetic latexes, such as CR and IR, are formulated with colophonium resins, which effectively perform the functions of particle stability and film forming.

Once the latex has been adequately stabilized, crosslinking agents are usually applied to bind the polymeric chains together and form a three-dimensional network that gives the material the desired flexibility and performance. The crosslinking process may involve the use of several crosslinking agents [27].

Vulcanization, in which crosslinking is carried out by means of sulfur bonds, is the most common technique [8]. Colloidal sulfur is often employed with NR, IR, and NBR latexes. Typically, 0.5 to 2.5 parts per hundred of rubber (phr) are used. Zinc oxide is utilized in the range of 4.0–5.0 phr for CR [26]. Carbamates in conjunction with thiazoles are ultra-fast accelerators for the crosslinking process. The latex mixture can alternatively be vulcanized by adding sulfur donors such as thiurams and thioureas as activators. Guanidines, or xanthates, also can be added [30].

Fillers, in particular calcium carbonate, are commonly used to reduce the cost of NR examination gloves [27]. The degree of reinforcement offered by a filler for a rubber glove depends on many factors. The most crucial aspect is to achieve a large filler–rubber interface, which only colloidal filler particles can offer. To avoid dispersibility and processability concerns, the particles must have a specific surface area between 6 and 400 m^2/cm^3 [31].

Medical gloves contain antioxidants that defend them against attack by oxygen while in storage. Surgical and examination gloves contain non-staining antioxidants such as phenolic antioxidants (styrenated and hindered phenols), which are sometimes combined with a secondary antioxidant [30].

Pigments and dyes are combined with gloves to achieve opacification and impart the desired hue to the product [27]. The use of pigments or UV absorbers can improve light fastness to prevent hardening of NR gloves when exposed to direct sunlight. Also, by adding so-called antiozonants, protection against ozone can be accomplished [30].

Pre-curing: After the latex dipping process, another drying phase takes place. In this stage, the curing process is partially carried out, which is called the pre-curing process. The compounded latex that has been deposited on the molds is allowed to acquire a certain wet gel strength before the leaching step [28].

Leaching: This stage is often referred to as "wet gel leaching." Once the latex mixture has dried, residual chemicals and proteins on the gloves surface are removed through immersion in tanks of hot water. The tanks are refilled periodically with fresh hot water [28]. The water immersion period ranges from 1 to 10 min, depending on the film width. Washing NR latex film in a weak aqueous alkaline solution, such as aqueous ammonia or aqueous potassium hydroxide solution, facilitates protein removal [26].

Curing: This process, also simply referred to as vulcanization, often involves a hot-air circulation blower. The lowest vulcanization temperature varies depending on the compounded latex. Normal ranges for NR and IR are 90–100 °C, for NBR 120–140 °C, and for CR 120–130 °C [26]. The rubber reaches its final strength upon leaving the vulcanization oven [28].

Surface treatment: The purpose of the treatment of the inner surface of gloves is to prevent sticking together, to facilitate donning, to ensure a smooth fit, and to provide comfort during use. Traditionally, powder was employed for this purpose. However, powder was associated with increased risks of irritation or hypersensitivity for both users and patients, especially in NR gloves. NR latex proteins, which cause allergies, adhere to the powder, and spread rapidly in the environment, increasing the prevalence of allergies. As a result, the use of powder is increasingly restricted by regulation. In several countries, such as the United States, Germany, and the United Kingdom, powder is prohibited [27,32]. As an alternative, other treatments can be applied, such as chlorination and polymeric coatings [33].

Powdered gloves are formed by dipping them in a slurry. This substance is also known as wet powder, and contains talc, silica, or crosslinked starch. For the chlorination process, the gloves are dipped in a solution containing chlorine. The reaction with the chlorine forms a thin film of chlorinated rubber on the glove surface. The chlorine solution is produced by pumping chlorine gas into the water or by combining hydrochloric acid with sodium hypochlorite [26]. Probably the most widely used method for producing powder-free NR gloves is chlorination. The double bonds of the polymer chains present in NR are highly prone to the addition of chlorine, which has the effect of stiffening and detackifying the rubber surface of the glove [28].

Regarding polymer coating, it is common practice to dip gloves in hydrogel, an aqueous dispersion based on acrylic or polyurethane diluted to the required concentration, silicone polymer, or a polymer blend [26]. Coatings can be classified into two categories: hydrogels and non-hydrogels. Hydrogel coatings are composed of substances that absorb water several times their weight, swell, and become slick so that gloves can be easily donned. Non-hydrogels are water-repellent, and the coating's topology matches the features of a powdered surface. Often, a dual strategy is employed: first, the donning side of the glove is coated, and then the grip side is chlorinated [28].

Stripping from molds: After surface treatment, the gloves undergo a drying process and are then demolded and packaged for sale [26].

1.4. Environmental Concerns Related to Medical Gloves

The global demand for rubber gloves keeps increasing despite the environmental problems related to their disposal [34]. Rubber gloves account for 24% of total medical solid waste [35]. Discarded NR gloves typically take at least two years to degrade in a natural environment. Many highly additivated and crosslinked commercial NR gloves require even longer to fully decompose in soil under ambient conditions [36].

The various stages of rubber glove production require multiple resources, including potable water, chemicals, energy, and electricity. Water is often used for the preparation of the compounded latex, as well as for cleaning, leaching, and cooling procedures. Heat is utilized in the drying and curing processes. Electricity is mainly used for lighting, pumping water, operating heavy machinery, and the treatment of liquid waste [37].

At each stage of the glove manufacturing process, there are material inflows and waste outflows. Contaminated rinse water flows can be said to occur throughout the washing and leaching stages. In operations involving heating or mechanical action, energy is consumed. Ovens fueled by liquefied petroleum gas (LPG) produce carbon dioxide emissions as well as energy losses. Gloves and packaging materials are also discarded downstream in the production process. This manufacturing technique has effects on the environment as well as human wellness [38].

It is important to note that sulfur is one of the most widely used crosslinking agents. The sulfur-based curing system (vulcanization) is harmful from the point of view of environmental and health problems. The emission of toxic sulfur-based gases can cause acid rain, which returns considerable quantities of sulfuric acid to the earth, destroying vegetation and degrading soil quality. In addition, gaseous sulfur compounds can induce irritation and inflammation of the respiratory system. Higher levels of sulfur dioxide can cause eye burns and be fatal to humans [39]. In addition, accelerators such as benzothiazoles, which are toxic to aquatic life, are used in the vulcanization process [40].

To counterbalance the disadvantages of the traditional sulfur process, alternative curing methods include metal ionic crosslinkers, organic peroxides, or physical methods such as UV and gamma rays. The basic mechanism underlying the functionality of the metal ion as a crosslinker is related to its charges. Sulfur forms covalent bonds between elastomer chains in vulcanization, and these sulfur bonds can be replaced by an ionic bond with a multivalent metal ion, resulting in a reduction in process time and energy consumption. The most common applications of metal ion crosslinking are NBR and CR gloves. As this method does not require initiators or crosslinking accelerators, the cost of materials is reduced [39].

In ultraviolet (UV) crosslinking, covalent bonds are generated via the UV-assisted thiol–ene reaction, which represents an unconventional method for the crosslinking of NR. It can be carried out at room temperature with short process times and without the use of hazardous chemicals. UV-crosslinked NR articles exhibit good skin compatibility and high tensile strength. Both the lattice density and Young's modulus have been found to increase with radiation intensity [41].

With respect to gamma ray crosslinking, research has shown that carboxylated NBR can be crosslinked (forming covalent bonds) through high-energy radiation, such as gamma rays or electron beams [41]. The advantages of this procedure include the absence of hazardous chemical residues, full control of the crosslinking density, and improved mechanical properties of the crosslinked material. Disadvantages include the large amount of energy required for the process, the fact that direct exposure of humans could cause cancer, and the lack of available technical data [42].

In recent years, the widespread usage of rubber and the resulting large amount of waste of this material has increased interest in this field, with the objective of applying bioremediation. NR can be degraded by bacteria and fungi, but the process is slow and even slower in gloves with higher crosslinking densities [35,43]. Linos et al. (2000) found that *Pseudomonas aeruginosa* AL98, a type of Gram-negative bacterium, was capable of disintegrating NR, in its natural form as NR latex concentrates or in its crosslinked forms as NR or IR gloves [44].

Although the biodegradation of NR has been widely investigated, progress in this field of study has been hampered by the difficult isolation of appropriate bacteria, extended cultivation periods, and the scarcity of genetic tools [45]. *Actinomycetes* have dominated the literature about the rupture of *cis*-1,4-polyisoprene among NR-degrading bacteria. The most prominent genera are *Streptomyces, Mycobacterium, Nocardia,* and *Gordonia* [46]. The three latter species directly attack the NR substrate, producing a biofilm and fusing with the polymer to induce cell surface degradation. The adherent group of bacteria has been implicated as much more efficient degraders of this substance than enzyme-secreting strains [47].

There is evidence that some NR glove additives limit microbial descomposing action. It has been demonstrated that the extraction of these inhibitory substances (antioxidants) using organic solvents promotes the proliferation of *Gordonia* and *Micromonospora* species. However, using chemical solvents to remove rubber inhibitors is not environmentally friendly, so an alternative via microbial action was studied. Due to the similarities between rubber additives and fungal degradable chemicals, the successful cleavage of antioxidants by white rot fungus has been reported [46].

An example of a plant for the recycling and remediation of NR by microbial action is shown in Figure 6. The waste NR is ground to promote further microbial attack. The ground rubber is then heated to denature the unstable compounds, while sterilizing the rubber to ensure the absence of pathogenic species that could inactivate or compete with the microorganisms used in the bioreactors [46].

After heating, a detoxification process is performed in which white rot fungi can be used to degrade the NR additives. Once the additives have been removed, a devulcanization process is performed with *Thiobacillus ferrooxidans* to break the sulfur bonds of the NR. The decomposition can be completed with potent degrader agents such as *Nocardia* sp. and

Gordonia polyisoprenivorans. Then, the lower-molecular weight molecules can be catabolized by *Streptomyces* sp. or *Xanthomonas* sp. Alternatively, the devulcanized NR can be filtered, cleaned, dried, and blended with fresh NR for reprocessing [46].

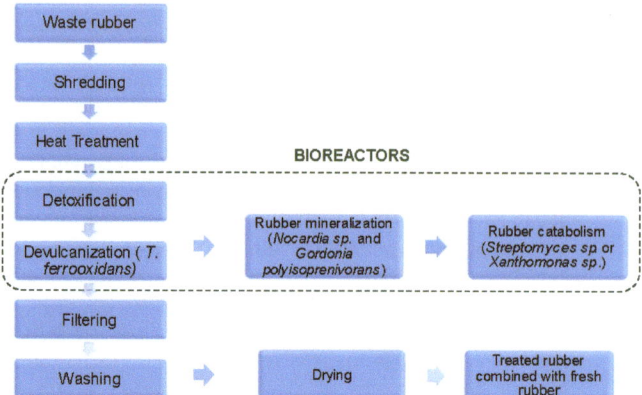

Figure 6. Recycling and remediation of NR through microbial action. Reprinted and adapted with permission from reference [46]. Copyright © 2013 Springer Nature.

2. Types of Medical Gloves

There is a variety of medical gloves based on the specific requirements of each application. Essentially, the two main types of medical gloves are examination gloves, used for normal medical check-ups and minor operations, and surgical gloves, used for operations [48]. Examination gloves are thin (50–150 μm) and ambidextrous. As they are usually for short-term use, they can be sterile or non-sterile depending on the risk to be handled. On the other hand, surgical gloves are always packed in a sterile bag in pairs, distinguishing the right hand from the left. These gloves are thicker than examination gloves (180–250 μm) as they are worn longer; it is advisable to change them every 90 min, or less if a perforation is detected [27].

The most common types of medical gloves (Figure 7) include those made of the following materials: natural rubber (NR), polyisoprene (IR), acrylonitrile butadiene rubber (NBR), chloroprene (CR), polyethylene (PE), and poly(vinyl chloride) (PVC) [32].

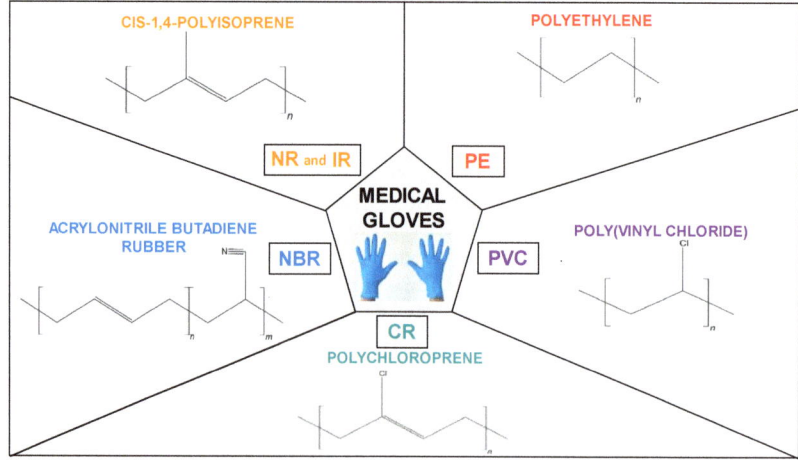

Figure 7. Chemical structure of common types of medical gloves.

2.1. Natural Rubber (NR)

Natural rubber (NR) is a key raw material that has modernized the world due to its wide functionality and excellent elastic properties. NR is present in the latex of more than 2000 plant species, including *Hevea* sp., *Castilla* sp., *Manihot* sp., *Guayule* sp., and *Taraxacum kok-saghyz* sp. [27,49]. Surprising examples, such as dandelions, are included. However, only one tree source, *Hevea brasiliensis*, is commercially significant [31].

Hevea brasiliensis NR latex is a colloidal system of *cis*-1,4-polyisoprene particles dispersed in an aqueous serum. The milky white sap consists of approximately 34% *cis*-1,4-polyisoprene, 2–3% protein, 0.1–0.5% sterol glycosides, 1.5–3.5% resins, 0.5–1.0% ash, 1.0–2.0% sugars, and 55–65% water [31,50]. The production of milky latex generated by the *Hevea brasiliensis* tree fluctuates between 19.8 g and 90.5 g per tree and per tap, using a half-spiral cut extraction method on the bark of the tree with alternating daily harvesting [42].

NR gloves, also known as latex gloves, are made of 90% to 95% NR and 5% to 10% compounding additives [30]. Thus, NR gloves are waterproof, and they exhibit excellent mechanical properties, such as high elasticity, tactility, and tension retention [39]. These gloves are excellent for delicate applications due to their extreme comfort and sensitivity [51]. The minimum and maximum operating temperatures are −51 °C and 104 °C, respectively. Most medical examination and surgical gloves are made of this material, which provides excellent barrier protection against microorganisms and infectious fluids [42]. A negative aspect of NR is the presence of impurities such as proteins, which have antimicrobial properties and play an important role in plant defense responses, but whose remaining presence in NR gloves causes allergies to a certain part of the exposed population [52,53]. Sensitization may occur with repeated exposure [54]. NR gloves typically have extractable protein (EP) levels ranging from 20 to 1000 µg/g. Despite this, EP can be removed through various leaching processes [1]. Once NR gloves were identified as a source of allergen exposure, awareness was raised, and risk reduction measures were implemented. The transition to powder-free, low-protein NR gloves and synthetic gloves corresponded with a decrease in the incidence of allergies [55].

2.2. Polyisoprene (IR)

Polyisoprene rubber (IR) is a synthetic rubber with the same chemical composition as NR and therefore shares similar properties. Shell Company was the first to commercialize IR in 1960 [56]. IR has a more uniform and lighter color than NR. IR also has a higher tensile and tear strength due to a narrower molecular weight dispersion. This material behaves like NR during processing and can be crosslinked using the same techniques [31]. Most synthetic surgical gloves are made of IR and are characterized by their high dexterity, sensitivity, absence of protein, and high level of wearer comfort [57]. IR contains 90–92% *cis*-1,4-polyisoprene, while NR contains approximately 99% of this configuration [58].

2.3. Acrylonitrile Butadiene Rubber (NBR)

Acrylonitrile butadiene rubber (NBR), also known as nitrile rubber, was patented in 1934 by the chemists Erich Konrad and Eduard Tschunkur of IG Farabenindustrie [59]. The acrylonitrile content (18% to 50%) in this material gives it higher hardness, higher resistance to oil and non-polar solvents, and better puncture and abrasion resistance compared to NR [31]. This material is used in various surgical and examination gloves. They are usually blue, purple, or black, and any needle puncture is evident [32]. NBR gloves have a longer shelf life than NR gloves [24]. NBR gloves are flexible, soft, and comfortable. However, they have drawbacks such as lower sensitivity and rougher texture than NR gloves [51]. Their pathogen protection and temperature tolerance are moderate, with lowest and maximum working temperatures of −34 °C and 121 °C, respectively [42]. NBR is one of the most widely used synthetic rubbers because of its lower cost compared to other synthetic rubbers [39,51].

2.4. Polychloroprene (CR)

Polychloroprene (CR) is a DuPont patented and registered product known as Neoprene® [42]. It is produced through emulsion polymerization of chloroprene [31]. This is one of the most frequently used synthetic rubbers for making gloves that are resistant to both temperature and aggressive chemicals. Its environmental resistance, thermal stability, and good oil resistance make it a standout in the glove sector [39].

CR gloves fit and feel like NR gloves. They are very comfortable and suitable for people sensitive to NR. These gloves are extremely durable and can stretch quickly while maintaining their original shape due to their high elasticity [60]. Their mechanical and flammability resistance are also superior to those of NBR gloves [42]. The minimum and maximum operating temperatures are $-25\ °C$ and $93\ °C$, respectively [61].

2.5. Polyethylene (PE)

Polyethylene (PE) is a polymer synthesized through polycondensation of ethylene. PE is malleable, flexible, and resistant to heat, electrical current, chemicals, and degradation [62]. Thin PE foils are welded together to create PE gloves available in various thicknesses and with textured surfaces. They have a wide range of applications, including non-sterile medical work, food handling, painting, and handling of electronic components. The protective effect depends more on the strength of the welded seams than on the inherent chemical resistance of the material [30].

2.6. Poly(vinyl Chloride) (PVC)

Polyvinyl chloride (PVC) is a synthetic rigid polymer that was converted into a flexible material by Waldo Semon at BFGoodrich in the 1920s. Flexible PVC is vinyl compounded with a plasticizer, which defines the properties of the final product [63]. Traditionally, phthalates have been added to PVC as plasticizers. These substances have been gradually replaced with less harmful substitutes such as adipates and vegetable oils [64]. PVC gloves, also known as vinyl gloves, are stiffer than NR gloves and have comparatively lower elastic modulus, tear strength, tensile strength, feel, and comfort, but on the plus side, they have no residual protein and are less expensive [27,65]. PVC gloves are usually transparent and fit loosely; they can be used in non-sterile environments and for handling non-hazardous materials and drugs [32]. PVC gloves are permeable; investigations into the permeability of gloves exposed to 13 chemotherapeutic drugs indicated that even after short-term applications, transfer to the wearer's skin occurs [51,66]. These gloves are easily worn out by use [67].

Table 2 summarizes the main advantages and disadvantages of the different types of medical gloves.

Table 2. Properties of main medical gloves.

Material	Advantages	Disadvantages
NR	Good resistance to alkali and acids. Comfortable, good fitting and feeling for hands. High elasticity and ability to adapt to shapes. High tear strength. Waterproof.	Permeable to several solvents. Poor resistance to chemicals. Possible allergies due to residual protein.
IR	Absence of allergy associated with proteins in NR gloves. Good elasticity and break resistance.	It is costly.

Table 2. Cont.

Material	Advantages	Disadvantages
NBR	Good alternative for people that are allergic to NR gloves. Resistance to various chemicals, especially oils, fuels, weak acids, caustics, and some organic solvents. Eligible for handling most food materials. Good resistance to mechanical stress.	It has a low level of sensitivity, which may restrict how well the hands adapt to and operate with the gloves. Low resistance to alcohols, amines, ketones, ester, ethers, concentrated acids, halogenated hydrocarbons, and aromatic hydrocarbons.
CR	Resistance to temperature and harsh chemicals. Mechanical and flammability resistance are superior to NBR gloves. CR gloves fit and feel like NR gloves. Appropriate for people allergic to NR.	It is costly.
PE	Can be used for food material. Inexpensive option.	Poor resistance and barrier protection.
PVC	It is cost-effective, since PVC is inexpensive. Good for those suffering from skin and chemical allergies as it is skin-friendly.	Due to plasticizer, not adequate for handling fatty food since there is the possibility of migration of the plasticizer into the food. Less stretch, comfort, and elongation than NR. Poor resistance to chemical degradation. High permeability to chemotherapy drugs.

3. Mechanical Properties of Medical Gloves

There are international requirements that must be followed for medical gloves to be suitable for their intended purpose. As an example, the ASTM standards for NR, NBR and CR rubber examination gloves are presented in Table 3.

Table 3. Mechanical properties of examination medical gloves according to ASTM standards.

Property of Examination Gloves	ASTM D3578—19 (NR) [68]				ASTM D6319—19 (NBR) [69]		ASTM D6977—19 (CR) [70]	
	Before Aging		After Aging		Before Aging	After Aging	Before Aging	After Aging
	Type I	Type II	Type I	Type II				
Minimum Tensile Strength (MPa)	18	14	14	14	14	14	14	14
Maximum Stress at 500% Elongation (MPa)	5.5	2.8	-	-	-	-	-	-
Minimum Ultimate Elongation (%)	650	650	500	500	500	400	500	400

The ASTM D3578 – 19 specification dictates the mechanical property values that NR examination gloves must reach. The appendix of the standard provides physical criteria for Type I and Type II gloves. This classification has been extended to provide customers with a greater selection of fit, feel, and comfort [68]. For NBR and CR examination gloves, the mechanical property values are dictated by ASTM D6319 – 19 [69] and ASTM D6977-19 [70], respectively.

For NR, NBR, and CR examination gloves, one of the following accelerated aging tests must be performed: (a) being exposed to 70 ± 2 °C for 166 ± 2 h or (b) 100 ± 2 °C for 22 ± 0.3 h. Aging tests are designed to demonstrate that the performance of the gloves will not deteriorate before the date of expiry. Accelerated aging testing is required since it is impracticable to conduct real-time aging tests prior to releasing these products onto the market. Under the standards' test conditions, gloves must be able to resist the deterioration caused by oxidative and thermal aging. Mechanical properties are expected to be altered over the lifespan of the product, so they are measured before and after the aging test to verify that gloves keep their physical integrity and protective capability [68–70].

The specifications for the mechanical properties of medical gloves according to European standards are addressed in EN 455-2:2015 (Medical gloves for single use. Part 2: Requirements and testing for physical properties). For accelerated aging, the gloves are heated in an oven at 70 ± 2 °C. The minimum force at break (before and after aging) for surgical gloves must be 9.0 N, for examination gloves except for thermoplastic materials 6.0 N, and for examination gloves made of thermoplastic materials (e.g., PVC, PE) 3.6 N [71].

Table 4 shows the mechanical properties of examples of gloves made of NR and NBR of KOSSAN Rubber Industries gloves published on the company website. It can be seen that the properties of the products meet the normative requirements [72].

Table 4. Examples of mechanical properties of KOSSAN medical gloves [72].

Property	Latex Examination Glove PS60Y		Nitrile Examination Glove CS30	
	Unaged	Aged	Unaged	Aged
Tensile Strength (MPa)	20–24	16–20	28–32	29–33
Ultimate Elongation (%)	700–740	600–640	500–540	460–500
Force at Break (N)	7.0–7.5	7.0–7.5	6.0–6.3	6.0–6.3

4. Prototypes of Medical Gloves with Performance-Enhancing Materials

The long-term viability of medical glove manufacturing processes is crucial from both a financial and environmental protection point of view. The use of performance-enhancing materials such as biomaterials, bio-fillers, biodegradable polymers, antimicrobial agents, etc. in conjunction with natural and synthetic rubbers could help to support the three pillars of sustainability in the environmental, social, and financial sectors.

Biomaterials such as bio-fillers help accelerate gloves' degradation after disposal. Thus, the extraction of bio-based chemicals and their incorporation into the polymeric matrix could lead the way in a new era in disposable glove manufacturing [73]. Food waste, terrestrial vegetation, and aquatic plants such as micro and macro algae could all be sources for these bio-based compounds [39]. Since the green market is growing dramatically each year, the introduction of biodegradable rubber gloves onto the market within the green technology sector would present an opportunity for manufacturing companies [36].

Antibacterial components have become prevalent in daily life, and the antibacterial properties of nanoparticles are rapidly being investigated and commercialized [39]. Despite being sterilized and separately packed, surgical gloves are exposed to germs when the package is opened [74].

The increasing number of antibiotic-resistant microorganisms has led to the search for new agents that can prevent the spread of pathogenic microorganisms. Antibacterial agents with the potential to be incorporated into natural or synthetic rubber gloves include biguanides such as chlorhexidine salts and poly(hexamethyl biguanide) (PHMB), quaternary ammonium salts such as benzalkonium chloride and benzethonium chloride, chlorinated phenols such as triclosan, essential oils such as farnesol, phenoxyethanol, octoxyglycerin, antifungal agents, iodine compounds, silver salts, some vegetable oil extracts, such as gentian violet, brilliant green, chitosan-based compounds, turmeric, and similar substances [4]. By covalently bonding the antibacterial agent to polymer surfaces, it is feasible to achieve an enduring effect, which leads to self-sterilized materials that may protect themselves from pathogens and contribute to the eradication of harmful microbes [75].

4.1. Biodegradable Green Gloves Containing Ascorbic Acid from Maleate Epoxidized Natural Rubber/Poly(vinyl Alcohol) Blend

Poly(vinyl alcohol) (PVA) is a biodegradable polymer that has been used as precursor material for the production of decomposable gloves, as it is non-toxic, physically and chemically resistant, and economically viable. Previous research has reported the improvement of biodegradability when PVA is combined with NR [76]. Ascorbic acid (L-ascorbic acid), also identified as vitamin C, has been shown in numerous studies to have antibacterial properties. It has been demonstrated that it inhibits the growth of *Helicobacter pylori*, *Campylobacter jejuni* [77], *Staphylococcus aureus*, *Enterococcus faecalis* [78], and *Mycobacterium tuberculosis* [79]. In vitro studies have demonstrated that L-ascorbic acid can improve the action of antibiotics like azithromycin [80] and levofloxacin [81,82].

Riyajan et al. studied maleate epoxidized natural rubber (MENR) and PVA (MENR/PVA) blends for producing a biodegradable glove with ascorbic acid (AA) encapsulated, represented in Figure 8. To produce MENR, under a nitrogen atmosphere and intensive stirring, 20 wt% NR latex was combined with 10% non-ionic surfactant. Then, formic acid and water were added to the previous mixture, which was held at 30 °C for 15 min. The temperature was increased to 70 °C and the reaction was completed after 5 h of stirring. Maleic anhydride (MA) in the presence of 10% Triton X-100 was then added to the resultant epoxidized natural rubber (ENR) latex at 80 °C and agitated for 3 h. The mixture was agitated for 15 min at 70 °C, after the addition of the free radical initiator potassium persulfate [83].

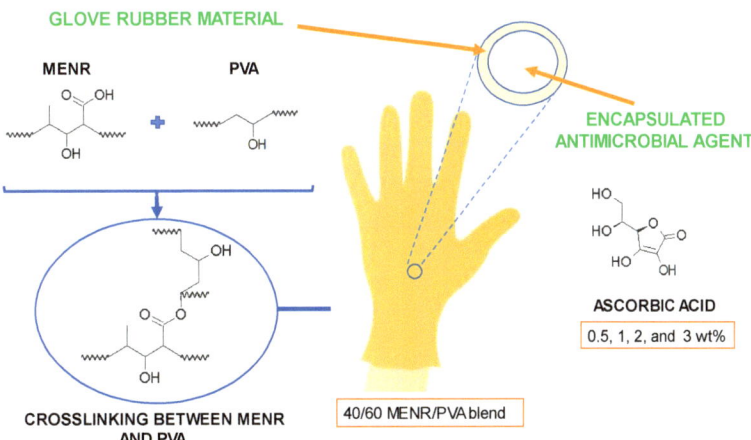

Figure 8. MENR/PVA blend glove with encapsulated AA. Graphic prepared by the authors based on reference information [83].

To prepare the MENR/PVA blends for the gloves, a 10 wt% PVA aqueous solution was combined, at 78 °C, with various MENR concentrations of 10, 20, 30, and 40% using magnetic agitation. Then, on glass plates, 80 g of the MENR/PVA blends was dehydrated at 30 °C for 3 days. The biodegradation of this material was examined by monitoring the weight loss of samples with different proportions of PVA and MENR in the blend. Samples of PVA alone and with MENR contents of 10, 20, 30, and 40% blended with the PVA were evaluated. The samples were weighed and then buried in soil, irrigated daily with water to maintain its moisture content, at ambient temperature. PVA alone exhibited the greatest biodegradation, due to the existence of hydroxyl groups in this compound. After 10 days buried in the soil, 50% of the PVA's weight had been lost, and after 40 days, it had decomposed completely. The biodegradation rate of the samples is reduced as the MENR proportion in the sample is increased, because crosslinking takes place. Nevertheless, all the blends decomposed properly in the natural environment through water-induced hydrolysis and enzymatic breakdown. After 90 days, at the end of the experiment, the samples containing 10 and 20% of MENR almost had a weight loss of 100%, and the samples with 30 and 40% of MENR lost around 75 and 60% of their weight, respectively [83].

The encapsulation of AA in a 40/60 MENR/PVA blend was explored in order to impart antibacterial activity to gloves. The encapsulation efficiency (EE) was 100, 99, 98.5, and 96%, respectively, for 0.5, 1, 2, and 3 wt% AA. The cumulative in vitro release of AA from the MENR/PVA blend films can be described as two distinct stages based on these data. The first 12 h are characterized by a burst release phase in which about 25, 33, 38, and 43% of the total AA was released from the MENR/PVA blends with 0.5, 1, 2, and 3 wt% AA, respectively. During this phase, AA was released via diffusion through the walls of the MENR/PVA blend. Up to 70 days, the release process is characterized by a more progressive release, accounting for approximately 100, 90, 75, and 64% of the total for 3, 2, 1, and 0.5 wt% AA, respectively. The initial burst release is caused by the leaching of AA near the capsule walls. As there is no polymer coating, the rate of matrix dissolution is quite rapid, and AA adjacent to the wall could promptly diffuse away [83].

It has been established that gloves manufactured with an MENR/PVA blend containing antimicrobial agent effectively prevent microbial transmission. Controlled and optimized release of AA from the MENR/PVA blend could play a significant role in the development of a medical glove [83].

4.2. NR Films/Gloves and Carboxylated-NBR (XNBR) Films Containing Sago Starch as Bio-Filler

The creation of effective bio-based products would aid in the prevention of environmental degradation. NR can be utilized as a matrix material in composite applications, where it is supplemented with bio-fillers to improve thermo-mechanical and barrier properties. In NR gloves, efforts to substitute ordinary calcium carbonate with bio-fillers such as polysaccharides, eggshell, and chitosan are frequently considered [36]. The advantages of employing bio-fillers over synthetic fillers are their renewability, abundance, and low cost; the negatives are comparatively weaker mechanical qualities. Because cellulose, chitin, and starch are hydrophilic, they are less compatible with the NR matrix. Achieving a homogenous filler–NR matrix mixing is difficult due to the different structural features of the components. Fillers with small particle size enhance the physical interaction with the matrix. Hence, the mechanical resistance, thermal stability, sorption, crystallinity, and biodegradability of the bio-fillers can be improved as result of their smaller size. On the other hand, the presence of hydroxyl groups in bio-fillers may result in low compatibility with NR [84].

A chemical treatment of the bio-filler can reduce the hydroxyl group content to improve compatibility, resulting in composites with higher strength and crystallinity. Further research is required to investigate the primary obstacles: inadequate hardness, moisture absorption, and suitability for outdoor and heavy-duty uses [85]. It is known that some bacteria and fungi are capable of degrading NR, despite the lengthy nature of the process [86]. The addition of polysaccharides to the NR system serves to enhance the action of microorganisms, facilitating degradation via enzymatic polysaccharide rupture and oxidation of the rubber backbone chain [87]. The polysaccharides are particularly favorable for the biodegradation process since they can be used as sustenance for microorganisms, hence promoting their proliferation and degradative action [36]. Starch is a typical polysaccharide used in biodegradable rubber films. It is made up of 70–80% amylopectin and 20–30% amylose [88].

Amylose content is a key criterion for its usage as a biodegradable material since it may provide nutrients to microorganisms, allowing them to begin the biodegradation activity [34]. When compared to other forms of starches, sago palm (*Metroxylon sagu*) starch has a greater amylose concentration (27%). To reach the required qualities of rubber films, starch must undergo a physical or chemical transformation. Acid hydrolysis may be used to chemically modify native sago starch (NSS) by inducing the creation of sulphate ester groups on the starch surface, which increases the interaction between the rubber matrix and the starch [89].

Daud et al. designed an experiment with sago starch to improve the biodegradability of NR and XNBR films. Sago starch with sulphate ester groups (AHSS) was obtained by treating NSS with aqueous sulfuric acid solution for 7 days at room temperature. The particle size of NSS was initially 1.233 µm, and it was lowered to 0.313 µm after the acid hydrolysis process. SEM micrographs of the NSS and AHSS are shown in Figure 9(a1) and Figure 9(a2), respectively. The surface of the AHSS particles is more porous, more rugged, and largely eroded than that of the NSS particles. In order to make an adequate comparison, unfilled NR, NSS-filled NR, AHSS-filled NR, unfilled XNBR, NSS-filled XNBR, and AHSS-filled XNBR films were prepared. To prepare the films, NR latex was mixed with compounding ingredients (with or without filler, depending on the case) and mechanically stirred for 1 h to obtain the NR compounded latex, which was then matured for 24 h at room temperature prior to the dipping process. For the prevulcanization procedure, the NR compounded latex was then heated to 80 °C and continuously stirred. XNBR latex was compounded similarly to NR latex, with the difference that the maturation period was 48 h. Prior to the dipping procedure, the prevulcanized compounded latexes were stirred for 15 min. Clean aluminum plates were dipped for 10 s in a coagulant bath, dried for 5 min, and left to cool at room temperature for 5 min before being dipped for 10 s in a latex dipping tank and cured at 100 °C. NR and XNBR were cured for 10 and 90 min, respectively [34].

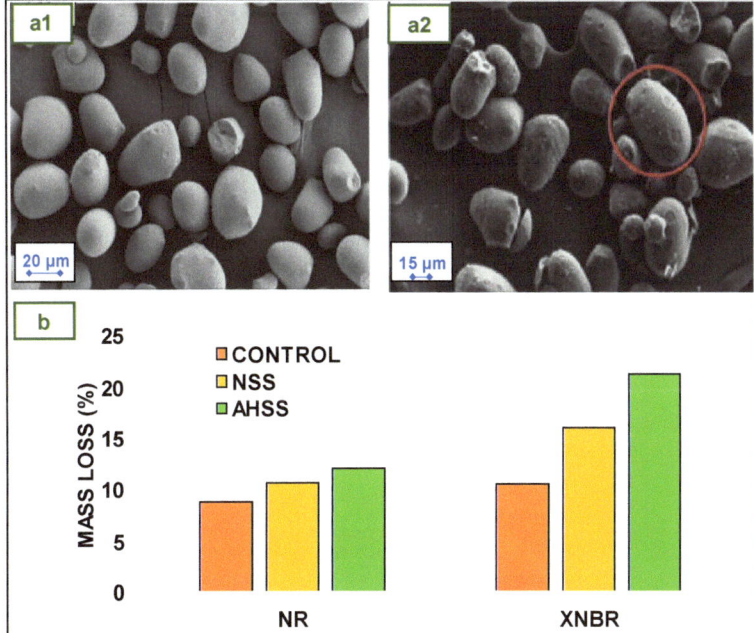

Figure 9. SEM images of (**a1**) NSS and (**a2**) AHSS. The red circle shows the porous surface of the starch particle after acid hydrolysis (**b**) Mass loss of NR and XNBR films (control, NSS-filled, and AHSS-filled) after 3 weeks. Reprinted and adapted with permission from reference [34]. Copyright © 2019 Elsevier.

Regarding mechanical behavior, in both cases, NR and XNBR unfilled films have the best properties. The poor interfacial bond between the hydrophilic sago starch and the hydrophobic rubbers resulted in a decrease in the tensile properties of films when NSS was added. Incorporating AHSS into the films improved the mechanical properties and swelling resistance of NR and XNBR compared to NSS. This distinction may be attributed to the superior compatibility of AHSS in NR and XNBR films compared to NSS in these same films. The low amorphous content, reduced particle size, and existence of sulphate ester groups contribute to the increased rubber–filler interaction between AHSS and the rubber matrix [34].

Figure 9b shows the mass loss of unfilled, NSS-filled, and AHSS-filled NR and XNBR films after 3 weeks of soil burial. The percentage of mass loss was highest for AHSS-filled NR films, followed by NSS and unfilled NR films [34].

The mass loss tendency of NR films is comparable to that of XNBR films. Both the unfilled NR and XNBR films experienced a lower mass loss. Compound additives, such as sulfur, are reported to inhibit the rate of biodegradation of rubber films. Incorporating sago starch, however, would encourage soil microorganisms to consume this bio-filler and secrete enzymes that can degrade rubber molecular chains [36]. The AHSS-filled NR and XNBR films showed significant mass loss. This could be accredited to the decrease in the amorphous section after acid hydrolysis of sago starch, which makes rubber and glycosidic chains more susceptible to attack by microorganisms [34].

Rahman et al. studied the degradation of gloves made from NR with sago starch as bio-filler from buried soil samples by a mixed culture containing starch-degrading bacteria as well as NR-degrading bacteria. The aim of the starch hydrolysis test was to confirm the presence of starch-degrading bacteria in the mixed culture. In this test, the evaluated bacteria were grown on agar plates containing starch. After incubation, an iodine indicator was added to the plates. Hence, when a few drops of potassium iodide solution were applied to the sample, the surface of the plate became blue-black because the reaction between starch and iodine produces polyiodide chains. The amylose in starch forms helices around which the iodine molecules are clustered. This blue-black color does not occur when starch is broken down or hydrolyzed into smaller carbohydrate units. Therefore, transparent, clear zones were formed next to the colonies that hydrolyze starch, while the other parts of the plate remained colored [90].

Figure 10 shows a clear zone in the iodine test on an agar plate that proved the starch hydrolyzation. Based on the biodegradation rate data, the presence of starch-degrading microorganisms as well as rubber-degrading bacteria was detected, which accelerated the biodegradation of sago-filled NR gloves by 53.68%, while the biodegradation rate for NR gloves (without filler) was lower, at around 50.31% [35].

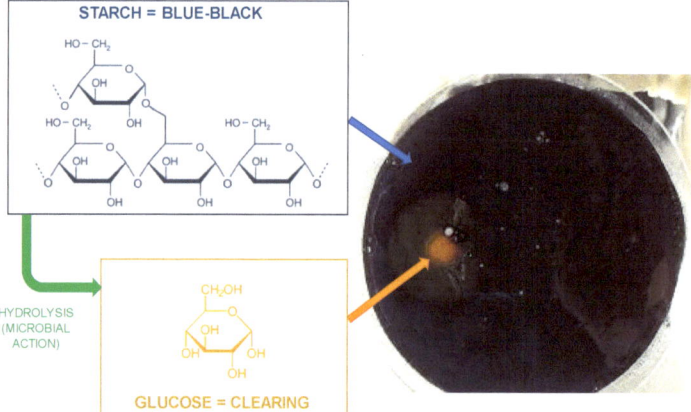

Figure 10. Starch hydrolysis test of the mixture culture. The blue and the orange arrows show the areas where the starch remains unchanged and where it has been hydrolyzed to glucose by microbial action, respectively. Reprinted and adapted with permission under a Creative Commons license (CC BY 3.0) from reference [35].

4.3. Mangosteen Peel as Antimicrobial Agent in NR Gloves

Xanthones are secondary metabolites found in plants, fungi, and lichens. They have been isolated in the pericarp area of the mangosteen, a typical fruit of the tropics. Xanthones have potent antioxidant, anticancer, anti-inflammatory, anti-allergic, antibacterial, antifungal, and antiviral properties [91]. In fact, the peel of mangosteen is a kind of hydrophobic biomaterial that can be used in medical care, cleaning products, skin care, and cosmetics. It can inhibit exposed cells such as *S. aureus*, *S. albus*, and *M. luteus*, as well as plant pathogenic fungus like *F. oxysporum f.* sp. *vasinfectum*, *A. tenuis*, and *D. oryzae*. It is also effective against *P. acnes* and *S. epidermidis* and can be used as an alternate therapy against acne. Furthermore, due to proven good properties, it can suppress cancer cells and has potential for both preventative and therapeutic purposes [92].

Moopayak and Tangboriboon used mangosteen peel as a bio-filler to produce NR medical gloves. The addition of mangosteen peel powder to NR formulation as a bio-filler can improve the antimicrobial properties of the gloves without sacrificing the softness, film thickness, and mechanical characteristics [93].

NR gloves with mangosteen peel powder have been obtained with good appearance, smooth, transparent, and thin, with good elongation, good tensile strength, no water leakage, and no skin toxicity. Comparing NR gloves with and without mangosteen peel, it was detected that the mechanical properties with the addition of the bio-filler were not only preserved, but slightly improved. The microstructure of the mangosteen peel used is presented Figure 11 [93].

Color	T_{degrad} (°C)	Particle size (µm)	Solubility in water (%)	Solubility in alcohol (%)	Pore diameter (Å)	Pore volume (cm³/g)
Reddish brown	110	24.777	Low	Excellent	10.970	0.000536

Figure 11. SEM (**a1**,**a2**) and FESEM (**b1**–**b4**) micrographs of mangosteen peel. The main physical properties of mangosteen peel powder are summarized below. Reprinted and adapted with permission from reference [93]. Copyright © 2020 John Wiley and Sons.

To prepare the NR gloves, a porcelain hand mold, concentrated NR latex, and compounding chemicals were used. The mold was washed, dried, and dipped for 3 s in coagulant. The coagulant-coated hand mold was then dipped into the NR for 15 s and dried at room temperature for 2–3 min. The NR compounded latex film was then cured for 30 min at 120 °C, allowed to dry, and demolded. The toxicity of gloves containing mangosteen peel was lower than that of gloves containing silver nitrate, which can impact human skin and should be used in the appropriate ratio to prevent microbial infections.

E. coli, *B. subtilis*, *S. aureus*, and *P. aeruginosa* were shown to be inhibited by mangosteen peel concentrations between 80 and 100 g/mL [93].

4.4. NR Films with Cellulose Nanocrystals as Reinforcing and Crosslinking Agent for Application in Gloves

Because of their elevated rigidity and reinforcing capacity, cellulose nanocrystals (CNCs) are a promising bio-filler. Typically, CNCs are obtained from renewable resources through acid hydrolysis, as is shown in Figure 12a [94–96]. CNCs are normally dispersed in NR latex without modification due to their great dispersibility in aqueous media, which is a result of their high content of hydroxyl groups [97]. However, it has been demonstrated that modifying the surface of CNCs enhances their reinforcement effect on NR. As a result of the hydrophobic–hydrophobic interaction between modified CNCs and NR, the tensile strength and the elongation at break increased significantly compared to unmodified CNCs. To ensure compatibility with the rubber while preserving the dispersion of CNCs aqueous media, it is crucial to strike a balance in the degree of modification of the CNCs [98].

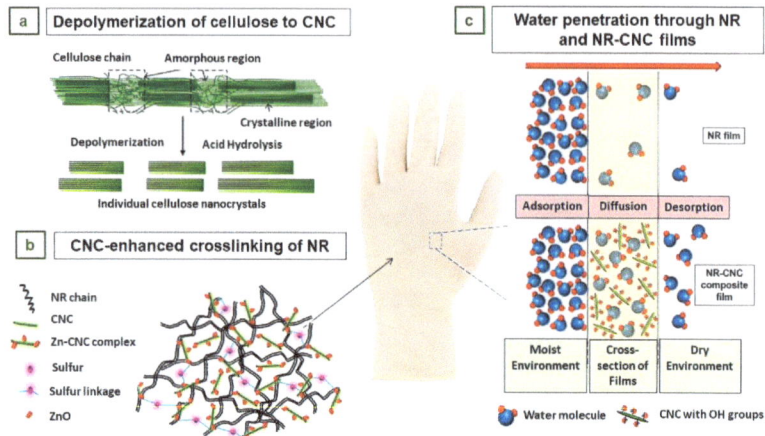

Figure 12. (**a**) Depolymerization of cellulose to nanocellulose (reprinted with permission under a Creative Commons license (CC BY 3.0) from reference [96]). (**b**) Illustration of the formation of a Zn–cellulose complex with CNC in the cross-linked NR matrix [99]. (**c**) Illustration of the proposed permeation mechanism through NR and NR–CNC nanocomposites and THF. (**b**,**c**) Reprinted and adapted with permission from reference [99]. Copyright © 2020 American Chemical Society.

Blanchard et al. studied the influence of CNCs on the reinforcing, crosslinking, and solvent barrier characteristics of lightly crosslinked NR films [99]. In nonpolar matrices, it is difficult to efficiently disperse CNCs due to their extensive surface area and their trend to form aggregates bonded together by hydrogen bonds. Therefore, for proper dispersion, it was necessary to prepare an aqueous colloidal suspension of CNCs [100].

As an initial step for experimentation, NR composite latex was prepared by predispersing the compound chemicals, including ZnO and sulfur, in water. This predispersion mixture was subsequently incorporated into NR formulations [99].

For NR-CNC films, CNCs were incorporated at concentrations of 0 (NR control), 0.5, 1.5, 3, and 5 phr. Dipping films were produced using glass substrates that were dipped in a coagulant solution for 10 s and then dried at 65 °C for 20 min. The substrate was then dipped for 40 s in the NR formulations and cured at 100 °C for 1 h. The cured films were then peeled off from the glass substrates and cured for an additional hour. The dynamic and tensile mechanical properties of these dipping films were analyzed. Increased crosslinking resulted in significant improvements in both tensile strength and modulus compared to the base NR control. The force required to break the films increased as film thickness decreased [99].

To prepare cast films, latex formulations containing 40 wt % total solids were cast on glass substrates to obtain dried NR films of 12 mm in thickness. The films were then cured at 100 °C for 1 h, peeled off, and post-cured for 1 h. The cast films were used to evaluate the impact of CNCs on morphology, crosslinking density, and barrier properties. The addition of CNCs resulted in an increase in the crosslinking density of the NR films. This was presumably attributed to increased dispersion of the crosslinking activator ZnO due to the development of a Zn–cellulose complex, with the CNCs acting as a dispersant (Figure 12b) [99].

The nanocomposite thin films had low permeability to nonpolar solvent vapors, such as tetrahydrofuran (THF), but high permeability to water vapor, as shown in Figure 12c. This ability of the material to reach or surpass NR strength at lower film thicknesses may allow for thinner gloves and for hand perspiration to pass through while functioning as a barrier to solvents. It may also lead to cost savings by reducing the use of NR. The findings of this investigation indicate that NR composite films produced using NR/CNCs have considerable potential for application as gloves [99].

4.5. NR and NBR Gloves Coated with Gardine Solution

Gardine solution is an innovative antiseptic dye with broad-spectrum antibacterial effects prepared by combining brilliant green with chlorhexidine. Brilliant green and chlorhexidine, when used independently, have been shown to have low antimicrobial efficacy, but when combined, they have a synergistic effect with significantly improved efficacy. Chlorhexidine is a non-toxic chemical widely used in low concentrations in mouthwash solutions along with other antiseptics [4]. Historically, brilliant green has been used as a topical anti-infective for skin lesions and is currently used in combination with gentian violet and proflavine hemisulfate in neonatal nurseries as a broad-spectrum antiseptic solution [101].

In the study conducted by Reitzel et al., NR and NBR gloves were impregnated with Gardine solution to create antimicrobial coating. The results indicated that Gardine-coated NR and NBR gloves were highly effective in reducing pathogenic contamination in the short term and long term. For the short-term exposure test, 1 cm^2 segments of NR and NBR coated and uncoated control gloves were exposed to 1.5×10^8 colony-forming units (cfu)/mL of methicillin-resistant *Staphylococcus aureus* (MRSA), vancomycin-resistant enterococci, multidrug-resistant (MDR) *E. coli*, MDR *Acinetobacter baumannii*, and *Candida albicans*. The segments were dried for 30 s, 10 min, 30 min, and 1 h, and then distributed on agar plates, which were incubated overnight at 37 °C, and growth was measured. All microorganisms tested were significantly reduced within 30 s and completely eliminated within 1 h when exposed to Gardine-coated NR gloves. Figure 13(a1) shows the complete kill within 30 s for *E coli* and Figure 13(a2) for MRSA [101].

For the long-term exposure test, MRSA and *E coli* were employed because they are biofilm-forming microorganisms typically found in hospital environments. Figure 13b shows that the average number of MRSA and *E. coli* colonies adhered to the surface of Gardine-coated gloves was significantly lower than that of control gloves. After 24 h, the adhesion of MRSA and *E. coli* to the surface of Gardine-coated NR gloves decreased by at least 95%. On the surface of Gardine-coated NBR gloves, there was an 80% reduction in MRSA and a 100% reduction (total kill) in *E coli* [101].

These coated gloves represent an alternative means of preventing the spread of invasive microbial pathogens. In terms of final cost, the Gardine impregnation process would be carried out during the manufacture of the gloves, reducing the costs associated with a separate additional manufacturing process. In addition, Gardine solution is made up of low-cost components. These antimicrobial gloves would be cost-effective based on material and production time estimates [101].

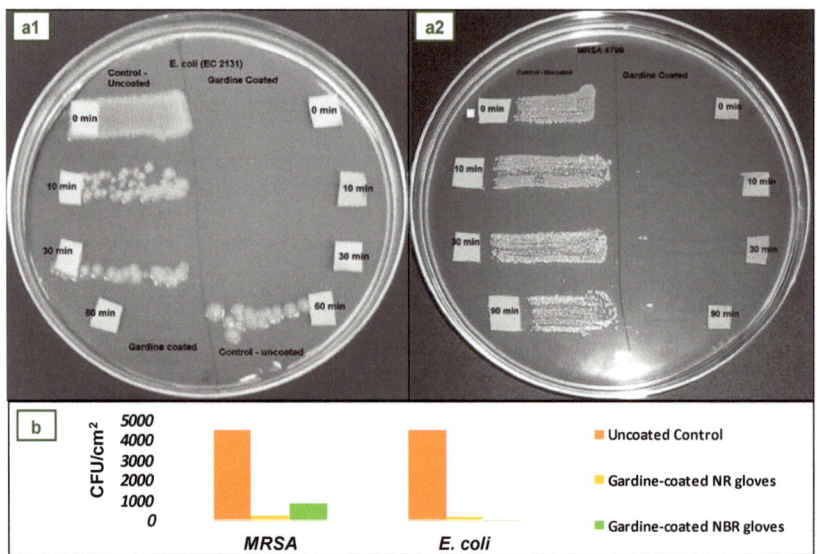

Figure 13. (**a1**) Brief exposure test of Gardine-coated gloves. (**a2**) Long-term exposure. (**b**) Mean colony counts recorded for all coated glove types after 24 h exposure to MRSA or *E. coli*. Reprinted and adapted with permission from reference [101]. Copyright © 2009 Elsevier.

4.6. NBR Gloves Coated with Poly(hexamethylene Biguanide) Hydrochloride

Poly(hexamethylene biguanide) hydrochloride (PHMB) is a positively charged polymer with antibacterial and antiviral activity [102]. It is effective against a wide range of pathogenic microorganisms, including Gram-negative bacteria, Gram-positive bacteria, and fungi [103]. Due to its strong and nonspecific interaction with negatively charged phospholipids in the cellular membranes of microorganisms, PHMB possesses a broad antibacterial spectrum [102]. PHMB has been utilized for decades with no reports of bacterial resistance [104]. It has been demonstrated to pose a minimal risk of skin sensitivity and a low toxicity risk to humans in general [102]. Moreover, PHMB has disinfectant and antiseptic properties [104], which makes it suitable for house cleaning, water sanitization, hygiene products, and wound treatment [103].

Leitgeb et al. conducted an in vitro examination of the antibacterial efficacy of a new non-sterile NBR medical glove coated with PHMB on its outer surface provided for Ansell Ltd. These gloves are intended for use during patient examinations to avoid microorganism cross-contamination across surfaces in healthcare environments. The study's goal was to evaluate the performance of NBR medical gloves, with and without antibacterial PHMB coating on the outside surface, (Figure 14a) made from the same formulation [105].

For this investigation, the quantity of bacteria recovered from a stainless-steel coupon after touching a pigskin substrate with both gloves was evaluated. Pigskin substrates were contaminated with suspensions containing 1×10^9 colony-forming units of *E. faecium* ATCC 51559, *E. coli* ATCC 25922, *K. pneumoniae* ATCC 4352, and *S. aureus* ATCC 33591. After impregnating sections of pigskin with bacterial suspensions, swatches of coated and uncoated (control) gloves were tightly pressed onto the inoculated pigskins. Immediately, a sterile weight was placed on the glove swatch and left in place for 1 min; then, the sample was placed in a sterile Petri plate with the exposed side facing up and left for 5 min at room temperature. The contaminated side of the glove swatch was then positioned on a sterile 40 mm diameter stainless steel coupon. The weight was immediately placed onto the test glove for 1 min. Separately, the contaminated pigskin, stainless steel coupon, and test glove swatch were placed in buffer solution and carefully vortexed [105].

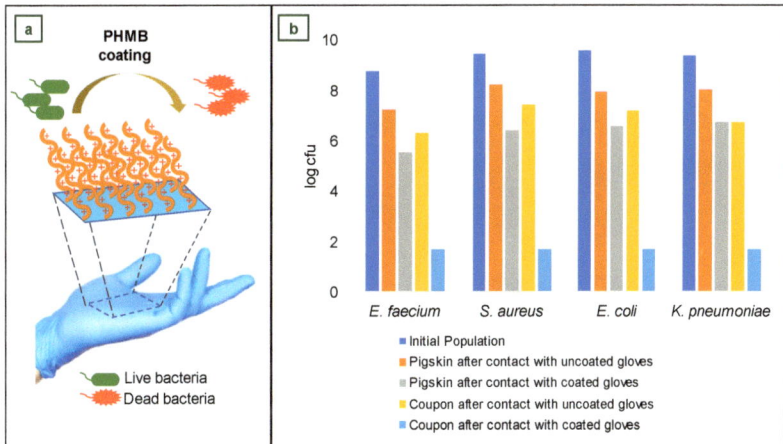

Figure 14. (**a**) Schematic illustration of coating (illustration prepared by the authors based on reference information [105]). (**b**) Pre- and post-exposure populations of challenge microorganisms following transfer procedures. Adapted with permission under a Creative Commons license (CC BY) from reference [105]. Copyright © 2013 Elsevier.

Bacterial extractions were carried out on the pigskin substrate, stainless steel coupons, and each glove swatch, and the difference between the coated and uncoated control gloves was analyzed (Figure 14b). In comparison to the non-coated control glove, the coated glove reduced *E. faecium* recovery by 4.63 log cfu, *E. coli* recovery by 5.48 log cfu, *K. pneumoniae* recovery by 5.03 log cfu, and *S. aureus* recovery by 5.72 log cfu. According to these findings, the use of antibacterial medical gloves may be an innovative method for preventing or limiting cross-contamination and, consequently, the indirect spread of infections in intensive care unit (ICU) settings [105].

4.7. NR Antimicrobial Three-Layer Glove

In some instances, external coating is not suggested for surgical gloves since it might create undesired side effects. There is a possibility of transferring the coating to the patient's tissues, cells, and organs during surgery. For these reasons, a three-layer glove is a good alternative for invasive procedures.

The three-layer antimicrobial coating method, used in surgical gloves, inserts antimicrobial chemicals between NR films. It is possible by triple-dipping the glove mold in NR compounded latex and antimicrobial solutions during the manufacturing process. Triclosan, nanocomposites, metal ion-based antimicrobial agents, vegetable oil surfactants, antiseptic dyes, chlorhexidine, gluconate, dodecyl dimethyl ammonium chloride salt, benzalkonium chloride, and similar antimicrobial agents might be incorporated in this manner [39].

Daeschlein et al. created a prototype of a new NR three-layer antibacterial surgical glove. Figure 15a shows a microscopic cross-sectional view of a droplet-like mixture of antimicrobial agents (chlorhexidine and quaternary ammonium salts) in the intermediate layer, while Figure 15b is a representation of the inner (I) and outer (O) surfaces adjacent to the rubber border layers. The antimicrobial agent is released from the interlayer upon penetration of the glove, resulting in deposition of the active antimicrobial agent at the site of damage or puncture. Because the antimicrobial agent droplets are trapped between two NR boundary layers, there is no continuous exposure of the material to the skin surface in the absence of lesions, hence lowering the possibility of sensitivity from extended contact [106].

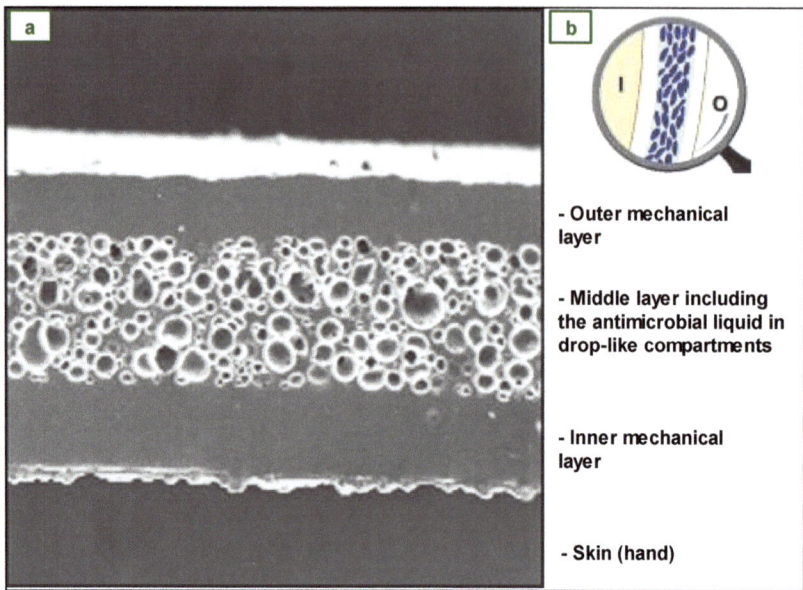

Figure 15. Three-layer NR glove with antimicrobial agent. (**a**) Cross-section micrograph. (**b**) Three-layer scheme. Reprinted with permission from reference [106]. Copyright © 2011 Elsevier.

4.8. NBR Antimicrobial Gloves Coated with Electrospun Trimethylated Chitosan (TMCh)-Loaded (PVA) Fibers

Usually, antimicrobial agents have been added to gloves through coatings. But after this treatment, the surface of the gloves tends to become smoother, and they tend to slip more when they are used. Thus, alternative coatings that make the surface of the glove rougher are needed. Ultrafine fibers, loaded with antibacterial agents, are one of the materials that solve this problem. Electrospinning is the method most often used to make these fibers because it provides the opportunity to conveniently control the fiber dimensions. This approach essentially utilizes an electric field to draw a polymer strand [107].

Chitosan is a highly biocompatible antibacterial agent composed of β-(1→4)-D-glucosamine and γ β-(1→4)-N-acetyl D-glucosamine units. Water-soluble chitosan derivatives such quaternized chitosan (QCh) and alkylated chitosan like trimethylated chitosan (TMCh) are alternatives to chitosan alone (usually only soluble in acidic media) for use as antibacterial agents in neutral pH conditions [107]. The presence of lipoteichoic acids, a significant component of the cell wall of Gram-positive bacteria, and lipopolysaccharide, of the outer membrane of Gram-negative bacteria, which provide a linkage for polycationic TMCh and disrupt the membrane functions, may explain the antibacterial capabilities of TMCh [108]. Normally, lipopolysaccharide and proteins are kept together by electrostatic interactions with divalent cations, which are essential for the outer membrane stability. Polycations compete with divalent metals such as Mg^{2+} and Ca^{2+} ions in the cell wall, hence compromising the cell wall integrity [109].

Vongsetskul et al. effectively coated NBR gloves with ultrathin electrospun PVA fibers loaded with TMCh. Using water as a solvent, solutions containing 4% w/v of TMCh mixed with 8% w/v of PVA and 2% w/v of TMCh mixed with 10% w/v of PVA were prepared. These solutions were subjected to the electrospinning process using a feed rate of the solutions of approximately 0.5 mL/h [107].

Different electrical voltage values were used (12, 14, 16, 18, and 20 kV) to analyze its effect on the morphological appearance of the produced fibers. As the applied voltage increased from 12 to 16 kV, the fibers became smoother and smaller. SEM studies revealed that the optimal conditions to produce uniform fibers (101 to 133 nm of diameter) were a voltage of 16 kV and solution of 4% w/v TMCh-8% w/v PVA [107].

For the surface roughness and wettability study, film-coated NBR gloves were prepared by dipping in a 4% w/v TMCh-8% w/v PVA solution and drying at room temperature. The surface roughness was increased from 429 to 511 µm^2 by coating electrospun fibers on the glove. The contact angle measurements of the NBR glove surface, TMCh-PVA film on the NBR glove surface, and TMCh-PVA electrospun fibers on the NBR glove surface were 80.1° ± 1.2°, 59.3° ± 8.9°, and 37.1° ± 2.7°, respectively. These values indicate that the hydrophilicity of the gloves increased when coated with TMCh-PVA films or TMCh-PVA fibers [107].

To evaluate the antimicrobial activity of the fiber-coated gloves, the agar plate method was used. *E. coli*, *P. aeruginosa*, *A. baumannii*, and *C. albicans* were tested. In the results of antimicrobial testing, a zone of growth inhibition against the tested microbes by the TMCh-PVA fiber-coated NBR gloves was observed, whereas no antimicrobial activity was observed for the PVA fiber-coated ones. In conclusion, NBR gloves coated with these TMCh)-loaded (PVA) fibers exhibited antibacterial properties against Gram-negative bacteria, including *E. coli*, *P. aeruginosa*, and *A. baumannii*, as well as yeast *Candida albicans*. Likewise, this coating on the external surface of the glove improved roughness and wettability, which would be advantageous for gripping and practical applications [107].

4.9. Antibacterial NR Films with Surface-Anchored QP4-VP for Application in Medical Gloves

Quaternary ammonium compounds (QACs) are cationic active biocides, which, in addition to their antibacterial action, are ideal for cleaning and deodorizing [4]. The mechanism of action of QACs against bacterial and viral phospholipid membranes is depicted in Figure 16a, where the red spheres represent positively charged nitrogen atoms. When bacteria encounter cationic ammonium agents, several processes take place: first, QACs connect to and insert themselves into the cell wall; then, they interact with the cytoplasmic membrane, releasing cytoplasmic material outside the membrane; and finally, they cause the cell wall to disintegrate via autolytic enzymes. In general, the loss and destruction of various sections of the bacteria results in their inactivation [4,110].

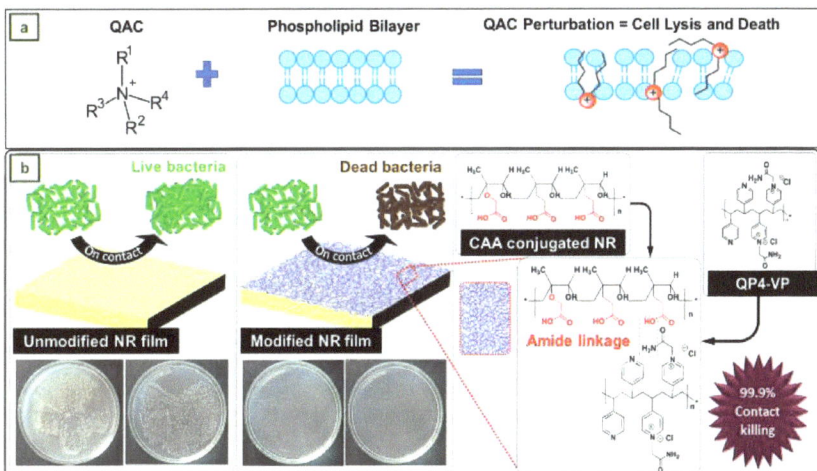

Figure 16. (**a**) Mode of action of QACs against both bacterial and viral phospholipid membranes (Reprinted with permission under standard ACS Author Choice/Editors' Choice usage agreement from reference [110]). (**b**) Antibacterial activity of QP-4VP-conjugated NR films vs. Control NR films. Reprinted with permission from reference [111]. Copyright © 2022 Elsevier.

In the work of Arakkal et al., NR films were converted into an effective antibacterial material (Figure 16b) through surface conjugation of quaternized poly(4-vinylpyridine) (QP4-VP) via an amide linkage bond using chloroacetic acid. The antibacterial action of poly(4-vinylpyridine) has been extensively examined and explored in ion exchange resins, but its low biocompatibility prevents its widespread application in biomedicine. However, it has also been shown that with the right choice of space groups and copolymerization, the hemolytic activities of the polyelectrolyte can be inhibited while maintaining antibacterial activity [111].

To evaluate the antimicrobial activity and stability of the NR films coated with a QP4-VP-conjugated surface layer, they were subjected to a leaching process in milli-Q water at 50 °C for 4 days. Subsequently, coated NR and leached coated NR films were exposed to *P. aeruginosa* and *A. baumannii* strains [111].

The results indicated that the microbial load of *P. aeruginosa* was reduced by 93.25% and 99.98% with the coated NR films and leached coated NR films, respectively. Similarly, the reduction in *A. baumannii* was 32.41% and 99.99%. The improved bacterial reduction rate confirmed that the leaching process at elevated temperatures allows the disoriented QP4-VP chains to organize efficiently, resulting in a higher conjugation density. This conjugation method could be used to develop similar antibacterial surfaces for various applications, such as medical gloves [111].

4.10. NR, NBR, and PE Medical Gloves with Blood-Repellent, Antibacterial, and Wound Healing Properties, Modified through Spraying Process

Medical blood-repellent gloves (MBRGs) were proposed by Zhuo et al., by means of treating the surface of conventional NR, NBR, and PE medical gloves with a novel procedure to achieve blood repellency and promote wound healing. This treatment was executed with a mist spray (MS), which was elaborated by mixing sodium citrate (SC), didecyldimethylammonium chloride (DDAC), and a silicon oil emulsion (SOE) containing aminoethylaminopropyl polydimethylsiloxane (AEAPS). It was intended that SC would combine with blood calcium ions to inhibit blood coagulation and glove adhesion, that AEAPS would be responsible for the hemophobicity and hydrophobicity of the treated gloves, and that DDAC, being a quaternary ammonium compound, would endow the gloves with antibacterial properties [112].

MBRGs were created by spraying MS onto the surface of commercial NR, NBR, or PE medical gloves and waiting for one minute. Experiments in vitro and in vivo demonstrated that these gloves are hemophobic and facilitate the healing of infected wounds. The antibacterial efficiency of MBRGs was tested against known bacteria strains. In vitro antimicrobial testing was performed with MS concentrations of 800, 400, 200, 100, and 50 g/mL. A solution of *S. aureus* or *E. coli* was added to each MS concentration and incubated first in tubes and then on agar plates. Phosphate-buffered saline (PBS) was used instead of MS in the control group. After 24 h, the antibacterial efficacy was assessed. MS showed outstanding activity against *S. aureus*, with an antibacterial rate close to 100% at a concentration of 50 μg/mL. In the case of *E. coli*, the antibacterial effect was close to 100% when the concentration was 200 μg/mL. The antibacterial activity of MS was also verified through the live/dead viability assay. In this study, *S. aureus* and *E. coli* were treated with MS. After treatment with MS, red fluorescence (dead bacteria) was clearly visible, whereas blue fluorescence (living bacteria) was nearly non-existent [112].

4.11. NR Gloves with SiO_2 and ZnO Hybrid Nanofillers

Silicon dioxide (SiO_2) or silica is a well-known reinforcing filler in the rubber industry and is commonly used to improve the physical and mechanical properties of NR [113]. To achieve the desired result of reinforcement, conventional silica fillers must be applied in high quantities. Studies have shown, however, that when the size of SiO_2 particles reaches the nanometer range, the nanoparticles can not only drastically reduce the filler content, but also provide superior reinforcement effects [114].

Zinc oxide (ZnO) is an n-type semiconducting particle with catalytic, electrical, and optical properties. This material has a broad UV absorption spectrum, good photostability, thermal stability, and biocompatibility. Due to the nano-size effect, ZnO has photocatalytic antibacterial properties when its size reaches the nanoscale. When ZnO is exposed to UV light, the photon energy is higher than the bandgap energy, which causes the valence band electrons to gain energy and migrate. As a result, many electron–hole pairs are generated on the nano-ZnO surface [115].

The holes (h^+) created on the surface of nano-ZnO generate reactive oxygen species (ROS) when they combine with water or oxygen from the air. These oxidative species adsorb onto the surface of the nanoparticles. Some studies have determined that the interaction between ROS and cells is the key antibacterial mechanism of nano-ZnO [115]. There are indications that antibacterial activity can be initiated not only by UV rays, but also by ambient light [116]. Furthermore, when microorganisms come into contact with nano-ZnO, the released Zn^{2+} and the sharp edges of ZnO nanoparticles can rupture their cell walls [117].

Mou et al. investigated the combination of the exceptional functional capabilities of nano-SiO_2 and nano-ZnO as fillers of NR to create medical gloves. The ZnO and SiO_2 nanoparticles used had an average particle size of about 78 nm and 65 nm, respectively. To evenly distribute the fillers in a composite nano-dispersion, the researchers created a high-speed and high-pressure nano-disperser. To obtain experimental glove samples using the dipping method, NR latex, composite nano-dispersion, and compounding chemicals were thoroughly combined. Initially, the cleaned and dried mold was dipped for 5 s into the previous mixture, then dried at 85 °C for 20 min, and then leached in water at 75 °C for 30 s. After hemming, the gloves were dried at 120 °C for 40 min. Finally, after demolding, they were placed in a drum drier and vulcanized at 120 °C for 20 min [114].

The results indicated that the uniform dispersion of nano-SiO_2 filler enhanced the amount of molecular chain entanglements in the NR, as depicted in Figure 17(a1,a2), making the material structure more compact and improving the barrier performance and aging resistance. The combination NR fillers with 1 phr of ZnO and 4.2 phr of SiO_2 reported the highest tensile strength of 32.6 MPa and elongation at break of 957% in mechanical properties tests. These results represented an interesting improvement in tensile strength and elongation at break compared to the unfilled NR sample, whose values were 27.8 MPa and 880%, respectively. Figure 17(b1,b2) shows scanning electron microscopy images at different magnifications. In addition, the elemental distribution of the LZ1S4.2 sample is presented in Figure 17(b3,b4) [114].

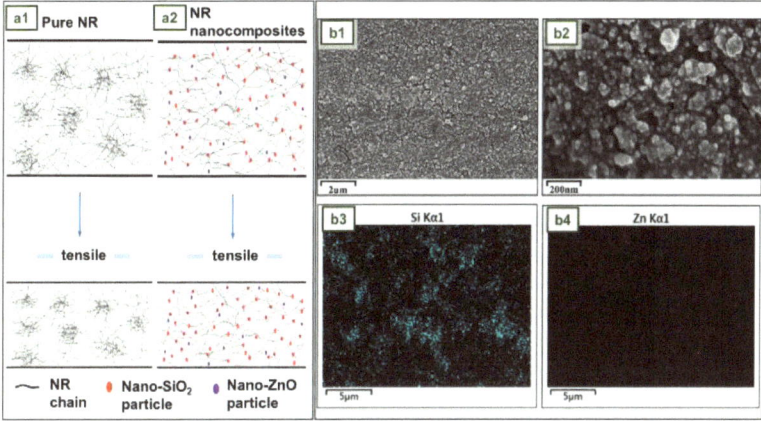

Figure 17. (**a1**,**a2**) Schematic diagram of nanoparticle enhancement mechanism. Scanning electron microscopy images of (**b1**,**b2**) LZ1S4.2 and element distribution of (**b3**) Si and (**b4**) Zn. Reprinted with permission from reference [114]. Copyright © 2022 John Wiley and Sons.

Strains of E. coli and S. aureus were chosen for antibacterial testing. The findings of the antibacterial activity of NR gloves are described in Table 5. The quantity of bacteria on the blank immediately after inoculation is given by U_0, whereas the quantity of bacteria on the blank and on the antibacterial sample after 24 h of incubation are U_t and At, respectively. The antibacterial activity R is equivalent to $log\ U_t - log\ A_t$. A value of R greater than 2 shows that the antimicrobial test is passed. After 24 h of culture, nano-ZnO-treated samples had virtually no bacteria. The R values for E. coli and S. aureus were more than 99.9% and greater than 5.2, respectively. Furthermore, the NR gloves containing hybrid nanofillers remained biocompatible. Therefore, this $NR/ZnO/SiO_2$ nanocomposite may have applications in the development of other NR products [114].

Table 5. Antibacterial test results of gloves [114].

Microorganisms Tested	U_0 (cfu/cm^2)	U_t (cfu/cm^2)	A_t (cfu/cm^2)	R	Antibacterial Rate (%)
E. coli	2.1×10^4	2.9×10^5	<0.6	>5.3	>99.9
S. aureus	2.1×10^4	2.3×10^5	<0.6	>5.2	>99.9

4.12. NR Antimicrobial Gloves Impregnated with Biosynthesized Silver Nanoparticles

Ionic silver (Ag^+) has long been recognized as an antibacterial metallic element capable of acting against bacteria such as E. coli [4,118]. To take advantage of silver ion activity, silver nitrate ($AgNO_3$), a solid powder with antiseptic qualities, is frequently utilized. It can be applied as a surface coating on various items to eliminate viral and bacterial cells [93]. The specific mechanism of silver antibacterial activity has not yet been fully understood. Among the variety of approaches, the main three of several pathways that determine the antibacterial activity of silver nanoparticles are the following: (1) irreversible bacterial cell membrane damage caused by direct contact; (2) production of reactive oxygen species (ROS); and (3) interaction with DNA and proteins [119]. Nanoparticle size plays an important role in antibacterial activity. It has been shown that the smaller the size of the nanoparticle, the greater its ability to penetrate bacteria. The nanoparticles attach to the bacterial cell wall, penetrate it, and cause damage and alterations in various metabolic pathways. It affects DNA replication and protein synthesis; due to oxidative stress, ROS generation occurs, which eventually leads to cell death [120].

Paosen et al. developed NR gloves coated with biosynthesized silver nanoparticles (AgNPs). The biosynthesis of AgNPs was performed using extract of *Eucalyptus citriodora* ethanolic leaf. NR gloves were cut into pieces, dipped into the AgNP solution, and dried. The elemental analysis of the coated gloves revealed that 24.8% wt silver was firmly adhered to the surface. Biofilms of S. aureus ATCC 25923, P. aeruginosa ATCC 27853, and *Candida albicans* ATCC 90028 were expected to develop on glove samples during 24 h of incubation [121].

Figure 18(a1–a6) shows the differences in the effect of staining microbial biofilms with uncoated gloves and AgNP-coated gloves. The results revealed that AgNP-coated gloves effectively removed S. aureus biofilms. Images from fluorescence microscopy were stained with a red fluorescent DNA-specific dye that only penetrates cells with damaged membranes and dead bacteria. The pictures revealed that P. aeruginosa cells were significantly less abundant and viable when cultured on AgNP-coated gloves compared to uncoated gloves, suggesting that AgNPs killed and inhibited microbial adhesion [121].

Figure 18(b1–b4) shows the scanning electron micrographs taken to examine the surface morphology of the bacterial adhesion by the mixed culture of P. aeruginosa, S. aureus, and C. albicans after 24 h of incubation. At magnifications of 5000 and 10,000, microbial cells on glove surfaces (indicated by red arrows) could be observed. Polymicrobial cells were fixed and colonized in the untreated gloves, whereas microbial cells were observed in tiny quantities on the coated surfaces. The viability of gloves coated with AgNPs was proven. These gloves presented significant antimicrobial activity, particularly against multidrug-resistant bacteria, and may be suitable for preventing or minimizing cross-contamination and indirect pathogen transmission in hospital settings [121].

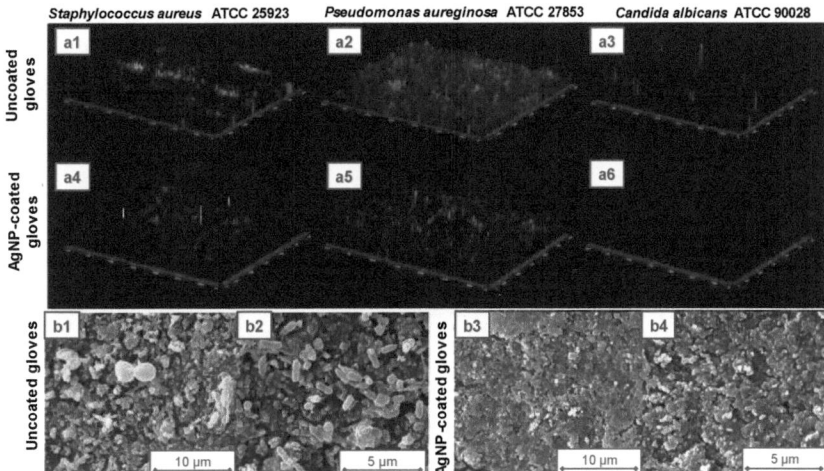

Figure 18. Fluorescence microscopy images of (**a1,a4**) *S. aureus* ATCC 25923, (**a2,a5**) *P. aeruginosa* ATCC 27853, and (**a3,a6**) *C. albicans* ATCC 90028 biofilms incubated with (**a1–a3**) uncoated gloves and (**a4–a6**) AgNP-coated gloves. Scanning electron microscopy (SEM) analysis. SEM micrograph of polymicrobial anti biofilm activity of (**b1,b2**) uncoated gloves and (**b3,b4**) AgNP-coated gloves. Reprinted with permission from reference [121]. Copyright © 2021 John Wiley and Sons.

4.13. NR Antimicrobial Gloves with Poly(dimethylsiloxane)-Copper Coating

Multiple studies have demonstrated that copper (Cu) possesses antimicrobial properties against *E. coli*, *L. monocytogenes*, *C. difficile*, yeasts, and viruses [122]. Cu-containing surfaces have been found to reduce environmental microbial contamination [123]. Cu has also been demonstrated to be a powerful antibacterial agent that can inhibit the growth of antibiotic-resistant bacteria such as MRSA, EMRSA-1, and EMRSA-16 [124,125].

In the research of Tripathy et al., an antimicrobial coating consisting of poly(dimethylsiloxane) (PDMS) combined with copper hydroxide nanowires (PDMS Cu) was developed. PDMS-Cu's antibacterial properties have been demonstrated to reduce the viability of a panel of multidrug-resistant clinical pathogens (*E. coli*, *S. aureus* (MRSA), and *K. pneumoniae*). The PDMS-Cu surface exhibited superior activity as an antimicrobial film compared to the control (a glass coverslip). The antibacterial effectiveness of the PDMS-Cu surface was also evaluated in a patient room alongside controls (glass coverslips and PDMS substrates). On this occasion, PDMS-Cu was found to have the lowest amount of attached bacterial colonies compared to the controls [126].

In addition to the previous tests, it was established that coating a stethoscope diaphragm with a thin layer of PDMS-Cu could inhibit the transfer of infections from one patient to another in a hospital setting. The possibility of coating commercially available NR gloves with a thin layer of PDMS-Cu provides compelling evidence and an attractive opportunity to introduce antibacterial gloves into hospitals to minimize the transmission of nosocomial infections [126].

Figure 19a presents a schematic explanation of the antibacterial behavior of the PDMS-Cu surface. Figure 19(b1,b2) shows bacterial colonies on chocolate agar plates after 2 and 4 h (respectively) of exposure to the environment in a patient room. Figure 19c shows the confocal microscopy of *E. coli* biofilm on coverslip and PDMS-Cu surfaces after 5 days of incubation in Luria broth (LB) culture media [126].

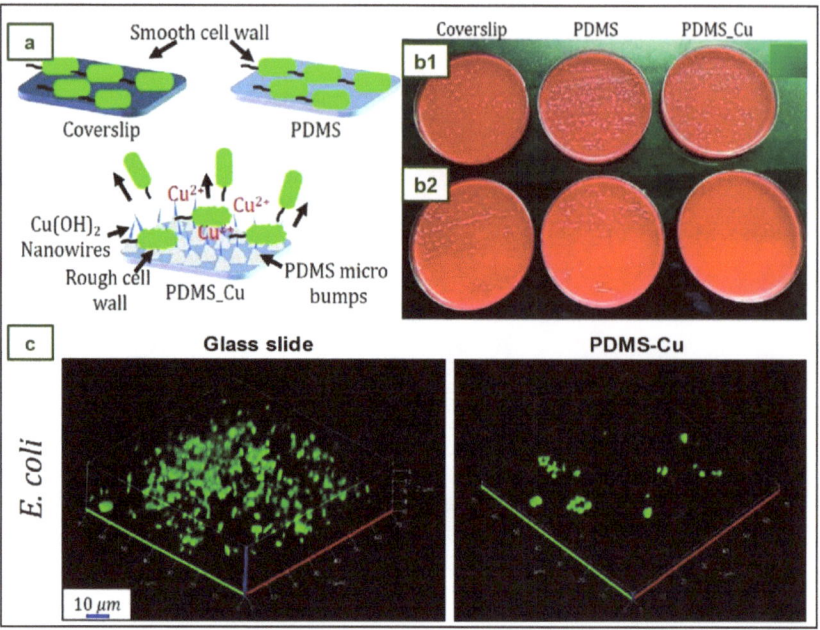

Figure 19. Images of the antibacterial activity of PDMS-Cu. Reprinted with permission from reference [126]. Copyright © 2018 American Chemical Society.

5. Conclusions

The sustainability of medical glove production processes is essential to support the three pillars of sustainability in the environmental, social, and financial sectors. In this sense, the use of performance-enhancing materials such as biomaterials, bio-fillers, biodegradable polymers, antimicrobial agents, etc. becomes a promising route to create new medical gloves with improved properties.

The integration of antiseptic substances or drugs with antimicrobial properties represent a viable option against drug-resistant bacteria and cross-contamination of pathogenic microorganisms and viruses. In this regard, ascorbic acid, biguanides, quaternary ammonium compounds, chitosan derivates, chlorhexidine, and Gardine solution have been successfully assayed. Furthermore, nanoparticles of metals, especially silver and copper, as well as metallic oxides such as ZnO and CuO have been effectively employed.

The use of low-cost bio-fillers such as sago starch, mangosteen peel, and cellulose nanocrystals can improve the biodegradability properties of gloves without sacrificing the softness, film thickness, and mechanical characteristics.

When scaling the knowledge obtained in the laboratory to industry, it is important to consider the ease of production and the profitability of the modifications. In this regard, processes that incorporate performance-enhancing materials directly into rubber formulations or that only require an extra dipping process may be the most attractive from an economic point of view. In summary, the optimization of crucial manufacturing parameters is necessary to obtain safe and high-quality gloves that meet regulatory criteria and are attractive to consumers and investors.

Author Contributions: Conceptualization, M.J.L., L.J.d.V., J.P. and L.F.; formal analysis, M.J.L. and L.F.; writing—original draft preparation, M.J.L.; writing—review and editing, M.J.L., J.P. and L.F. All authors have read and agreed to the published version of the manuscript.

Funding: This work was funded by the Generalitat de Catalunya (Grant: 2021SGR-01042).

Acknowledgments: M.J.L. acknowledges the Departament de Universitats i Recerca de la Generalitat de Catalunya i del Fons Social Europeu for the financial support of an FI-UPC predoctoral grant (2022 FI_B 00709). M.J.L., L.J.d.V., J.P. and L.F. acknowledge the group eb-POLICOM/Polímers i Compòsits Ecològics i Biodegradables, a research group of the Generalitat de Catalunya (Grant 2021 SGR 01042).

Conflicts of Interest: The authors declare no conflict of interest.

References

1. Mazón, L.; Orriols, R.M. Sanitary Gloves Management. Adequate Protection-Effectiveness and Environmental Responsibility. *Rev. Asoc. Española Espec. Med. Trab.* **2018**, *27*, 175–181.
2. Delves, P.J.; Martin, S.J.; Burton, D.R.; Roitt, I.M. *Roitt's Essential Immunology*, 13th ed.; John Wiley and Sons: Hoboken, NJ, USA, 2017; ISBN 9781118415771.
3. Ford, C.; Park, L.J. How to Apply and Remove Gloves. *Br. J. Nurs.* **2019**, *28*, 26–28. [CrossRef]
4. Babadi, A.A.; Bagheri, S.; Hamid, S.B.A. Progress on Antimicrobial Surgical Gloves: A Review. *Rubber Chem. Technol.* **2016**, *89*, 117–125. [CrossRef]
5. Ellis, H. Surgical Gloves. *J. Perioper. Pract.* **2010**, *20*, 219–220. [CrossRef]
6. Ellis, H. Evolution of the Surgical Glove. *J. Am. Coll. Surg.* **2008**, *207*, 948–950. [CrossRef]
7. Carraher, C.E. *Introduction to Polymer Chemistry*, 4th ed.; CRC Press: Boca Raton, FL, USA, 2017; ISBN 978-1-4987-3761-6.
8. Ikeda, Y. Understanding Network Control by Vulcanization for Sulfur Cross-Linked Natural Rubber (NR). *Chem. Manuf. Appl. Nat. Rubber* **2014**, *2014*, 119–134. [CrossRef]
9. Boyd, D.A. Sulfur and Its Role In Modern Materials Science. *Angew. Chem.* **2016**, *128*, 15486–15502. [CrossRef] [PubMed]
10. Hill, D.M. *The Science and Technology of Latex Dipping*, 1st ed.; Smithers Rapra: Shrewsbury, UK, 2018; ISBN 9781909030053.
11. von Manteuffel, W.Z. Rubber Gloves in the Practice of Surgery. *Cent. Surg.* **1897**, *24*, 553–556.
12. Palma, S. How the World's Largest Maker of Rubber Gloves Is Coping with Covid | Financial Times. Available online: https://www.ft.com/content/359047c2-89fb-11ea-a109-483c62d17528 (accessed on 10 October 2022).
13. Kim Man, M.M. Glove Industry Spikes during Covid-19 Pandemic: A Case Study of Comfort Gloves Berhad (CGB). *Int. Bus. Res.* **2021**, *14*, 105. [CrossRef]
14. Ugalmugle, S.; Swain, R. Medical Gloves Market Size by Product, by Form, by Application, by Usage, by Sterility, by Distribution Channel, by End-Use, COVID-19 Impact Analysis, Regional Outlook, Growth Potential, Price Trends, Competitive Market Share & Forecast, 2021–2027. Available online: https://www.gminsights.com/industry-analysis/medical-gloves-market (accessed on 4 July 2022).
15. Grand View Research Disposable Gloves Market Size Report, 2022–2030. Available online: https://www.grandviewresearch.com/industry-analysis/disposable-gloves-market (accessed on 9 March 2023).
16. Eurostat EU Trade since 2015 of COVID-19 Medical Supplies by Categories. Available online: https://appsso.eurostat.ec.europa.eu/nui/submitViewTableAction.do (accessed on 8 June 2022).
17. Kanjanavisut, K. COVID-19 Increased Global Demand for Medical Glove. EIC Indicates That Malaysia Gains More from Export Than Thailand. Available online: https://www.scbeic.com/en/detail/file/product/6857/fnyoncnigx/EIC-Note_rubber-glove_EN_20200601.pdf (accessed on 29 April 2023).
18. Tey, C. MARGMA Expects Global Glove Demand to Resume Growth Next Year, after 19% Drop | The Edge Markets. Available online: https://www.theedgemarkets.com/article/margma-expects-global-glove-demand-resume-growth-next-year-after-19-drop (accessed on 7 October 2022).
19. MarketWatch Disposable Gloves Market 2022 Recent Developments, Size, Share, Growth Strategies, Segment by Type, Region and Future Forecast 2028—MarketWatch. Available online: https://www.marketwatch.com/press-release/disposable-gloves-market-2022-recent-developments-size-share-growth-strategies-segment-by-type-region-and-future-forecast-2028-2022-08-01 (accessed on 7 October 2022).
20. Lim, J. Top Glove Makes the Highest Quarterly Profit among Malaysia's Top 10 Companies | The Edge Markets. Available online: https://www.theedgemarkets.com/article/top-glove-makes-highest-quarterly-profit-among-malaysias-top-10-companies (accessed on 10 October 2022).
21. Wong, E.L. Top Glove Posts Record Net Profit of RM348m in 3Q, Declares 10 Sen Dividend | The Edge Markets. Available online: https://www.theedgemarkets.com/article/top-glove-posts-record-3q-net-profit-rm348m-declares-10-sen-dividend (accessed on 10 October 2022).
22. Syafiqah, S. Comfort Gloves Quarterly Net Profit Declines on Two Fronts | The Edge Markets. Available online: https://www.theedgemarkets.com/article/comfort-gloves-quarterly-net-profit-declines-two-fronts (accessed on 10 October 2022).
23. Kuala Lumpur Stock Exchange (KLSE) COMFORT (2127): Quarterly Results for Last 10 Financial Years | I3investor. Available online: https://klse.i3investor.com/web/stock/financial-quarter/2127 (accessed on 10 October 2022).
24. Sajeev, S.; Chandra, G. Disposable Gloves Market Size, Share, Trends & Industry Analysis 2023. Available online: https://www.alliedmarketresearch.com/disposable-gloves-market (accessed on 19 May 2020).
25. Maximize Market Research Nitrile Gloves Market: Global Industry Analysis and Forecast (2022–2029). Available online: https://www.maximizemarketresearch.com/market-report/nitrile-gloves-market/126531/ (accessed on 7 October 2022).

26. Akabane, T. Production Method & Market Trend of Rubber Gloves. *Int. Polym. Sci. Technol.* **2016**, *43*, 369–373. [CrossRef]
27. Crépy, M.-N.; Hoerner, P. Gloves: Types, Materials, and Manufacturing. In *Protective Gloves for Occupational Use*; CRC Press: Boca Raton, FL, USA, 2022; pp. 17–44, ISBN 9781003126874.
28. Yip, E.; Cacioli, P. The Manufacture of Gloves from Natural Rubber Latex. *J. Allergy Clin. Immunol.* **2002**, *110*, S3–S14. [CrossRef]
29. Wang, M.; Morris, M. Advances in Fillers for the Rubber Industry. In *Rubber Technologist's Handbook*; White, J., De, S.K., Naskar, K., Eds.; Smithers Rapra Technology Limited: Shawbury, UK, 2009; Volume 2, pp. 189–214, ISBN 978-1-84735-100-5.
30. Mellström, G.A.; Boman, A. Gloves: Types, Materials, and Manufacturing. In *Protective Gloves for Occupational Use*; Wahlberg, J.E., Boman, A., Estlander, T., Maibach, H.I., Eds.; CRC Press: Boca Raton, FL, USA, 2005; pp. 15–28, ISBN 9780367393854.
31. Simpson, R.B. *Rubber Basics*; Rapra Technology Limited: Shrewsbury, UK, 2002; ISBN 185957307X.
32. Srinivasan, S. Powdered Gloves: Time to Bid Adieu. *J. Postgrad. Med.* **2018**, *64*, 68. [CrossRef] [PubMed]
33. Preece, D.; Hong Ng, T.; Tong, H.K.; Lewis, R.; Carré, M.J. The Effects of Chlorination, Thickness, and Moisture on Glove Donning Efficiency. *Ergonomics* **2021**, *64*, 1205–1216. [CrossRef]
34. Daud, S.; You, Y.S.; Azura, A.R. The Effect of Acid Hydrolyzed Sago Starch on Mechanical Properties of Natural Rubber and Carboxylated Nitrile Butadiene Rubber Latex. *Mater. Today Proc.* **2019**, *17*, 1047–1055. [CrossRef]
35. Rahman, M.F.A.; Rusli, A.; Adzami, N.S.; Azura, A.R. Studies on the Influence of Mixed Culture from Buried Soil Sample for Biodegradation of Sago Starch Filled Natural Rubber Latex Gloves. *IOP Conf. Ser. Mater. Sci. Eng.* **2019**, *548*, 012018. [CrossRef]
36. Misman, M.A.; Azura, A.R. Overview on the Potential of Biodegradable Natural Rubber Latex Gloves for Commercialization. *Adv. Mat. Res.* **2013**, *844*, 486–489. [CrossRef]
37. Jawjit, W.; Pavasant, P.; Kroeze, C. Evaluating Environmental Performance of Concentrated Latex Production in Thailand. *J. Clean. Prod.* **2015**, *98*, 84–91. [CrossRef]
38. Rattanapan, C.; Suksaroj, T.; Ounsaneha, W. Development of Eco-Efficiency Indicators for Rubber Glove Product by Material Flow Analysis. *Procedia Soc. Behav. Sci.* **2012**, *40*, 99–106. [CrossRef]
39. Yew, G.Y.; Tham, T.C.; Show, P.L.; Ho, Y.C.; Ong, S.K.; Law, C.L.; Song, C.; Chang, J.S. Unlocking the Secret of Bio-Additive Components in Rubber Compounding in Processing Quality Nitrile Glove. *Appl. Biochem. Biotechnol.* **2020**, *191*, 1–28. [CrossRef]
40. Kloepfer, A.; Jekel, M.; Reemtsma, T. Occurrence, Sources, and Fate of Benzothiazoles in Municipal Wastewater Treatment Plants. *Environ. Sci. Technol.* **2005**, *39*, 3792–3798. [CrossRef]
41. Lenko, D.; Schlögl, S.; Temel, A.; Schaller, R.; Holzner, A.; Kern, W. Dual Crosslinking of Carboxylated Nitrile Butadiene Rubber Latex Employing the Thiol-Ene Photoreaction. *J. Appl. Polym. Sci.* **2013**, *129*, 2735–2743. [CrossRef]
42. Yew, G.Y.; Tham, T.C.; Law, C.L.; Chu, D.T.; Ogino, C.; Show, P.L. Emerging Crosslinking Techniques for Glove Manufacturers with Improved Nitrile Glove Properties and Reduced Allergic Risks. *Mater. Today Commun.* **2019**, *19*, 39–50. [CrossRef]
43. Nik Yahya, N.Z.; Zulkepli, N.N.; Ismail, H.; Ting, S.S.; Abdullah, M.M.A.B.; Kamarudin, H.; Hamzah, R. Properties of Natural Rubber/Styrene Butadiene Rubber/Recycled Nitrile Glove (NR/SBR/RNBRg) Blends: The Effects of Recycled Nitrile Glove (RNBRg) Particle Sizes. *Key Eng. Mater.* **2016**, *673*, 151–160. [CrossRef]
44. Linos, A.; Reichelt, R.; Keller, U.; Steinbüchel, A. A Gram-Negative Bacterium, Identified as Pseudomonas Aeruginosa AL98, Is a Potent Degrader of Natural Rubber and Synthetic Cis-1,4-Polyisoprene. *FEMS Microbiol. Lett.* **2000**, *182*, 155–161. [CrossRef]
45. Rose, K.; Steinbüchel, A. Biodegradation of Natural Rubber and Related Compounds: Recent Insights into a Hardly Understood Catabolic Capability of Microorganisms. *Appl. Environ. Microbiol.* **2005**, *71*, 2803–2812. [CrossRef]
46. Chengalroyen, M.D.; Dabbs, E.R. The Biodegradation of Latex Rubber: A Minireview. *J. Polym. Environ.* **2013**, *21*, 874–880. [CrossRef]
47. Arenskötter, M.; Baumeister, D.; Berekaa, M.; Pötter, G.; Kroppenstedt, R.M.; Linos, A.; Steinbüchel, A. Taxonomic Characterization of Two Rubber Degrading Bacteria Belonging to the Species Gordonia Polyisoprenivorans and Analysis of Hyper Variable Regions of 16S RDNA Sequences. *FEMS Microbiol. Lett.* **2001**, *205*, 277–282. [CrossRef]
48. Preece, D.; Lewis, R.; Carré, M.J. A Critical Review of the Assessment of Medical Gloves. *Tribol.-Mater. Surf. Interfaces* **2021**, *15*, 10–19. [CrossRef]
49. Mooibroek, H.; Cornish, K. Alternative Sources of Natural Rubber. *Appl. Microbiol. Biotechnol.* **2000**, *53*, 355–365. [CrossRef]
50. Cacioli, P. Introduction to Latex and the Rubber Industry. *Rev. Fr. Allergol. Immunol. Clin.* **1997**, *37*, 1173–1176. [CrossRef]
51. Das, D.; Nag, S.; Naskar, H.; Acharya, S.; Bakchi, S.; Ali, S.S.; Roy, R.B.; Tudu, B.; Bandyopadhyay, R. Personal Protective Equipment for COVID-19: A Comprehensive Review. In *Healthcare Informatics for Fighting COVID-19 and Future Epidemics. EAI/Springer Innovations in Communication and Computing*; Garg, L., Chakraborty, C., Mahmoudi, S., Sohmen, V.S., Eds.; Springer: Cham, Switzerland, 2022; pp. 141–154, ISBN 978-3-030-72751-2.
52. Hänninen, A.R.; Mikkola, J.H.; Kalkkinen, N.; Turjanmaa, K.; Ylitalo, L.; Reunala, T.; Palosuo, T. Increased Allergen Production in Turnip (Brassica Rapa) by Treatments Activating Defense Mechanisms. *J. Allergy Clin. Immunol.* **1999**, *104*, 194–201. [CrossRef] [PubMed]
53. Mercurio, J. Creating a Latex-Safe Perioperative Environment. *OR Nurse* **2011**, *5*, 18–25. [CrossRef]
54. Charous, B.L.; Tarlo, S.M.; Charous, M.A.; Kelly, K. Natural Rubber Latex Allergy in the Occupational Setting. *Methods* **2002**, *27*, 15–21. [CrossRef] [PubMed]
55. Liberatore, K.; Kelly, K.J. Latex Allergy Risks Live On. *J. Allergy Clin. Immunol. Pract.* **2018**, *6*, 1877–1878. [CrossRef]
56. McMillan, F.M. *The Chain Straighteners*, 1st ed.; The MacMillan Press Ltd.: New York, NY, USA, 1979; ISBN 978-1-349-04432-0.

57. Zetune, K.; Dombrowski, R.; Day, J.; Wagner, N.J. Puncture and/or Cut Resistant Glove Having Maximized Dexterity, Tactility, and Comfort. U.S. Patent 13/639,740, 6 June 2013.
58. Cuillo, P.A.; Hewitt, N. *The Rubber Formulary*; Noyes Publications: Norwich, NY, USA; William Andrew Publishing, LLC: Norwich, NY, USA, 1999; ISBN 0-8155-1434-4.
59. Konrad, E.; Tschunkur, E. Rubber like Masses from Butadiene Hydrocarbons and Polymerizable Nitrils. U.S. Patent 1,973,000, 11 September 1934.
60. SmartPractice Polychloroprene Gloves vs Nitrile Gloves: What You Need to Know » SmartPractice Blog. Available online: https://blog.smartpractice.com/polychloroprene-gloves-vs-nitrile-gloves-what-you-need-to-know/ (accessed on 13 March 2023).
61. Wypych, G. *Handbook of Polymers*, 2nd ed.; ChemTec Publishing: Toronto, ON, Canada, 2016; ISBN 9781895198928.
62. Sugiura, K.; Sugiura, M.; Shiraki, R.; Hayakawa, R.; Shamoto, M.; Sasaki, K.; Itoh, A. Contact Urticaria Due to Polyethylene Gloves. *Contact Dermat.* **2002**, *46*, 262–266. [CrossRef] [PubMed]
63. Carroll, W.F.; Johnson, R.W.; Moore, S.S.; Paradis, R.A. Poly(Vinyl Chloride). In *Applied Plastics Engineering Handbook: Processing, Materials, and Applications*, 2nd ed.; William Andrew Publishing: Norwich, NY, USA, 2017; pp. 73–89, ISBN 9780323390408.
64. Tapia-Fuentes, J.; Cruz-Salas, A.; Álvarez-Zeferino, C.; Martínez-Salvador, C.; Pérez-Aragón, B.; Vázquez-Morillas, A. Bioplastics in Personal Protective Equipment. *Biodegrad. Mater. Appl.* **2022**, 173–210. [CrossRef]
65. Gnaneswaran, V.; Mudhunuri, B.; Bishu, R.R. A Study of Latex and Vinyl Gloves: Performance versus Allergy Protection Properties. *Int. J. Ind. Ergon.* **2008**, *38*, 171–181. [CrossRef]
66. Wallemacq, P.E.; Capron, A.; Vanbinst, R.; Boeckmans, E.; Gillard, J.; Favier, B. Permeability of 13 Different Gloves to 13 Cytotoxic Agents under Controlled Dynamic Conditions. *Am. J. Health-Syst. Pharm.* **2006**, *63*, 547–556. [CrossRef]
67. Landeck, L.; Gonzalez, E.; Koch, O.M. Handling Chemotherapy Drugs—Do Medical Gloves Really Protect? *Int. J. Cancer* **2015**, *137*, 1800–1805. [CrossRef]
68. *ASTM D3578-19*; Standard Specification for Rubber Examination Gloves. ASTM International: West Conshohocken, PA, USA, 2019; pp. 1–5. [CrossRef]
69. *ASTM D6319-19*; Standard Specification for Nitrile Examination Gloves for Medical Application. ASTM International: West Conshohocken, PA, USA, 2019; pp. 1–4. [CrossRef]
70. *ASTM D6977-19*; Standard Specification for Polychloroprene Examination Gloves for Medical Application. ASTM International: West Conshohocken, PA, USA, 2019; pp. 1–4. [CrossRef]
71. *EN 455-2:2015*; Medical Gloves for Single Use—Part 2: Requirements and Testing for Physical Properties. European Committee for Standardization: Brussels, Belgium, 2015; pp. 1–13.
72. Kossan Rubber Industries Bhd Rubber Disposable Gloves. Available online: https://kossan.com.my/products/gloves/healthcare.html (accessed on 18 October 2022).
73. Jayathilaka, I.; Ariyadasa, T.U.; Egodage, S.M. Powdered Corn Grain and Cornflour on Properties of Natural Rubber Latex Vulcanizates: Effect of Filler Loading. In Proceedings of the MERCon 2018—4th International Multidisciplinary Moratuwa Engineering Research Conference, Moratuwa, Sri Lanka, 30 May–1 June 2018; pp. 235–240. [CrossRef]
74. McDonnell, G.; Russell, A.D. Antiseptics and Disinfectants: Activity, Action, and Resistance. *Clin. Microbiol. Rev.* **1999**, *12*, 147–179. [CrossRef]
75. Kenawy, E.-R. Electrospun Polymer Nanofibers with Antimicrobial Activities. In *Polymeric Materials with Antimicrobial Activity: From Synthesis to Applications*; Muñoz-Bonilla, A., Cerrada, M.L., Fernández-García, M., Eds.; Royal Society of Chemistry: London, UK, 2013; pp. 208–223.
76. Riyajan, S.A.; Chaiponban, S.; Tanbumrung, K. Investigation of the Preparation and Physical Properties of a Novel Semi-Interpenetrating Polymer Network Based on Epoxised NR and PVA Using Maleic Acid as the Crosslinking Agent. *Chem. Eng. J.* **2009**, *153*, 199–205. [CrossRef]
77. Zhang, H.M.; Wakisaka, N.; Maeda, O.; Yamamoto, T. Vitamin C Inhibits the Growth of a Bacterial Risk Factor for Gastric Carcinoma: Helicobacter Pylori. *Cancer* **2000**, *80*, 1897–1903. [CrossRef]
78. Sánchez-Najera, R.I.; Nakagoshi-Cepeda, S.; Martínez-Sanmiguel, J.J.; Hernandez-Delgadillo, R.; Cabral-Romero, C. Ascorbic Acid On Oral Microbial Growth and Biofilm Formation. *Pharma Innov. J.* **2013**, *2*, 103–109.
79. Vilchèze, C.; Hartman, T.; Weinrick, B.; Jacobs, W.R. Mycobacterium Tuberculosis Is Extraordinarily Sensitive to Killing by a Vitamin C-Induced Fenton Reaction. *Nat. Commun.* **2013**, *4*, 1–10. [CrossRef]
80. Biswas, S.; Thomas, N.; Mandal, A.; Mullick, A.; Chandra, D.; Mukherjee, S.; Sett, S.; Kumar Mitra, A. In Vitro Analysis of Antibacterial Activity of Vitamin C alone and in Combination with Antibiotics on Gram Positive Rod isolated from Soil of a Dumping Site of Kolkata. *Int. J. Pharm. Biol. Sci.* **2013**, *3*, 101–110.
81. El-Gebaly, E.; Essam, T.; Hashem, S.; El-Baky, R.A. Effect of Levofloxacin and Vitamin C on Bacterial Adherence and Preformed Biofilm on Urethral Catheter Surfaces. *J. Microb. Biochem. Technol.* **2012**, *4*, 131–136. [CrossRef]
82. Verghese, R.; Mathew, S.; David, A. Antimicrobial Activity of Vitamin C Demonstrated on Uropathogenic Escherichia Coli and Klebsiella Pneumoniae. *J. Curr. Res. Sci. Med.* **2017**, *3*, 88. [CrossRef]
83. Riyajan, S. ad Biodegradable Green Glove Containing Ascorbic Acid from Maleated Epoxidized Natural Rubber/Poly(Vinyl Alcohol) Blend: Preparation and Physical Properties. *J. Polym. Environ.* **2022**, *30*, 1141–1150. [CrossRef]
84. Mohammed, L.; Ansari, M.N.M.; Pua, G.; Jawaid, M.; Islam, M.S. A Review on Natural Fiber Reinforced Polymer Composite and Its Applications. *Int. J. Polym. Sci.* **2015**, *2015*, 243947. [CrossRef]

85. Thomas, S.K.; Parameswaranpillai, J.; Krishnasamy, S.; Begum, P.M.S.; Nandi, D.; Siengchin, S.; George, J.J.; Hameed, N.; Salim, N.V.; Sienkiewicz, N. A Comprehensive Review on Cellulose, Chitin, and Starch as Fillers in Natural Rubber Biocomposites. *Carbohydr. Polym. Technol. Appl.* **2021**, *2*, 100095. [CrossRef]
86. Bode, H.B.; Kerkhoff, K.; Jendrossek, D. Bacterial Degradation of Natural and Synthetic Rubber. *Biomacromolecules* **2001**, *2*, 295–303. [CrossRef]
87. Afiq, M.M.; Azura, A.R. Effect of Sago Starch Loadings on Soil Decomposition of Natural Rubber Latex (NRL) Composite Films Mechanical Properties. *Int. Biodeterior. Biodegradation* **2013**, *85*, 139–149. [CrossRef]
88. Taghvaei-Ganjali, S.; Motiee, F.; Shakeri, E.; Abbasian, A. Effect of Amylose/Amylopectin Ratio on Physico-Mechanical Properties of Rubber Compounds Filled by Starch. *J. Appl. Chem. Res.* **2010**, *4*, 53–60.
89. Wei, B.; Xu, X.; Jin, Z.; Tian, Y. Surface Chemical Compositions and Dispersity of Starch Nanocrystals Formed by Sulfuric and Hydrochloric Acid Hydrolysis. *PLoS ONE* **2014**, *9*, e86024. [CrossRef] [PubMed]
90. Abiola, C.; Oyetayo, V.O. Isolation and Biochemical Characterization of Microorganisms Associated with the Fermentation of Kersting's Groundnut (Macrotyloma Geocarpum). *Res. J. Microbiol.* **2016**, *11*, 47–55. [CrossRef]
91. Riscoe, M.; Kelly, J.X.; Winter, R. Xanthones as Antimalarial Agents: Discovery, Mode of Action, and Optimization. *Curr. Med. Chem.* **2005**, *12*, 2539–2549. [CrossRef]
92. Pedraza-Chaverri, J.; Cárdenas-Rodríguez, N.; Orozco-Ibarra, M.; Pérez-Rojas, J.M. Medicinal Properties of Mangosteen (Garcinia Mangostana). *Food Chem. Toxicol.* **2008**, *46*, 3227–3239. [CrossRef]
93. Moopayak, W.; Tangboriboon, N. Mangosteen Peel and Seed as Antimicrobial and Drug Delivery in Rubber Products. *J. Appl. Polym. Sci.* **2020**, *137*, 49119. [CrossRef]
94. Cao, L.; Yuan, D.; Fu, X.; Chen, Y. Green Method to Reinforce Natural Rubber with Tunicate Cellulose Nanocrystals via One-Pot Reaction. *Cellulose* **2018**, *25*, 4551–4563. [CrossRef]
95. Ding, C.; Matharu, A.S. Recent Developments on Biobased Curing Agents: A Review of Their Preparation and Use. *ACS Sustain. Chem. Eng.* **2014**, *2*, 2217–2236. [CrossRef]
96. Lee, H.V.; Hamid, S.B.A.; Zain, S.K. Conversion of Lignocellulosic Biomass to Nanocellulose: Structure and Chemical Process. *Sci. World J.* **2014**, *2014*, 631013. [CrossRef]
97. Panchal, P.; Ogunsona, E.; Mekonnen, T. Trends in Advanced Functional Material Applications of Nanocellulose. *Processes* **2018**, *7*, 10. [CrossRef]
98. Parambath Kanoth, B.; Claudino, M.; Johansson, M.; Berglund, L.A.; Zhou, Q. Biocomposites from Natural Rubber: Synergistic Effects of Functionalized Cellulose Nanocrystals as Both Reinforcing and Cross-Linking Agents via Free-Radical Thiol-Ene Chemistry. *ACS Appl. Mater. Interfaces* **2015**, *7*, 16303–16310. [CrossRef] [PubMed]
99. Blanchard, R.; Ogunsona, E.O.; Hojabr, S.; Berry, R.; Mekonnen, T.H. Synergistic Cross-Linking and Reinforcing Enhancement of Rubber Latex with Cellulose Nanocrystals for Glove Applications. *ACS Appl. Polym. Mater.* **2020**, *2*, 887–898. [CrossRef]
100. Thakore, S. Nanosized Cellulose Derivatives as Green Reinforcing Agents at Higher Loadings in Natural Rubber. *J. Appl. Polym. Sci.* **2014**, *131*, 1–7. [CrossRef]
101. Reitzel, R.A.; Dvorak, T.L.; Hachem, R.Y.; Fang, X.; Jiang, Y.; Raad, I. Efficacy of Novel Antimicrobial Gloves Impregnated with Antiseptic Dyes in Preventing the Adherence of Multidrug-Resistant Nosocomial Pathogens. *Am. J. Infect. Control.* **2009**, *37*, 294–300. [CrossRef] [PubMed]
102. Koburger, T.; Hübner, N.O.; Braun, M.; Siebert, J.; Kramer, A. Standardized Comparison of Antiseptic Efficacy of Triclosan, PVP-Iodine, Octenidine Dihydrochloride, Polyhexanide and Chlorhexidine Digluconate. *J. Antimicrob. Chemother.* **2010**, *65*, 1712–1719. [CrossRef] [PubMed]
103. Peng, J.; Liu, P.; Peng, W.; Sun, J.; Dong, X.; Ma, Z.; Gan, D.; Liu, P.; Shen, J. Poly(Hexamethylene Biguanide) (PHMB) as High-Efficiency Antibacterial Coating for Titanium Substrates. *J. Hazard. Mater.* **2021**, *411*, 125110. [CrossRef]
104. Sowlati-Hashjin, S.; Karttunen, M.; Carbone, P. Insights into the Polyhexamethylene Biguanide (PHMB) Mechanism of Action on Bacterial Membrane and DNA: A Molecular Dynamics Study. *J. Phys. Chem. B* **2020**, *124*, 4487–4497. [CrossRef]
105. Leitgeb, J.; Schuster, R.; Eng, A.H.; Yee, B.N.; Teh, Y.P.; Dosch, V.; Assadian, O. In-Vitro Experimental Evaluation of Skin-to-Surface Recovery of Four Bacterial Species by Antibacterial and Non-Antibacterial Medical Examination Gloves. *Antimicrob. Resist. Infect. Control.* **2013**, *2*, 27. [CrossRef]
106. Daeschlein, G.; Kramer, A.; Arnold, A.; Ladwig, A.; Seabrook, G.R.; Edmiston, C.E. Evaluation of an Innovative Antimicrobial Surgical Glove Technology to Reduce the Risk of Microbial Passage Following Intraoperative Perforation. *Am. J. Infect. Control.* **2011**, *39*, 98–103. [CrossRef]
107. Vongsetskul, T.; Wongsomboon, P.; Sunintaboon, P.; Tantimavanich, S.; Tangboriboonrat, P. Antimicrobial Nitrile Gloves Coated by Electrospun Trimethylated Chitosan-Loaded Polyvinyl Alcohol Ultrafine Fibers. *Polym. Bull.* **2015**, *72*, 2285–2296. [CrossRef]
108. Raafat, D.; Von Bargen, K.; Haas, A.; Sahl, H.G. Insights into the Mode of Action of Chitosan as an Antibacterial Compound. *Appl. Environ. Microbiol.* **2008**, *74*, 3764–3773. [CrossRef]
109. Kong, M.; Chen, X.G.; Xing, K.; Park, H.J. Antimicrobial Properties of Chitosan and Mode of Action: A State of the Art Review. *Int. J. Food Microbiol.* **2010**, *144*, 51–63. [CrossRef]
110. Schrank, C.L.; Minbiole, K.P.C.; Wuest, W.M. Are Quaternary Ammonium Compounds, the Workhorse Disinfectants, Effective against Severe Acute Respiratory Syndrome-Coronavirus-2? *ACS Infect. Dis.* **2020**, *6*, 1553–1557. [CrossRef] [PubMed]

111. Arakkal, A.; Rathinam, P.; Sirajunnisa, P.; Gopinathan, H.; Vengellur, A.; Bhat, S.G.; Sailaja, G.S. Antibacterial Natural Rubber Latex Films with Surface-Anchored Quaternary Poly(4-Vinylpyridine) Polyelectrolyte. *React. Funct. Polym.* **2022**, *172*, 105190. [CrossRef]
112. Zhuo, Y.; Cheng, X.; Fang, H.; Zhang, Y.; Wang, B.; Jia, S.; Li, W.; Yang, X.; Zhang, Y.; Wang, X. Medical Gloves Modified by a One-Minute Spraying Process with Blood-Repellent, Antibacterial and Wound-Healing Abilities. *Biomater. Sci.* **2022**, *10*, 939–946. [CrossRef]
113. Chen, Y.; Peng, Z.; Kong, L.X.; Huang, M.F.; Li, P.W. Natural Rubber Nanocomposite Reinforced with Nano Silica. *Polym. Eng. Sci.* **2008**, *48*, 1674–1677. [CrossRef]
114. Mou, W.; Li, J.; Fu, X.; Huang, C.; Chen, L.; Liu, Y. SiO_2 and ZnO Hybrid Nanofillers Modified Natural Rubber Latex: Excellent Mechanical and Antibacterial Properties. *Polym. Eng. Sci.* **2022**, *62*, 3110–3120. [CrossRef]
115. Jalal, R.; Goharshadi, E.K.; Abareshi, M.; Moosavi, M.; Yousefi, A.; Nancarrow, P. ZnO Nanofluids: Green Synthesis, Characterization, and Antibacterial Activity. *Mater. Chem. Phys.* **2010**, *121*, 198–201. [CrossRef]
116. Raghupathi, K.R.; Koodali, R.T.; Manna, A.C. Size-Dependent Bacterial Growth Inhibition and Mechanism of Antibacterial Activity of Zinc Oxide Nanoparticles. *Langmuir* **2011**, *27*, 4020–4028. [CrossRef] [PubMed]
117. Leung, Y.H.; Chan, C.M.N.; Ng, A.M.C.; Chan, H.T.; Chiang, M.W.L.; Djurišić, A.B.; Ng, Y.H.; Jim, W.Y.; Guo, M.Y.; Leung, F.C.C.; et al. Antibacterial Activity of ZnO Nanoparticles with a Modified Surface under Ambient Illumination. *Nanotechnology* **2012**, *23*, 475703. [CrossRef] [PubMed]
118. Turner, R.J. Is Silver the Ultimate Antimicrobial Bullet? *Antibiotics* **2018**, *7*, 112. [CrossRef]
119. Bedlovičová, Z.; Salayová, A. Green-Synthesized Silver Nanoparticles and Their Potential for Antibacterial Applications. *Bact. Pathog. Antibact. Control.* **2018**, *12*, 73–94. [CrossRef]
120. More, P.R.; Pandit, S.; De Filippis, A.; Franci, G.; Mijakovic, I.; Galdiero, M. Silver Nanoparticles: Bactericidal and Mechanistic Approach against Drug Resistant Pathogens. *Microorganisms* **2023**, *11*, 369. [CrossRef]
121. Paosen, S.; Lethongkam, S.; Wunnoo, S.; Lehman, N.; Kalkornsurapranee, E.; Septama, A.W.; Voravuthikunchai, S.P. Prevention of Nosocomial Transmission and Biofilm Formation on Novel Biocompatible Antimicrobial Gloves Impregnated with Biosynthesized Silver Nanoparticles Synthesized Using Eucalyptus Citriodora Leaf Extract. *Biotechnol. J.* **2021**, *16*, e2100030. [CrossRef]
122. Airey, P.; Verran, J. Potential Use of Copper as a Hygienic Surface; Problems Associated with Cumulative Soiling and Cleaning. *J. Hosp. Infect.* **2007**, *67*, 271–277. [CrossRef]
123. Casey, A.L.; Adams, D.; Karpanen, T.J.; Lambert, P.A.; Cookson, B.D.; Nightingale, P.; Miruszenko, L.; Shillam, R.; Christian, P.; Elliott, T.S.J. Role of Copper in Reducing Hospital Environment. *J. Hosp. Infect.* **2010**, *74*, 72–77. [CrossRef]
124. Noyce, J.O.; Michels, H.; Keevil, C.W. Potential Use of Copper Surfaces to Reduce Survival of Epidemic Meticillin-Resistant Staphylococcus Aureus in the Healthcare Environment. *J. Hosp. Infect.* **2006**, *63*, 289–297. [CrossRef]
125. Arendsen, L.P.; Thakar, R.; Sultan, A.H. The Use of Copper as an Antimicrobial Agent in Health Care, Including Obstetrics and Gynecology. *Clin. Microbiol. Rev.* **2019**, *32*, 1–28. [CrossRef]
126. Tripathy, A.; Kumar, A.; Chowdhury, A.R.; Karmakar, K.; Purighalla, S.; Sambandamurthy, V.; Chakravortty, D.; Sen, P. A Nanowire-Based Flexible Antibacterial Surface Reduces the Viability of Drug-Resistant Nosocomial Pathogens. *ACS Appl. Nano Mater.* **2018**, *1*, 2678–2688. [CrossRef]

Disclaimer/Publisher's Note: The statements, opinions and data contained in all publications are solely those of the individual author(s) and contributor(s) and not of MDPI and/or the editor(s). MDPI and/or the editor(s) disclaim responsibility for any injury to people or property resulting from any ideas, methods, instructions or products referred to in the content.

Article

3D Filaments Based on Polyhydroxy Butyrate—Micronized Bacterial Cellulose for Tissue Engineering Applications

Matheus F. Celestino [1], Lais R. Lima [2], Marina Fontes [1,3], Igor T. S. Batista [1], Daniella R. Mulinari [4], Alessandra Dametto [3], Raphael A. Rattes [1], André C. Amaral [1], Rosana M. N. Assunção [5], Clovis A. Ribeiro [6], Guillermo R. Castro [7,*] and Hernane S. Barud [1,*]

1. Biopolymers and Biomaterials Group, Postgraduate Program in Biotechnology, University of Araraquara (UNIARA), Araraquara 14801-320, SP, Brazil; itsbatista@gmail.com (I.T.S.B.); acamaral@uniara.edu.br (A.C.A.)
2. Institute of Chemistry, University of São Paulo (USP), São Carlos 13566-590, SP, Brazil; laisroncalho@gmail.com
3. Biosmart Nanotechnology LTDA, Araraquara 14808-162, SP, Brazil
4. Department of Mechanics and Energy, State University of Rio de Janeiro (UEJR), Rio de Janeiro 20550-900, RJ, Brazil
5. Faculty of Integrated Sciences of Pontal (FACIP), Federal University of Uberlandia (UFU), Pontal Campus, Ituiutaba 38304-402, MG, Brazil
6. Institute of Chemistry, São Paulo State University (UNESP), Araraquara 14800-900, SP, Brazil
7. Nanomedicine Research Unit (Nanomed), Center for Natural and Human Sciences, Federal University of ABC (UFABC), Santo André 09210-580, SP, Brazil
* Correspondence: guillermo.castro@ufabc.edu.br (G.R.C.); hsbarud@uniara.edu.br (H.S.B.)

Abstract: In this work, scaffolds based on poly(hydroxybutyrate) (PHB) and micronized bacterial cellulose (BC) were produced through 3D printing. Filaments for the printing were obtained by varying the percentage of micronized BC (0.25, 0.50, 1.00, and 2.00%) inserted in relation to the PHB matrix. Despite the varying concentrations of BC, the biocomposite filaments predominantly contained PHB functional groups, as Fourier transform infrared spectroscopy (FTIR) demonstrated. Thermogravimetric analyses (i.e., TG and DTG) of the filaments showed that the peak temperature (T_{peak}) of PHB degradation decreased as the concentration of BC increased, with the lowest being 248 °C, referring to the biocomposite filament PHB/2.0% BC, which has the highest concentration of BC. Although there was a variation in the thermal behavior of the filaments, it was not significant enough to make printing impossible, considering that the PHB melting temperature was 170 °C. Biological assays indicated the non-cytotoxicity of scaffolds and the provision of cell anchorage sites. The results obtained in this research open up new paths for the application of this innovation in tissue engineering.

Keywords: 3D printing; micronized bacterial cellulose; poly(hydroxybutyrate); tissue engineering; scaffolds

1. Introduction

Three-dimensional (3D) printing, also referred to as additive manufacturing (AM), has gained significant traction in the industry due to its ability to achieve mass customization and bring intricate designs to life, surpassing the limitations of traditional manufacturing methods [1]. Among the several 3D printing technologies, fused deposition modeling (FDM) stands out as the market's most widely used method of obtaining scaffolds [2].

The FDM technique was introduced and commercialized by the Stratasys corporation in the United States during the early 1990s. Since then, FDM has been employed to produce several materials, ranging from polymers and metal powder to ceramics and composites. FDM has gained significant popularity in biomaterial research due to its affordability, compact size, ability to create intricate structures, and lack of organic solvents [3]. In

FDM, materials are melted and deposited layer by layer onto a print bed, following a programmed pattern [4].

Polymers are the most used class of materials in additive manufacturing due to their high availability, diversity of applications, and low cost [5]. Nowadays, polymers from natural origins are preferred for 3D printing for tissue engineering because of their biocompatibility, in addition to being renewable, non-toxic, biodegradable, sustainable, and ecologically correct [6].

Currently, thermoplastics such as poly(lactic acid) (PLA) and acrylonitrile butadiene styrene (ABS) dominate the FDM materials market [7]. Due to its non-biodegradability and limited cell integration, ABS is not the preferred choice for fabricating tissue engineering scaffolds [8]. PLA is an example of a biodegradable and bioresorbable thermoplastic biopolymer that has been successfully applied in regenerative medicine and tissue engineering, being widely used for the construction of scaffolds via 3D FDM printing [9–12]. In addition, PLA gels can cause irritation via polymer hydrolysis in the surrounding tissue in which they are applied because the *pKa* of lactic acid is 3.86.

A class of biopolymers that has gained prominence in recent years is polyhydroxyalkanoates (PHAs). PHAs are synthesized directly from bacterial metabolism occurring at low concentrations of nitrogen, phosphorus, oxygen, or magnesium and an excess of carbon, representing an advantage compared to PLA. Other advantages are their complete biodegradability and the multiplicity of their structures [5,9]. Poly(hydroxybutyrate) (PHB) is an extensively investigated member of the polyhydroxyalkanoate (PHA) family. It is a biodegradable biopolymer with piezoelectric properties, rendering it suitable for various biomedical applications. PHB demonstrates high biocompatibility with different cell types, including osteoblasts, epithelial cells, and chondrocytes. Its biocompatibility is attributed to the presence of low molecular weight PHB in the body and the natural occurrence of its degradation product, 3-hydroxybutyric acid, which serves as a natural metabolite in organs like the brain, heart, and lungs. These inherent characteristics, coupled with its ability to promote bone growth, favorable mechanical properties, and cost-effectiveness, position PHB as a highly promising material for medical applications [13,14].

Although it has desirable characteristics for a wide spectrum of applications, PHB has some shortcomings that limit its use for producing the scaffolds applied in tissue engineering, especially if the FDM technique is used; these shortcomings include hydrophobicity, a low degradation rate, fragility, and contamination via pyrogenic compounds and thermal instability [13,14]. To address the limitations of PHB, the incorporation of PHB into composites with other biopolymers has been extensively explored [15–17]. In particular, cellulose has emerged as a commonly used biopolymer for this purpose. In a study by da Silva Moura et al. (2019) [18], treated coconut fibers containing approximately 40–60% cellulose were employed as reinforcing agents in PHB composites. Including these fibers resulted in enhanced mechanical properties, improved thermal stability, and an increased modulus of elasticity without additional additives. Similarly, Barud et al. (2011) [19] utilized bacterial cellulose (BC) to fabricate composites, which exhibited superior mechanical properties compared to the individual polymers, mainly in terms of tensile strength, elongation to rupture, and Young's modulus.

BC is composed of β-D-glucopyranose units joined together by β-1,4-glycosidic bonds arranged in a ribbon-like network of fibrils less than 100 nm in length and 2–4 nm in diameter. Unlike vegetal cellulose, the BC network is free of lignin, hemicellulose, and other constituents of lignocellulosic materials [20]. In addition to the composition, BC differs from celluloses from other sources due to its high degree of purity and polymerization (up to 8000), crystallinity (70–80%), high water content (up to 99%), physical and mechanical resistance, flexibility, and high biocompatibility [21]. BC is a non-cytotoxic, non-genotoxic, biodegradable, and biocompatible biomaterial [22,23]. This interest is reflected by the variety of works in the literature involving bacterial cellulose, many of them focused on medical applications and, more specifically, in the tissue engineering sector [24–27].

The high applicability of BC is not only due to its unique characteristics but also to the forms and structures employed. BC can be used in the form of a membrane, either wet or dry, in the form of cellulose nanomaterials (CNM), such as cellulose nanocrystals (CNC), cellulose nanofibers (CNF), obtained through the rupture of amorphous domains or from the simple separation of fiber bundles, and as micronized cellulose particles (CMP), obtained through treatments capable of weakening supramolecular interactions through a process called cellulose activation [28]. However, there are challenges to be faced when producing high-quality BC (and its derivatives), such as the reduction of operating costs and the viability of large-scale production. For this reason, there has been a growing interest in the search for solutions, such as developing new bioreactor projects and process automation. Furthermore, alternative raw materials (especially waste) have been explored as a promising option [23,29].

MELO et al. (2020) [30] demonstrated the circular economy for BC by recycling its waste from commercial wound dressings to obtain CNC and develop sustainable and biodegradable packaging. An additional example of the circular economy in action in the production chain of BC and its derivatives is the replacement of hazardous chemical processes and/or reagents commonly used in BC pre-treatment and hydrolysis to obtain CNM and CMP through processes and/or sustainable reagents. Swelling agents such as dimethylsulfoxide (DMSO) and dimethylformamide (DMF) are often used to activate cellulose and obtain micronized particles. However, these reagents pose risks to human health and are not ecologically friendly, in addition to requiring high financial and energy costs for production, purification, collection, recycling, and disposal. Therefore, there is a growing search for sustainable alternatives, such as mechanochemical processes, to obtain these structures [31].

In this context, aiming at the concept of upcycling and circular economy, the present work developed a biocomposite product with high added value, such as scaffolds. Micronized BC was obtained from industrial dressing residues through the mechanochemical process, and was used as a reinforcing agent in the development of biocomposite filaments based on a PHB matrix. The manufactured scaffolds were physicochemically and morphologically characterized, and in vitro cytotoxicity assays were performed to confirm the viability for tissue engineering applications.

2. Materials and Methods

2.1. Materials

The PHB, Biocycle 1000 (Mn 147.596 g mol^{-1}, Mw 376 g mol^{-1}, and PDI 2.5), was supplied by PHB Industrial S/A, São Paulo. Bacterial cellulose residues were provided by the company BioSmart Nanotechnology LDTA. The micronizes were obtained using a Polymix® PX-IG 2000 Impact Grinder ball milling from the company Kinematica. The mechanochemical process was applied for 20 min at frequencies of 10 Hz, 20 Hz, and 30 Hz.

To obtain the biocomposites, micronized BC at 20 Hz was used. Both PHB and BC biopolymers were submitted to a thermokinetic mixer (MH-50H, 48 A), with speed maintained at 5250 rpm, for 1 min. The amount of micronized BC inserted into the PHB varied between 0.25%, 0.50%, 1.00%, and 2.00% (wt%). The mixtures obtained were ground in a granulating mill (Plastimax, 3.7 kW) and dried at 50 °C for 2 h. The biocomposites were extruded in a mini extruder (Weellzoom, model B Desktop, Guangdong Province, Guangzhou, China) to obtain the filaments. The temperature in processing the filaments was 165 °C, and the extrusion speed was 85 mm min^{-1}.

2.2. Methods

X-ray diffraction (XRD) measurements were performed on an XRD-6000 diffractometer (Shimadzu). The patterns of micronized BC were recorded using Cu-Kα radiation (λ = 1.5406 Å) at 40 kV and 40 mA in the 2θ region from 10 to 60°. Segal and peak deconvolution methods were applied to diffractograms to analyze the influence of micronization on crystallinity indices (CI). The pseudovoight 1 function was used to determine the crystalline portion from peak deconvolution. The particle size of BC was determined using the Anton-Paar PSA 1190 LD particle size analyzer. The results were reported for D10, D50, and D90, which are the volume diameters of the particles at 10%, 50%, and 90% cumulative volume, respectively. The surface area was measured using the Anton-Paar NOVA touch BET Surface Area and Pore Size Analyzers, with nitrogen as the adsorbate. The degree of polymerization (DP) of BC was determined using the transparent Cannon-Fenske-type viscometer n. 150, according to TAPPI standard T 230om-94: Viscosity of pulp (capillary viscometer method), 2013. Time measurements were performed in triplicate and the average obtained was used to calculate the degree of polymerization following the calculations presented by [32].

BC surface and filament cross-section morphologies were visualized in a JEOL 7500F electron microscope at 2.00 kV after being fixed onto stubs using a carbon film and coated with a thin carbon layer. Thermal analysis of both BC and PHB/BC biocomposite filaments was performed using TA Instruments SDT Q600 equipment. The samples were heated in an alumina crucible from 30 to 600 °C within an atmosphere of N_2 flowing at 100 mL min^{-1}. Fourier Transform Infrared Spectroscopy was performed using a Bruker-Vertex 70 spectrophotometer in attenuated total reflection (ATR) mode. The BC and PHB/BC biocomposites spectra were obtained by accumulating 64 scans with a 2 cm^{-1} resolution in the range of 4000–650 cm^{-1}.

For the 3D printing of the scaffolds, the digital design was taken from the "Thingiverse" file bank. Design sizing and slicing were performed using Ultimaker Cura 4.0 software. The scaffolds were sized $10 \times 10 \times 5$ mm on the x, y, and z axes, respectively, and the part fill was set to 50%. Printing was performed using a CREALITY Ender-3 3D printer, with a maximum speed of 180 mm s^{-1}, 0.4 mm nozzle, structure in anodized aluminum, and printing area of $220 \times 220 \times 250$ mm. The temperature used in the process was 185 °C.

For PHB/BC biocomposite filament cell viability assays, L929 cells were cultured at 1×10^4 cells/well in a 96-well culture plate. The cell viability was performed via MTT (3-(4,5-dimethylthiazol-2-yl)-2,5-diphenyltetrazolium bromide) assay, according to International Organization of Standardization (ISO) protocols for the biological evaluation of medical devices [33,34]. Thus, filaments from polymer blending between PHB and BC named PHB/BC 0.25%, PHB/BC 0.50%, PHB/BC 1.00%, and PHB/BC 2.00% were cut into 6 cm size and incubated with 3 mL of DMEM extraction medium at 37 °C for 24 h. Subsequently, the cell culture medium was removed, and 100 µL of each extraction media was added in contact with monolayer culture for 24 h in a CO_2 incubator. Afterward, the extract media were removed, and the wells were washed three times with PBS 1X. Next, 100 µL of the MTT solution (1 mg mL^{-1}) was added to each test well. Further, the microplate was incubated at 37 °C for 3 h. After the MTT was removed and 50 µL of isopropanol added to each well, the optical density was read at a wavelength of 570 nm using a spectrophotometer reader (SoftMax® Pro 5). The assay was performed in triplicate. Cell viability was defined as the absorbance ratio from the sample to the absorbance measured for negative control (survive control) and represented as a mean value ± standard deviation [35].

To confirm the PHB/BC-based biocomposite as a scaffold, cell adhesion assays were performed. PHB/BC scaffolds were seeded with 5×10^4 cells/well in a 24-well culture plate containing 2 mL of DMEM supplemented with 10% FBS and 100 U/mL penicillin-streptomycin and maintained at 37 °C in a CO_2 incubator for 3 days to study cell attachment. The cultured cells on the scaffolds were then fixed with glutaraldehyde 0.25% for 0.5 h and then washed three times with PBS 1X. Afterward, the scaffolds were dehydrated with ethanol series and air-dried. Scanning Electron Microscopy (SEM) micrographs were captured under a JEOL JSM-6510 operating at 5 kV. All specimens were sputter coated with gold (Balzer, SCD 050) [36].

3. Results

3.1. Bacterial Cellulose Micronized

The micronized BC resulting from the mechanochemical process at three different grinding frequencies (10, 20, and 30 Hz) had a powdery appearance and a slightly yellowish color when micronized at 30 Hz. This coloration is possibly due to the rise in the temperature of the process, consequent to the increase in frequency and, therefore, the increase in energy caused by the greater collision of the sample in the ball mill (Figure 1A). Scanning Electron Microscopy (SEM) showed that despite the milling at frequencies 10 and 20 Hz, the BC fibers were maintained compared to the BC before the mechanochemical process. Figure 1B shows a representative image of micronized BC at 20 Hz and Figure 1C confirms that the fibers morphology was maintained. It was also observed that increasing the grinding frequency to 30 Hz disfigured the cellulose fibers, forming agglomerates and exposing more of the amorphous portion of the cellulose.

The mechanochemical method known as ball milling is widely disseminated in the literature and used to obtain many micronized products [37–40]. This technique is used to fragment the BC into microparticles through mechanical collisions that break the hydrogen bonds, which are responsible for most of the intra- and intermolecular bonds and, consequently, for the three-dimensional crystalline structure of this biopolymer [31,41,42].

For the development of biocomposite filaments, the micronized particles must present an adequate balance between the characteristics of crystalline and amorphous cellulose structures since BC is used as a reinforcing agent. Still, it also needs to be degraded by the organism as cells differentiate and tissue regenerates.

Crystallinity indices (*CI*) were determined by Segal and peak deconvolution methods. The equation used for the Segal method was:

$$CI = \frac{(I_{200} - I_{am})}{I_{200}}$$

where *CI* is the relative crystallinity index, I_{200} is the maximum intensity (in arbitrary units) of the peak referring to the plane (200), and I_{am} is the diffraction intensity of the amorphous portion at $2\theta = 18°$.

The pseudovoight 1 function was used to determine the crystalline portion via peak deconvolution. Through this function, it was possible to determine the peaks referring to the crystalline portion and the amorphous portion. Consequently, the amounts of crystalline and amorphous material in the micronized BC samples were determined. This method considers the contributions of amorphous and crystalline cellulose for the entire spectrum; therefore, this technique has greater precision for determining the *CI* than the Segal method [43]. The *CI* of micronized BC samples and BC residue, calculated using the diffractograms in Figure 1D, are shown in Table 1.

It is observed that the Segal method provides higher *CI* values than the peak deconvolution method, except for the micronized *CI* at 30 Hz. These values portray an overestimation of the *CI* as previously reported [43], and reinforce the results obtained by [42,44]. It is also noted that the BC micronized at 10 Hz maintained most of its crystalline structure. In comparison, the BC micronized at 30 Hz showed the dominance of the amorphous portion, probably due to the disruption of the intra- and intermolecular bonds. The BC micronized at 20 Hz presented a greater balance between the portions, which is favorable for obtaining the biocomposite.

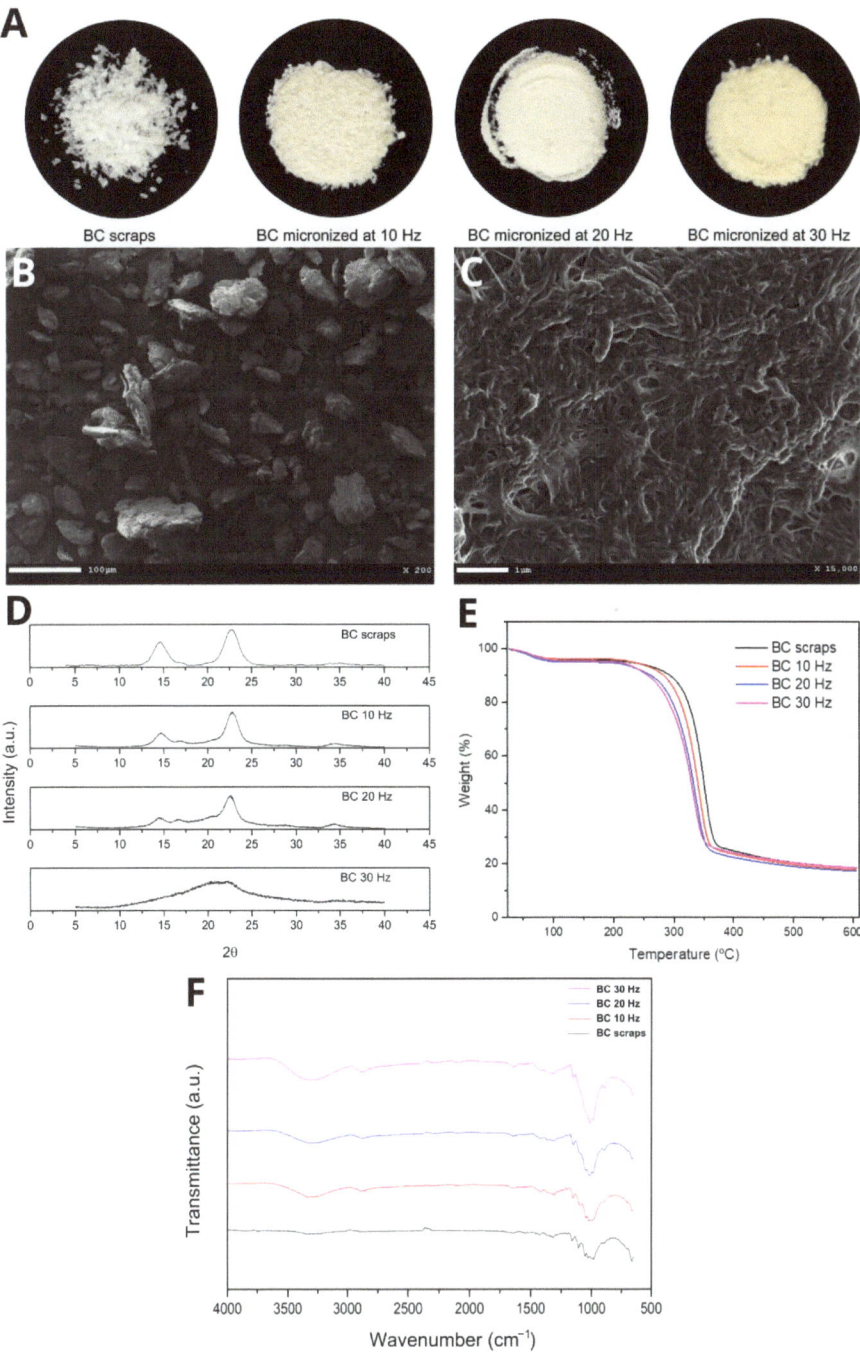

Figure 1. (**A**) Images of BC scraps and micronized BC at 10, 20, and 30 Hz, respectively; (**B**,**C**) representative SEM images of micronized BC at 20 Hz; (**D**) DRX curves, (**E**) TGA curves and (**F**) FTIR spectra of BC and micronized BC scraps at 10, 20, and 30 Hz, respectively.

Table 1. Crystallinity indices (*CI*) determined by Segal and peak deconvolution methods for bacterial cellulose (BC) scraps and BC micronized at 10, 20, and 30 Hz, respectively.

Method	Bacterial Cellulose			
	Scraps	10 Hz	20 Hz	30 Hz
Segal	0.99	0.89	0.81	0.27
Peak deconvolution	0.99	0.80	0.69	0.32

The TGA curves (Figure 1E) show two events with considerable mass losses observed in all micronized BC samples. The first event between 50 and 150 °C (≈5% initial mass loss) is attributed to water loss. The second event between 250 and 400 °C (with a loss of ≈75% of the initial mass) corresponds to the thermal degradation of cellulose, i.e., the processes of depolymerization, dehydration, and decomposition of the glycosidic units followed by the formation of carbonaceous residues [45]. Cellulose micronization breaks the hydrogen bonds that maintain the three-dimensional crystalline structure, causing the cellulose chains to depolymerize and become amorphous, which corroborates the XRD results. Increasing the frequency of the mechanochemical process increased the entropy of the system, consequently leading to a decrease in decomposition temperatures [31]. That is, the depolymerization and amorphization phenomena are incremented with the increase in the milling frequency; therefore, the T_{peak}, T_{onset}, and T_{offset} temperatures are lower, and mass loss events are anticipated.

In the FTIR spectrum for BC scraps and their micronized counterparts (Figure 1F), the presence of significant absorption bands at the same wavelength was observed, indicating similarity between the functional groups, i.e., there was no evidence of the formation of new bonds or changes in chemical structure. It was also observed that as the grinding frequency of the scraps increased, the transmittance of the bands decreased, which was very evident in the bands at 3300 cm^{-1} referring to OH stretching, and at 1140–1015 cm^{-1} referring to deformation CO [46]. This can be explained by the decrease in the crystallinity of BC, while micronization breaks hydrogen bonds [47]. This effect was also observed by [48–50] in their respective works. The other absorption characteristic bands for BC are: 2880 cm^{-1}—CH stretching of alkanes and asymmetric stretching CH$_2$; 1645 cm^{-1}—OH strain; 1420 cm^{-1}—CH$_2$ deformation; 1370 cm^{-1}—OH deformation [46].

The results of the granulometric analysis (Table 2) demonstrated an inversely proportional relationship between the grinding frequency and the average diameter of the micronized BC. As the frequency in the mechanochemical process increased, the average particle diameter decreased. This relationship can be observed in the three measurement ranges, named D10, D50, and D90, which represent 10, 50, and 90% of the total volume of particles, respectively. Ball milling reduced the mean particle diameter, but standard deviation variations in percentiles were observed. At the frequency of 10 Hz, particles were reduced, but not all of them were homogeneous; therefore, the observed standard deviation was higher than the deviations of the other frequencies used. At the frequency of 30 Hz, an increase in standard deviation was also observed in the deviation at 20 Hz. This result is due to the agglomerative phenomena that occur in more energetic processes in which the portion amorphous part of the material is gradually more exposed [47,51–53]. In this case, the phenomena of breakage and aggregation co-occur.

Table 2. Mean size and standard deviation of bacterial cellulose (BC) micronized at 10, 20, and 30 Hz.

BC Samples	Average Diameter (μm)		
	D10	D50	D90
10 Hz	33.3 ± 6.4	109.9 ± 18.6	276.1 ± 87.5
20 Hz	7.8 ± 0.7	53.6 ± 3.5	123.8 ± 12.7
30 Hz	6.4 ± 1.8	41.5 ± 12.6	74.9 ± 24.4

The average diameter of the micronized BC and its respective standard deviations are fundamental parameters for its application in filament formation. The sizes directly interfere with the ability to obtain a homogeneous biocomposite and the ability of this material to be extruded uniformly without breaking, preventing clogging of the extruder heads and the 3D printer.

The micronized BC at 10 Hz showed a high diameter and standard deviation compared to the others, which could become an obstacle to obtaining and printing the filament. The micronized filaments at 20 and 30 Hz had similar diameters, but the standard deviation of the micronized filaments at 20 Hz was smaller; therefore, the micronized ones at 20 Hz could provide better homogeneity to the filaments.

In the analysis of the determination of the surface area, it was expected that the increase in the grinding frequency correlated with an increase in the observed surface area, following the results of the granulometric analysis and the results obtained by [44]. This relationship was observed with the frequencies 10 and 20 Hz. However, at the frequency of 30 Hz, the surface area decreased compared to the milling performed at lower frequencies (Table 3), possibly due to the aggregation of micronized products due to the high level of energy transferred in the process.

Table 3. The surface area of bacterial cellulose (BC) micronized at 10, 20, and 30 Hz.

BC Samples	Surface Area ($m^2\ g^{-1}$)
10 Hz	1.47
20 Hz	1.59
30 Hz	1.20

The degree of polymerization (DP) refers to the number of repeated structural units observed in the constitution of a macromolecule; that is, the number of monomers that constitute the polymer. As the frequency used in the micronization process increased, bond breaking also increased and, consequently, the DP decreased (Table 4). These results corroborate the literature, in which micronization processes tend to reduce the DP of polymers due to the breaking of the bonds that unite the monomers [54,55], in the case of BC, corresponding to the β-1,4 glycosidic bonds.

Table 4. Degree of polymerization (DP) of bacterial cellulose (BC) scraps and their micronization at 10, 20, and 30 Hz.

Samples	Average Flow Time (s)	Degree of Polymerization (DP)
Solvent + water	31	-
Scraps	977.33	1576.28
BC 10 Hz	798.33	1387.87
BC 20 Hz	380.33	845.12
BC 30 Hz	83	207.97

3.2. PHB/BC Biocomposite Filaments

The filaments obtained after extrusion were homogeneous, and no cracks, flaws, burns, or BC agglomerates were observed. All filaments showed relative fracture resistance, but the filaments without BC, with 0.25 and 0.50% BC were more malleable and flexible than the others.

Figure 2 shows the TGA and DTG curves of the biocomposite filaments. Two mass loss events were observed in all samples. In the first event, between 50 °C and 120 °C, a small mass loss corresponds to the vaporization of residual moisture. In the second event, between 170 °C and 350 °C, the mass loss is significant, as complete degradation of PHB occurs, mainly due to the β-cleavage of the PHB chains in C=O and C–O bonds, which facilitates the formation of crotonic acid, dimeric, trimeric, and tetrameric volatiles [19,56,57].

PRADHAN et al. (2017) [56] also observed that the PHB synthesized from the fermentation of hydrolysates rich in hexose had greater crystallinity and resistance to thermal degradation. In addition, it is possible to observe in the TGA and DTG curves that the filaments with higher concentrations of BC showed a reduction in thermal resistance. The biocomposite filament PHB/2.00% BC, with the highest concentration of micronized BC, showed the lowest thermal resistance. This reduction in thermal resistance is related to the decrease in crystallinity of the filament as the micronized material was added. The amorphous portion of BC is exposed with the micronization of cellulose and consequent breakage of hydrogen bonds, and the three-dimensional crystalline structure responsible for thermal resistance is lost.

Considering that the filaments obtained are intended for application in scaffolds, the observed decrease in thermal stability does not represent a problem for application in 3D printing because the average melting temperature (T_m) of PHB (temperature used in printing) is 170 °C and the lowest T_{onset} observed was 235 °C.

The DSC curves of the filaments (Figure 3) show two peaks of endothermic events at approximately 180 °C and 290 °C, which correspond to the melting and decomposition of PHB, respectively [19,56,58]. The DSC peak between 250 and 300 °C refers to the thermal decomposition of PHB. What is being observed is that the thermal stability of PHB decreases with the addition of micronized BC. Considering only PHB, the temperature of thermal decomposition depends on the number average molecular weight, Mn, and the weight average molecular weight, MW, and polydispersibility, PDI. PHB has a high PDI (2.5), which can contribute to decreased thermal stability. The literature presents the T_m of PHB close to 170 °C, but the T_m of PHB and its composites depend on many factors, such as morphology and particle size, crystallization kinetics, and the composition process [58]. PRADHAN et al. (2017) [56] obtained PHB through *Bacillus megaterium* and *Cupriavidus necator* with T_m observed at 175 °C and 176 °C, respectively; these values were also higher than the standard literature presents.

In Figure 3, TGA/DTG curves of PHB/BC, the T_{onset} and T_{offset} values were indicated. The increase in BC up to 2.0% in the PHB/BC composite decreased the maximum de-composition temperature (MDT) from 290 °C to 248 °C (ΔT = 42 °C). However, BC concentrations up to 0.5% do not show drastic changes in the MDT, but high BC concentrations than 0.5% decrease the MDT in −25.1 °C/BC (%) (Figure S1, Supplementary Material) with consequent partial loss of PHB/BC composite stability.

Comparative analysis of TGA/DTGA with DSC shows some differences in the maximum endothermic temperatures compared with the MDT. However, it is possible to observe at the beginning of the reaction T_{onset}, similar values in the DSC and TGA/DTG curves. Additionally, the trends of temperature decrease with the increase in BC content are similar in Figures 2 and 3. Also, it is relevant to consider that during the processes of sample heating and cooling, TGA instruments are measuring the mass changes with the temperature; meanwhile, DSC measures the shifts in energy absorption or release related to the temperature variations.

In fact, TGA/DTG and DSC were obtained using the same equipment simultaneously and separated in Figures 2 and 3 to facilitate the visualization of the events.

Figure 4 shows the main bands observed in the PHB/BC biocomposite filaments. Referring to PHB: 1000–1300 cm^{-1}—C-O elongation of the ester group; 1455 cm^{-1}—asymmetric bending of -CH_2 or -CH_3; 1718 cm^{-1}—ester C=O elongation; 1271 cm^{-1}—CH group; 2853 cm^{-1}, 2926 cm^{-1} and 2981 cm^{-1}—C-H elongation [59,60]. Regarding the micronized BC, the bands are different compared to the PHB. However, these vibrations were not observed in the spectra of the biocomposite filaments that remained with most of the individual characteristics of the PHB, probably because of the low sensitivity of the technique. This demonstrates that even when using micronized BC as a reinforcing agent, maintaining the PHB chemical groups in most of the biocomposite material was possible.

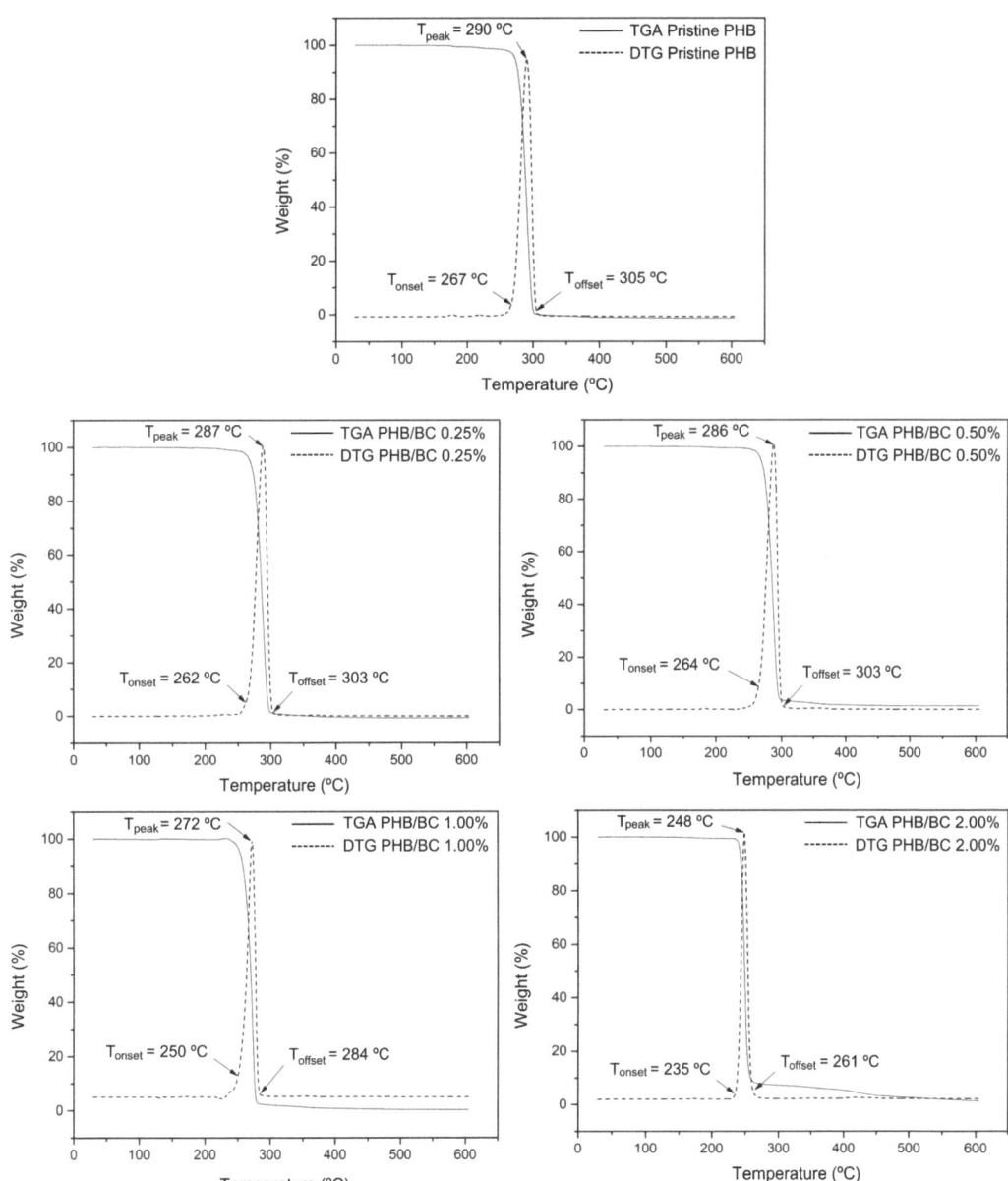

Figure 2. TGA and DTG curves of pristine PHB and PHB/BC-based biocomposite filaments containing 0.25, 0.50, 1.00, and 2.00% BC, respectively.

The average diameter of the filaments was 1.60 ± 0.04 mm, as determined by the images obtained by SEM (Figure 5). It was also possible to observe that the cross-section contained an entirely uniform, compact, and bubble-free surface, indicating high homogeneity, an essential characteristic for filaments used in 3D printing.

The pristine PHB filaments and the PHB/BC-based biocomposites containing 0.25, 0.50, and 1.00% BC were printed homogeneously without clogging the printer head or breaking the filament during printing. Figure 6 shows a representative image comparing pristine PHB and PHB/0.50% BC-based scaffolds. Filaments with 2.00% of micronized BC caused clogging of the 0.4 mm printer nozzle diameter. This clogging probably occurred due to the cellulose particles aggregation on the surface of the filament and the consequent change in the viscosity of the material, as previously reported [61].

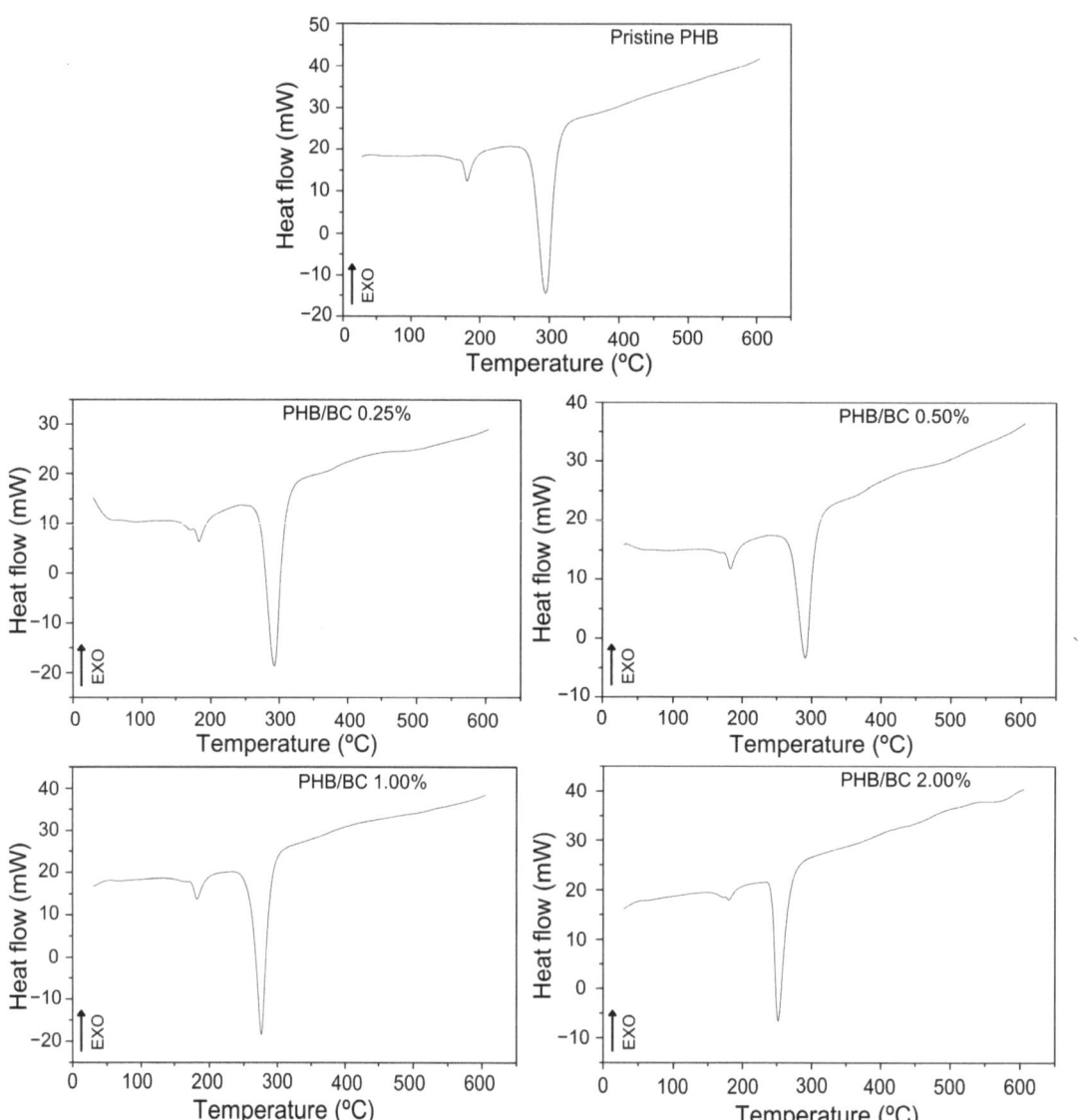

Figure 3. DSC curves of pristine PHB and PHB/BC-based biocomposite filaments containing 0.25, 0.50, 1.00, and 2.00% BC, respectively.

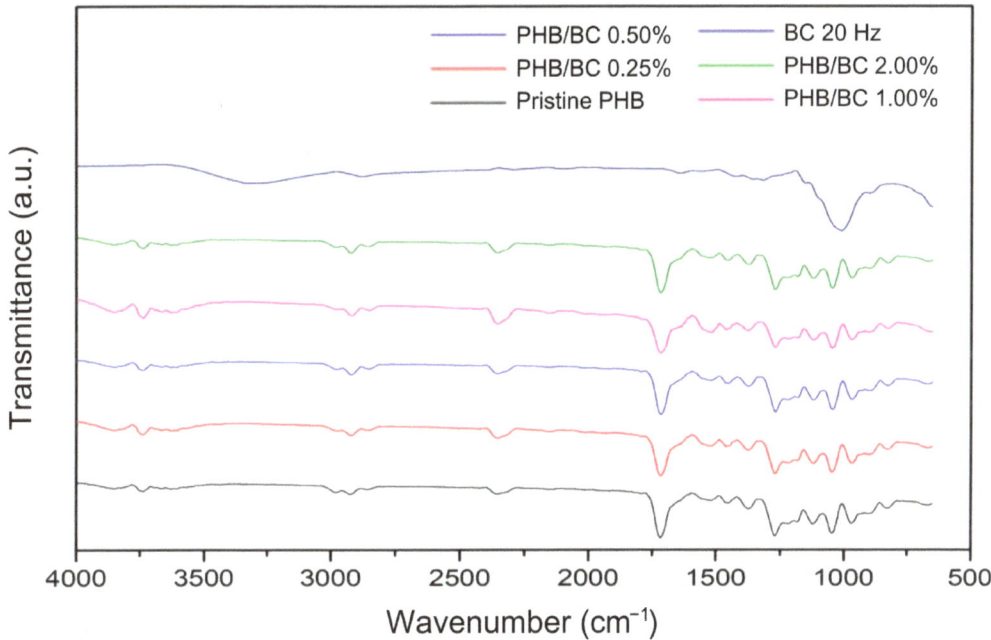

Figure 4. Fourier Transform Infrared Spectroscopy (FTIR) of micronized bacterial cellulose (BC) at 20 Hz, pristine poly(hydroxybutyrate) (PHB) filaments, and PHB/BC-based biocomposite filaments.

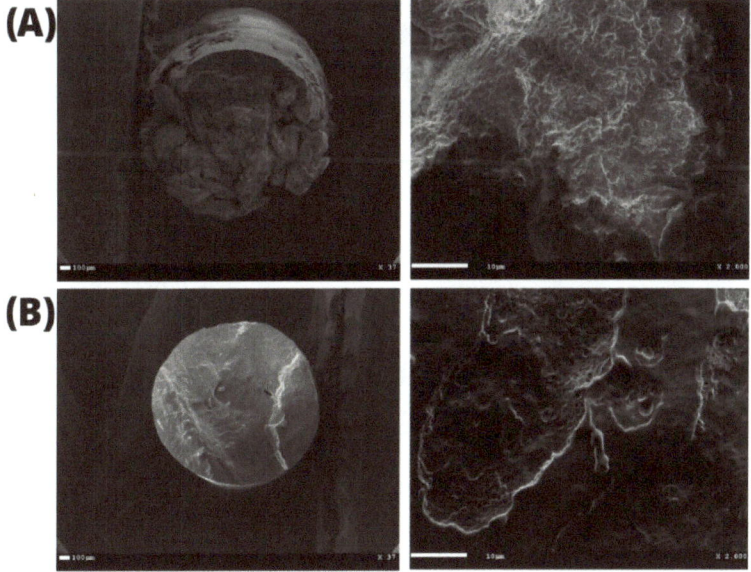

Figure 5. *Cont.*

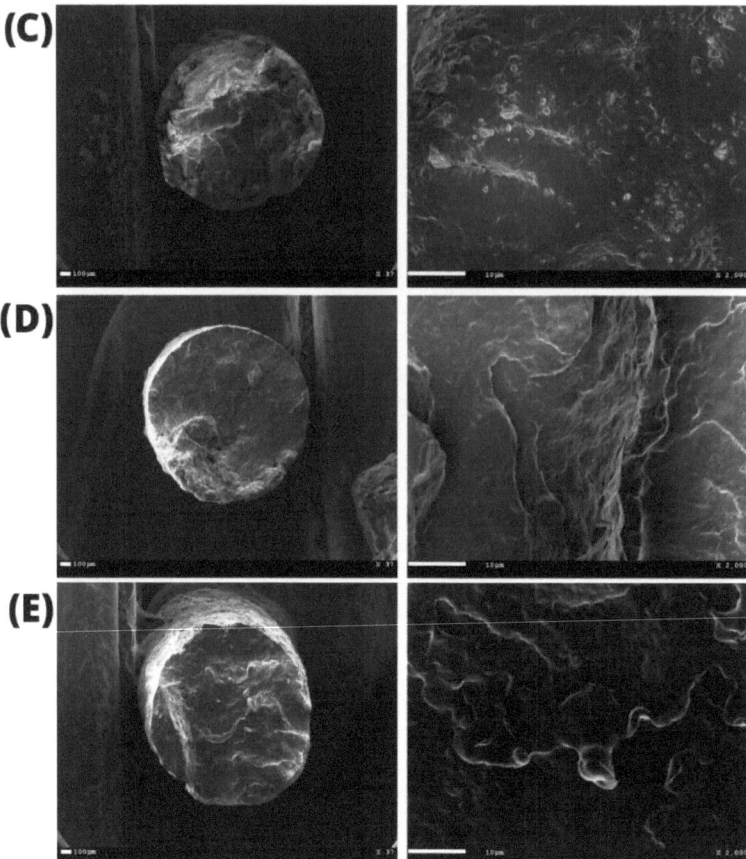

Figure 5. Scanning Electron Microscopy (SEM) of the printed filaments: (**A**) pristine PHB and PHB/BC-based biocomposites with 0.25, 0.50, 1.00, 2.00% BC ((**B–E**), respectively).

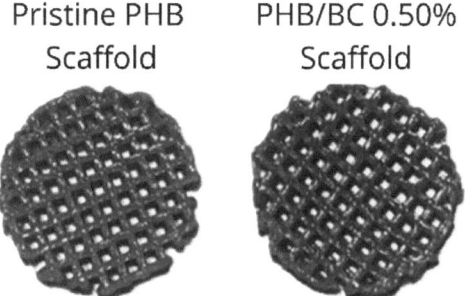

Figure 6. Scaffolds printed using pristine Poly(3-hydroxybutyrate) (PHB, **left**) and a representative image of PHB/0.50% BC-based biocomposite filaments (**right**).

3.3. Biological In Vitro Assays

The metabolic viability of L929 cells exposed to PHB/BC-based biocomposites was determined via the MTT assay after 24 h of exposure. The results showed that all samples were safe, resulting in cell viability higher than 70%, similar to control (Figure 7). This result confirmed previous ones that demonstrated no toxicity of biocomposites based on

PHB/BC, and PHB blended with different biopolymers, including PLA and Hydroxyapatite (HA) [62,63]. Furthermore, the PHB/BC-based biocomposites proved to support 3T3-L1 preadipocyte proliferation and induced a positive effect on osteoblast differentiation in vivo.

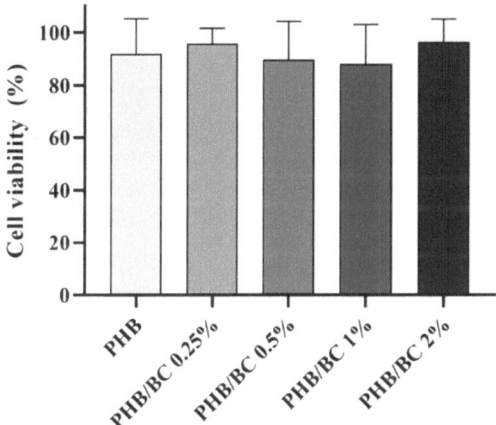

Figure 7. L929 viability assayed with the MTT assay revealed the using liquid extracts derived from PHB/BC-based biocomposites after 24 h of incubation.

Regarding cell adhesion assay, the cell–scaffold interaction was examined following the cultivation of L929 cells on PHB/BC scaffolds. Following a three-day culture, SEM images were obtained. Figure 8A–D demonstrates a representative sample PHB/BC 0.50%. It is possible to observe that the blend greatly supports cell adhesion, indicating the non-cytotoxicity of the scaffolds and the provision of cell anchorage sites.

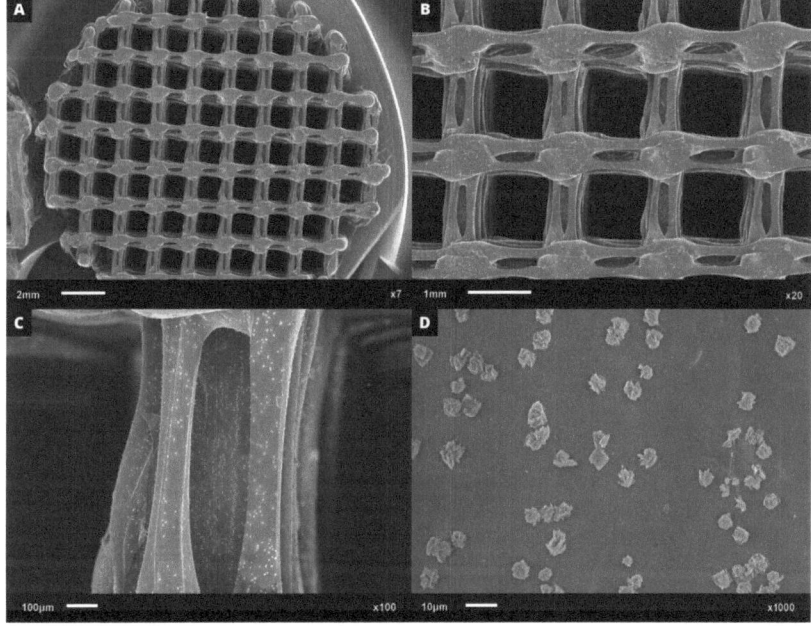

Figure 8. SEM microphotographs with approximation of: (**A**) 7×; (**B**) 20×, (**C**) 100×, and (**D**) 1000×, proving the L929 cell adhesion on the surface of PHB/BC 0.50% scaffolds after 3 days culture.

4. Conclusions

The PHB/BC-based scaffolds were obtained through 3D printing using filaments containing micronized BC. Among the different mechanochemical processes, BC micronized at 20 Hz showed the best properties in terms of granulometry and homogeneity for obtaining biocomposite filaments. In addition, BC micronized at 20 Hz presented a balance between the crystalline and amorphous portions verified via XRD analysis. This is an important characteristic, since BC performs the function of a reinforcing agent but can simultaneously be degraded as tissue regeneration occurs. TGA, FTIR, and SEM measurements confirmed that there is a reduction in the crystalline portion of BC, and that micronized BC at 20 Hz maintained the structure of the fibers and exhibited an insignificant decrease in thermal stability, considering the melting point of PHB for application in 3D printing. The production process of the biocomposite filaments resulted in homogeneous structures based on PHB/BC without cracks or breaks, which allows them to be used in 3D printing. The physical–chemical characterization, cytotoxicity, and cell adhesion evaluation carried out in the present study are sufficient to preliminarily validate that biocomposite filaments have the potential to be applied in tissue engineering. Obtaining these filaments also stands out, as BC industrial scraps were used. These new processes and products add value to them, as they are aligned with the concepts of circular economy, upcycling, and sustainability.

Supplementary Materials: The following supporting information can be downloaded at: https://www.mdpi.com/article/10.3390/jfb14090464/s1, Figure S1: Changes of the maximum decomposition temperature of PHB/BC related to the BC composition. Data obtained from Figure 2.

Author Contributions: Conceptualization, M.F.C. and L.R.L.; methodology, M.F.C., M.F., I.T.S.B. and R.A.R.; software, M.F.C., L.R.L. and M.F.; validation, L.R.L., M.F, D.R.M., C.A.R., G.R.C. and H.S.B.; formal analysis, M.F.C., L.R.L., C.A.R. and G.R.C.; investigation, M.F.C.; resources, D.R.M., A.C.A., R.M.N.A., G.R.C. and H.S.B.; data curation, M.F.C., L.R.L. and M.F.; writing—original draft preparation, L.R.L.; writing—review and editing, L.R.L., M.F, D.R.M., A.D., A.C.A., R.M.N.A., C.A.R., G.R.C. and H.S.B.; visualization, L.R.L.; supervision, D.R.M. and H.S.B.; project administration, D.R.M. and H.S.B.; funding acquisition, D.R.M. and H.S.B. All authors have read and agreed to the published version of the manuscript.

Funding: This research was funded by Rio de Janeiro State Foundation to Support Research (FAPERJ), grant number E26/211.829/2021; National Council for Scientific and Technological Development (CNPq), grant number 309614/2021-0, INCT-INFO; and São Paulo Research Foundation (FAPESP), grant number 2013/07276-1.

Data Availability Statement: The data presented in this study are available on request from the corresponding authors.

Acknowledgments: The authors thank the National Institutes of Science and Technology (INCT)—Science Institute for Polysaccharides and National Institute of Photonics (INFO), São Paulo; and the Anton Paar company for the opportunity to use the PSA 1190 LD Particle Size Analyzer and BET Surface Area and Pore Size Analyzers.

Conflicts of Interest: The authors declare no conflict of interest.

References

1. Yu, I.K.M.; Chan, O.Y.; Zhang, Q.; Wang, L.; Wong, K.-H.; Tsang, D.C.W. Upcycling of Spent Tea Leaves and Spent Coffee Grounds into Sustainable 3D-Printing Materials: Natural Plasticization and Low-Energy Fabrication. *ACS Sustain. Chem. Eng.* **2023**, *11*, 6230–6240. [CrossRef]
2. Krueger, L.; Miles, J.A.; Popat, A. 3D Printing Hybrid Materials Using Fused Deposition Modelling for Solid Oral Dosage Forms. *J. Control. Release* **2022**, *351*, 444–455. [CrossRef]
3. Winarso, R.; Anggoro, P.W.; Ismail, R.; Jamari, J.; Bayuseno, A.P. Application of Fused Deposition Modeling (FDM) on Bone Scaffold Manufacturing Process: A Review. *Heliyon* **2022**, *8*, e11701. [CrossRef]
4. Rafiee, M.; Farahani, R.D.; Therriault, D. Multi-Material 3D and 4D Printing: A Survey. *Adv. Sci.* **2020**, *7*, 1902307. [CrossRef] [PubMed]
5. Mehrpouya, M.; Vahabi, H.; Barletta, M.; Laheurte, P.; Langlois, V. Additive Manufacturing of Polyhydroxyalkanoates (PHAs) Biopolymers: Materials, Printing Techniques, and Applications. *Mater. Sci. Eng. C* **2021**, *127*, 112216. [CrossRef]

6. Udayakumar, G.P.; Muthusamy, S.; Selvaganesh, B.; Sivarajasekar, N.; Rambabu, K.; Banat, F.; Sivamani, S.; Sivakumar, N.; Hosseini-Bandegharaei, A.; Show, P.L. Biopolymers and Composites: Properties, Characterization and Their Applications in Food, Medical and Pharmaceutical Industries. *J. Environ. Chem. Eng.* **2021**, *9*, 105322. [CrossRef]
7. Tandon, S.; Kacker, R.; Singh, S.K. Correlations on Average Tensile Strength of 3D-Printed Acrylonitrile Butadiene Styrene, Polylactic Acid, and Polylactic Acid + Carbon Fiber Specimens. *Adv. Eng. Mater.* **2023**, *25*, 2201413. [CrossRef]
8. Alagoz, A.S.; Hasirci, V. 3D Printing of Polymeric Tissue Engineering Scaffolds Using Open-Source Fused Deposition Modeling. *Emergent Mater.* **2020**, *3*, 429–439. [CrossRef]
9. Jiang, L.; Zhang, J. Biodegradable and Biobased Polymers. In *Applied Plastics Engineering Handbook Processing, Materials, and Applications*, 2nd ed.; Plastics Design Library, Kutz, M., Eds.; William Andrew Publishing: Norwich, NY, USA, 2017; pp. 127–143. ISBN 978-0-323-39040-8. [CrossRef]
10. Matos, B.D.M.; Rocha, V.; da Silva, E.J.; Moro, F.H.; Bottene, A.C.; Ribeiro, C.A.; dos Santos Dias, D.; Antonio, S.G.; do Amaral, A.C.; Cruz, S.A.; et al. Evaluation of Commercially Available Polylactic Acid (PLA) Filaments for 3D Printing Applications. *J. Therm. Anal. Calorim.* **2019**, *137*, 555–562. [CrossRef]
11. Barud, H.S.; Machado, L.G.; Zaldivar, M.P.; Iemma, M.R.d.C.; Cruz, S.A.; Rangel, E.C.; Nassar, E.J. Polylactic Acid Scaffolds Obtained by 3D Printing and Modified by Oxygen Plasma. *Rev. Bras. Multidiscip.* **2020**, *23*, 97–106. [CrossRef]
12. Moro, F.H.; de Carvalho, R.A.; Barud, H.d.S.; Amaral, A.C.; da Silva, E.J. 3D Printer Nozzle Modification to Obtain Scaffolds for Use in Regenerative Medicine. *Res. Soc. Dev.* **2022**, *11*, e58111629472. [CrossRef]
13. Parvizifard, M.; Karbasi, S. Physical, Mechanical and Biological Performance of PHB-Chitosan/MWCNTs Nanocomposite Coating Deposited on Bioglass Based Scaffold: Potential Application in Bone Tissue Engineering. *Int. J. Biol. Macromol.* **2020**, *152*, 645–662. [CrossRef]
14. Soleymani Eil Bakhtiari, S.; Karbasi, S.; Toloue, E.B. Modified Poly(3-Hydroxybutyrate)-Based Scaffolds in Tissue Engineering Applications: A Review. *Int. J. Biol. Macromol.* **2021**, *166*, 986–998. [CrossRef] [PubMed]
15. Turco, R.; Santagata, G.; Corrado, I.; Pezzella, C.; Di Serio, M. In Vivo and Post-Synthesis Strategies to Enhance the Properties of PHB-Based Materials: A Review. *Front. Bioeng. Biotechnol.* **2021**, *8*, 619266. [CrossRef]
16. Briassoulis, D.; Tserotas, P.; Athanasoulia, I.-G. Alternative Optimization Routes for Improving the Performance of Poly(3-Hydroxybutyrate) (PHB) Based Plastics. *J. Clean. Prod.* **2021**, *318*, 128555. [CrossRef]
17. Vayshbeyn, L.I.; Mastalygina, E.E.; Olkhov, A.A.; Podzorova, M. V Poly(Lactic Acid)-Based Blends: A Comprehensive Review. *Appl. Sci.* **2023**, *13*, 5148. [CrossRef]
18. da Silva Moura, A.; Demori, R.; Leão, R.M.; Crescente Frankenberg, C.L.; Campomanes Santana, R.M. The Influence of the Coconut Fiber Treated as Reinforcement in PHB (Polyhydroxybutyrate) Composites. *Mater. Today Commun.* **2019**, *18*, 191–198. [CrossRef]
19. Barud, H.S.; Souza, J.L.; Santos, D.B.; Crespi, M.S.; Ribeiro, C.A.; Messaddeq, Y.; Ribeiro, S.J.L. Bacterial Cellulose/Poly(3-Hydroxybutyrate) Composite Membranes. *Carbohydr. Polym.* **2011**, *83*, 1279–1284. [CrossRef]
20. Barud, H.S.; Regiani, T.; Marques, R.F.C.; Lustri, W.R.; Messaddeq, Y.; Ribeiro, S.J.L. Antimicrobial Bacterial Cellulose-Silver Nanoparticles Composite Membranes. *J. Nanomater.* **2011**, *2011*, 721631. [CrossRef]
21. Lima, L.R.; Conte, G.V.; Brandão, L.R.; Sábio, R.M.; de Menezes, A.S.; Resende, F.A.; Caiut, J.M.A.; Ribeiro, S.J.L.; Otoni, C.G.; Alcântara, A.C.S.; et al. Fabrication of Noncytotoxic Functional Siloxane-Coated Bacterial Cellulose Nanocrystals. *ACS Appl. Polym. Mater.* **2022**, *4*, 2306–2313. [CrossRef]
22. Barud, H.d.S.; Cavicchioli, M.; Amaral, T.S.D.; Junior, O.B.d.O.; Santos, D.M.; Petersen, A.L.d.O.A.; Celes, F.; Borges, V.M.; de Oliveira, C.I.; de Oliveira, P.F.; et al. Preparation and Characterization of a Bacterial Cellulose/Silk Fibroin Sponge Scaffold for Tissue Regeneration. *Carbohydr. Polym.* **2015**, *128*, 41–51. [CrossRef] [PubMed]
23. Seddiqi, H.; Oliaei, E.; Honarkar, H.; Jin, J.; Geonzon, L.C.; Bacabac, R.G.; Klein-Nulend, J. Cellulose and Its Derivatives: Towards Biomedical Applications. *Cellulose* **2021**, *28*, 1893–1931. [CrossRef]
24. Rasouli, M.; Soleimani, M.; Hosseinzadeh, S.; Ranjbari, J. Bacterial Cellulose as Potential Dressing and Scaffold Material: Toward Improving the Antibacterial and Cell Adhesion Properties. *J. Polym. Environ.* **2023**, *10*, 1–20. [CrossRef]
25. Gorgieva, S. Bacterial Cellulose as a Versatile Platform for Research and Development of Biomedical Materials. *Processes* **2020**, *8*, 624. [CrossRef]
26. Islam, S.U.; Ul-Islam, M.; Ahsan, H.; Ahmed, M.B.; Shehzad, A.; Fatima, A.; Sonn, J.K.; Lee, Y.S. Potential Applications of Bacterial Cellulose and Its Composites for Cancer Treatment. *Int. J. Biol. Macromol.* **2021**, *168*, 301–309. [CrossRef]
27. Raut, M.P.; Asare, E.; Syed Mohamed, S.M.; Amadi, E.N.; Roy, I. Bacterial Cellulose-Based Blends and Composites: Versatile Biomaterials for Tissue Engineering Applications. *Int. J. Mol. Sci.* **2023**, *24*, 986. [CrossRef]
28. Kuhnt, T.; Camarero-Espinosa, S. Additive Manufacturing of Nanocellulose Based Scaffolds for Tissue Engineering: Beyond a Reinforcement Filler. *Carbohydr. Polym.* **2021**, *252*, 117159. [CrossRef]
29. Klemm, D.; Cranston, E.D.; Fischer, D.; Gama, M.; Kedzior, S.A.; Kralisch, D.; Kramer, F.; Kondo, T.; Lindström, T.; Nietzsche, S.; et al. Nanocellulose as a Natural Source for Groundbreaking Applications in Materials Science: Today's State. *Mater. Today* **2018**, *21*, 720–748. [CrossRef]
30. Melo, P.T.S.; Otoni, C.G.; Barud, H.S.; Aouada, F.A.; de Moura, M.R. Upcycling Microbial Cellulose Scraps into Nanowhiskers with Engineered Performance as Fillers in All-Cellulose Composites. *ACS Appl. Mater. Interfaces* **2020**, *12*, 46661–46666. [CrossRef]

31. Huang, L.; Wu, Q.; Wang, Q.; Wolcott, M. Mechanical Activation and Characterization of Micronized Cellulose Particles from Pulp Fiber. *Ind. Crops Prod.* **2019**, *141*, 111750. [CrossRef]
32. Andritsou, V.; de Melo, E.M.; Tsouko, E.; Ladakis, D.; Maragkoudaki, S.; Koutinas, A.A.; Matharu, A.S. Synthesis and Characterization of Bacterial Cellulose from Citrus-Based Sustainable Resources. *ACS Omega* **2018**, *3*, 10365–10373. [CrossRef]
33. *ISO 10993-5*; Biological Evaluation of Medical Devices. Part 5: Tests for Cytotoxicity: In Vitro Methods, 3ed. ISO: Geneva, Switzerland, 2009.
34. *ISO 10993-12*; Biological Evaluation of Medical Devices. Part 12: Sample Preparation and Reference Materials, 4ed. ISO: Geneva, Switzerland, 2012.
35. Mosmann, T. Rapid Colorimetric Assay for Cellular Growth and Survival: Application to Proliferation and Cytotoxicity Assays. *J. Immunol. Methods* **1983**, *65*, 55–63. [CrossRef]
36. Paschoalin, R.T.; Traldi, B.; Aydin, G.; Oliveira, J.E.; Rütten, S.; Mattoso, L.H.C.; Zenke, M.; Sechi, A. Solution Blow Spinning Fibres: New Immunologically Inert Substrates for the Analysis of Cell Adhesion and Motility. *Acta Biomater.* **2017**, *51*, 161–174. [CrossRef]
37. Reynes, J.F.; Isoni, V.; García, F. Tinkering with Mechanochemical Tools for Scale Up. *Angew. Chem.* **2023**. First published online: 28 April 2023. e202300819. [CrossRef]
38. Martínez, L.M.; Cruz-Angeles, J.; Vázquez-Dávila, M.; Martínez, E.; Cabada, P.; Navarrete-Bernal, C.; Cortez, F. Mechanical Activation by Ball Milling as a Strategy to Prepare Highly Soluble Pharmaceutical Formulations in the Form of Co-Amorphous, Co-Crystals, or Polymorphs. *Pharmaceutics* **2022**, *14*, 2003. [CrossRef]
39. Feng, X.; Lin, X.; Deng, K.; Yang, H.; Yan, C. Facile Ball Milling Preparation of Flame-Retardant Polymer Materials: An Overview. *Molecules* **2023**, *28*, 5090. [CrossRef] [PubMed]
40. Joy, J.; Krishnamoorthy, A.; Tanna, A.; Kamathe, V.; Nagar, R.; Srinivasan, S. Recent Developments on the Synthesis of Nanocomposite Materials via Ball Milling Approach for Energy Storage Applications. *Appl. Sci.* **2022**, *12*, 9312. [CrossRef]
41. Mattonai, M.; Pawcenis, D.; Del Seppia, S.; Łojewska, J.; Ribechini, E. Effect of Ball-Milling on Crystallinity Index, Degree of Polymerization and Thermal Stability of Cellulose. *Bioresour. Technol.* **2018**, *270*, 270–277. [CrossRef]
42. Ling, Z.; Wang, T.; Makarem, M.; Santiago Cintrón, M.; Cheng, H.N.; Kang, X.; Bacher, M.; Potthast, A.; Rosenau, T.; King, H.; et al. Effects of Ball Milling on the Structure of Cotton Cellulose. *Cellulose* **2019**, *26*, 305–328. [CrossRef]
43. Park, S.; Baker, J.O.; Himmel, M.E.; Parilla, P.A.; Johnson, D.K. Cellulose Crystallinity Index: Measurement Techniques and Their Impact on Interpreting Cellulase Performance. *Biotechnol. Biofuels* **2010**, *3*, 10. [CrossRef]
44. Lan, L.; Chen, H.; Lee, D.; Xu, S.; Skillen, N.; Tedstone, A.; Robertson, P.; Garforth, A.; Daly, H.; Hardacre, C.; et al. Effect of Ball-Milling Pretreatment of Cellulose on Its Photoreforming for H2 Production. *ACS Sustain. Chem. Eng.* **2022**, *10*, 4862–4871. [CrossRef]
45. Lima, L.R.; Santos, D.B.; Santos, M.V.; Barud, H.S.; Henrique, M.A.; Pasquini, D.; Pecoraro, E.; Ribeiro, S.J.L. Cellulose Nanocrystals from Bacterial Cellulose. *Quim. Nova* **2015**, *38*, 1140–1147. [CrossRef]
46. Drozd, R.; Rakoczy, R.; Konopacki, M.; Frąckowiak, A.; Fijałkowski, K. Evaluation of Usefulness of 2DCorr Technique in Assessing Physicochemical Properties of Bacterial Cellulose. *Carbohydr. Polym.* **2017**, *161*, 208–218. [CrossRef] [PubMed]
47. Zheng, Y.; Fu, Z.; Li, D.; Wu, M. Effects of Ball Milling Processes on the Microstructure and Rheological Properties of Microcrystalline Cellulose as a Sustainable Polymer Additive. *Materials* **2018**, *11*, 1057. [CrossRef]
48. Avolio, R.; Bonadies, I.; Capitani, D.; Errico, M.E.; Gentile, G.; Avella, M. A Multitechnique Approach to Assess the Effect of Ball Milling on Cellulose. *Carbohydr. Polym.* **2012**, *87*, 265–273. [CrossRef]
49. Tang, L.; Huang, B.; Yang, N.; Li, T.; Lu, Q.; Lin, W.; Chen, X. Organic Solvent-Free and Efficient Manufacture of Functionalized Cellulose Nanocrystals via One-Pot Tandem Reactions. *Green Chem.* **2013**, *15*, 2369–2373. [CrossRef]
50. Kano, F.S.; de Souza, A.G.; Rosa, D. dos S. Variation of the Milling Conditions in the Obtaining of Nanocellulose from the Paper Sludge. *Rev. Mater.* **2019**, *24*, e12406. [CrossRef]
51. Van Craeyveld, V.; Holopainen, U.; Selinheimo, E.; Poutanen, K.; Delcour, J.A.; Courtin, C.M. Extensive Dry Ball Milling of Wheat and Rye Bran Leads to in Situ Production of Arabinoxylan Oligosaccharides through Nanoscale Fragmentation. *J. Agric. Food Chem.* **2009**, *57*, 8467–8473. [CrossRef]
52. Niemi, P.; Faulds, C.B.; Sibakov, J.; Holopainen, U.; Poutanen, K.; Buchert, J. Effect of a Milling Pre-Treatment on the Enzymatic Hydrolysis of Carbohydrates in Brewer's Spent Grain. *Bioresour. Technol.* **2012**, *116*, 155–160. [CrossRef]
53. Gao, C.; Xiao, W.; Ji, G.; Zhang, Y.; Cao, Y.; Han, L. Regularity and Mechanism of Wheat Straw Properties Change in Ball Milling Process at Cellular Scale. *Bioresour. Technol.* **2017**, *241*, 214–219. [CrossRef]
54. Ji, G.; Gao, C.; Xiao, W.; Han, L. Mechanical Fragmentation of Corncob at Different Plant Scales: Impact and Mechanism on Microstructure Features and Enzymatic Hydrolysis. *Bioresour. Technol.* **2016**, *205*, 159–165. [CrossRef] [PubMed]
55. Liu, H.; Chen, X.; Ji, G.; Yu, H.; Gao, C.; Han, L.; Xiao, W. Mechanochemical Deconstruction of Lignocellulosic Cell Wall Polymers with Ball-Milling. *Bioresour. Technol.* **2019**, *286*, 121364. [CrossRef] [PubMed]
56. Pradhan, S.; Borah, A.J.; Poddar, M.K.; Dikshit, P.K.; Rohidas, L.; Moholkar, V.S. Microbial Production, Ultrasound-Assisted Extraction and Characterization of Biopolymer Polyhydroxybutyrate (PHB) from Terrestrial (*P. hysterophorus*) and Aquatic (*E. crassipes*) Invasive Weeds. *Bioresour. Technol.* **2017**, *242*, 304–310. [CrossRef]
57. Vahabi, H.; Michely, L.; Moradkhani, G.; Akbari, V.; Cochez, M.; Vagner, C.; Renard, E.; Saeb, M.R.; Langlois, V. Thermal Stability and Flammability Behavior of Poly(3-Hydroxybutyrate) (PHB) Based Composites. *Materials* **2019**, *12*, 2239. [CrossRef] [PubMed]

58. Reis, K.C.; Pereira, L.; Melo, I.C.N.A.; Marconcini, J.M.; Trugilho, P.F.; Tonoli, G.H.D. Particles of Coffee Wastes as Reinforcement in Polyhydroxybutyrate (PHB) Based Composites. *Mater. Res.* **2015**, *18*, 546–552. [CrossRef]
59. Sindhu, R.; Ammu, B.; Binod, P.; Deepthi, S.K.; Ramachandran, K.B.; Soccol, C.R.; Pandey, A. Production and Characterization of Poly-3-Hydroxybutyrate from Crude Glycerol by Bacillus Sphaericus NII 0838 and Improving Its Thermal Properties by Blending with Other Polymers. *Brazilian Arch. Biol. Technol.* **2011**, *54*, 783–794. [CrossRef]
60. Ramezani, M.; Amoozegar, M.A.; Ventosa, A. Screening and Comparative Assay of Poly-Hydroxyalkanoates Produced by Bacteria Isolated from the Gavkhooni Wetland in Iran and Evaluation of Poly-β-Hydroxybutyrate Production by Halotolerant Bacterium Oceanimonas Sp. GK1. *Ann. Microbiol.* **2015**, *65*, 517–526. [CrossRef]
61. Zanini, N.; Carneiro, E.; Menezes, L.; Barud, H.; Mulinari, D. Palm Fibers Residues from Agro-Industries as Reinforcement in Biopolymer Filaments for 3D-Printed Scaffolds. *Fibers Polym.* **2021**, *22*, 2689–2699. [CrossRef]
62. Codreanu, A.; Balta, C.; Herman, H.; Cotoraci, C.; Mihali, C.V.; Zurbau, N.; Zaharia, C.; Rapa, M.; Stanescu, P.; Radu, I.-C.; et al. Bacterial Cellulose-Modified Polyhydroxyalkanoates Scaffolds Promotes Bone Formation in Critical Size Calvarial Defects in Mice. *Materials* **2020**, *13*, 1433. [CrossRef]
63. Kohan, M.; Lancoš, S.; Schnitzer, M.; Živčák, J.; Hudák, R. Analysis of PLA/PHB Biopolymer Material with Admixture of Hydroxyapatite and Tricalcium Phosphate for Clinical Use. *Polymers* **2022**, *14*, 5357. [CrossRef]

Disclaimer/Publisher's Note: The statements, opinions and data contained in all publications are solely those of the individual author(s) and contributor(s) and not of MDPI and/or the editor(s). MDPI and/or the editor(s) disclaim responsibility for any injury to people or property resulting from any ideas, methods, instructions or products referred to in the content.

Review

Properties, Production, and Recycling of Regenerated Cellulose Fibers: Special Medical Applications

Sandra Varnaitė-Žuravliova * and Julija Baltušnikaitė-Guzaitienė

Department of Textile Technologies, Center for Physical Sciences and Technology, Demokratų Str. 53, LT-48485 Kaunas, Lithuania; julija.baltusnikaite@ftmc.lt
* Correspondence: sandra.varnaite.zuravliova@ftmc.lt

Abstract: Regenerated cellulose fibers are a highly adaptable biomaterial with numerous medical applications owing to their inherent biocompatibility, biodegradability, and robust mechanical properties. In the domain of wound care, regenerated cellulose fibers facilitate a moist environment conducive to healing, minimize infection risk, and adapt to wound topographies, making it ideal for different types of dressings. In tissue engineering, cellulose scaffolds provide a matrix for cell attachment and proliferation, supporting the development of artificial skin, cartilage, and other tissues. Furthermore, regenerated cellulose fibers, used as absorbable sutures, degrade within the body, eliminating the need for removal and proving advantageous for internal suturing. The medical textile industry relies heavily on regenerated cellulose fibers because of their unique properties that make them suitable for various applications, including wound care, surgical garments, and diagnostic materials. Regenerated cellulose fibers are produced by dissolving cellulose from natural sources and reconstituting it into fiber form, which can be customized for specific medical uses. This paper will explore the various types, properties, and applications of regenerated cellulose fibers in medical contexts, alongside an examination of its manufacturing processes and technologies, as well as associated challenges.

Keywords: regenerated cellulose fibers; medical; viscose; lyocell; modal; recycling

1. Introduction

Civilizations dating as early as the ancient Egyptians used natural materials such as flax, silk, hemp, and cotton to create sutures and bandages for wound healing. Some natural fibers are still in use today because of their intricate and versatile structures, favorable biocompatibility, and abundance.

In recent years, the definition of "textile" has broadened, necessitating an emphasis on the increasing significance of diverse textile materials for applications in technology, industry, construction, architecture, agriculture, and medicine [1–3]. The medical textile segment is regarded as the most rapidly expanding sector within the technical textile industry [4–6]. Textile materials, in conjunction with scientific methodologies, are extensively utilized in medical and surgical applications due to their durability, versatility, convenience, and antimicrobial properties [7]. These polymers should be dissolvable or meltable for extrusion, and their molecular chains should be linear, long, flexible, and capable of orientation and crystallization.

Medical textiles are engineered for various medical uses, including both internal and external applications, with the capability of being implantable or non-implantable. They are utilized in biological structures for the assessment, treatment, enhancement, or regeneration of tissues, organs, or physiological functions. Examples include plasters, dressings, bandages, and compression garments [8].

Ensuring a sustainable world from the textile perspective has motivated the usage of sustainable and biodegradable materials as alternatives.

1.1. Textile Fibers for Medical Applications

Polysaccharide-based fibers, particularly cellulose, are widely used in medical applications such as wound dressings and hemostats due to their thrombogenic properties [9]. As a natural biopolymer, cellulose is an integral component of plant cell walls and is synthesized by specific microbes, making it the most abundant organic substance on Earth. This abundance, coupled with properties like biocompatibility, biodegradability, and renewability, makes cellulose an attractive choice for designing biomaterials in biomedical fields.

Cellulose, derived from various natural sources, such as cotton and woody biomass, varies in content; cotton has up to 90% cellulose, making it nearly pure, while woody biomass contains about 40% to 50% cellulose [10]. Its structural strength comes from β-1,4-glycosidic linkages that form long chains of D-glucose units. These properties are harnessed in the creating of advanced healthcare products like aerogels, hydrogels, films, and fillers [11,12].

Incorporating cellulose into medicated textiles and pharmaceutical products fosters the development of new healthcare approaches, aligning with the goal of reducing the medical sector's impact by substituting non-biodegradable waste with sustainable, green polymers. This aligns with the wider movement towards sustainability in material science, emphasizing the critical role of naturally occurring substances in modern applications.

The regeneration process transforms the cellulose chain conformation from cellulose I to cellulose II, resulting in a structure with more amorphous regions and enhanced crystallinity [13]. This change also facilitates extensive modifications in regenerated cellulose (RC) products, including hydrogels, aerogels, cryogels, xerogels, membranes, thin films, and fibers [14]. The most commonly used structures in medical textiles are fibers [15–17]. Fibers are the primary materials for the textile industry, which vary in type based on their origin and chemical composition, as natural and chemical (man-made and synthetic). Regenerated cellulose fibers are paramount in the medical textile industry due to their exceptional biocompatibility, biodegradability, and mechanical properties. These fibers, derived from natural cellulose, undergo various chemical processes to transform their structure, making them suitable for diverse medical applications.

Cellulose can be regenerated to produce films/membranes, hydrogels/aerogels, filaments/fibers, microspheres/beads, bioplastics, etc., which show potential applications in textiles, biomedicine, energy storage, packaging, etc. Importantly, these cellulose-based materials can be biodegraded in soil and oceans, reducing environmental pollution [18]. The cellulose solvents, dissolving mechanism, and strategies for constructing the regenerated cellulose functional materials with high strength and performances, together with the current achievements and urgent challenges, are summarized, and some perspectives are also proposed.

There are five conventional types of regenerated cellulose or cellulose-derived fibers: viscose; lyocell; cupro; acetate; and modal. These fibers have been widely recognized for their biocompatibility, moisture retention capabilities, and biodegradability, making them suitable candidates for various medical applications, including wound care and tissue scaffolding. Recent innovations have focused on improving their structural properties, antimicrobial functionality, and integration with bioactive agents. These developments have opened new avenues for more efficient and sustainable healthcare solutions. The advantages of regenerated cellulose fibers in medical applications compared to synthetic fibers (like polypropylene and polyester) and natural fibers (like cotton or silk) are presented in Table 1.

Table 1. The advantages of regenerated cellulose fibers, used for medical applications, compared to synthetic and natural fibers.

Property	Regenerated Cellulose Fibers (Viscose, Lyocell, Modal, and Acetate)	Synthetic Fibers (Polypropylene and Polyester)	Natural Fibers (Cotton and Silk)	Scientific References
Cost-effectiveness	Moderate cost, relatively inexpensive production due to large-scale manufacturing.	Low cost, very cheap due to petrochemical origin and established production lines.	High cost for premium fibers like silk; cotton can vary in price depending on the quality.	[19–21]
Biodegradability	Highly biodegradable (especially lyocell and viscose); reduces environmental impact in disposable medical products.	Non-biodegradable; contributes to long-term environmental pollution (plastic waste).	Biodegradable, but cotton production has a significant environmental footprint (e.g., water usage).	[19–21]
Biocompatibility	Excellent biocompatibility due to cellulose origin; minimal risk of allergic reactions or irritation.	Variable; some synthetic fibers may cause skin irritation or inflammation, especially in sensitive patients.	High biocompatibility; natural fibers like silk are hypoallergenic, but cotton can sometimes be harsh on sensitive skin.	[21–23]
Absorbency	Excellent moisture absorption (e.g., lyocell and viscose); ideal for wound care and surgical dressings.	Poor moisture absorption; tends to repel liquids, which is useful in some protective textiles but not for wound care.	High absorbency (especially cotton); suitable for certain medical textiles but not as specialized as cellulose fibers.	[21–23]
Sterilization compatibility	Can be sterilized via autoclaving, gamma radiation, and other methods without losing structural integrity (except acetate).	Excellent for sterilization; resistant to degradation by most sterilization methods.	Natural fibers like cotton can be sterilized but may lose structural integrity or shrink over time. Silk is more delicate in sterilization.	[19,20,22]
Environmental sustainability	High sustainability (especially lyocell with closed-loop processing); lower chemical and energy inputs than synthetic fibers.	Poor sustainability due to petrochemical base and non-renewable resources used in production.	Mixed; cotton production is water-intensive, but silk is more environmentally friendly in small-scale production.	[19,20,23]
Patient comfort	Soft, breathable, and skin-friendly; suitable for long-term wear in medical garments and wound dressings.	Less breathable and can cause skin irritation; often uncomfortable for long-term skin contact (e.g., hospital gowns).	High comfort for silk, but cotton can sometimes cause friction on sensitive or healing skin.	[21–23]
Applications in the medical field	Used in wound dressings, surgical swabs, hygiene products, and biodegradable implants due to biocompatibility and absorbency.	Used in protective clothing, surgical masks, and other disposable medical products, but less suitable for direct skin contact.	Used in traditional bandages, sutures (silk), and some medical textiles, but less common in advanced medical applications.	[20,21,23]
Durability	Moderate durability; can be engineered for strength in specific medical applications (e.g., lyocell scaffolds).	High durability; long-lasting and tear-resistant, ideal for protective gear like surgical drapes.	Variable; silk is strong but delicate, while cotton is less durable in clinical use due to wear and tear.	[19,20,22]
Antimicrobial functionalization	Can be functionalized with antimicrobial agents (e.g., silver and iodine) for advanced wound care applications.	Often treated with antimicrobial agents, but additives may leach over time or be toxic.	Silk has natural antibacterial properties, while cotton may require treatment to gain antimicrobial properties.	[19,21,22]

The microscopic views of regenerated cellulose fibers and some common synthetic fibers are presented in Figure 1. More microscopic views of different fibers may be found in [24].

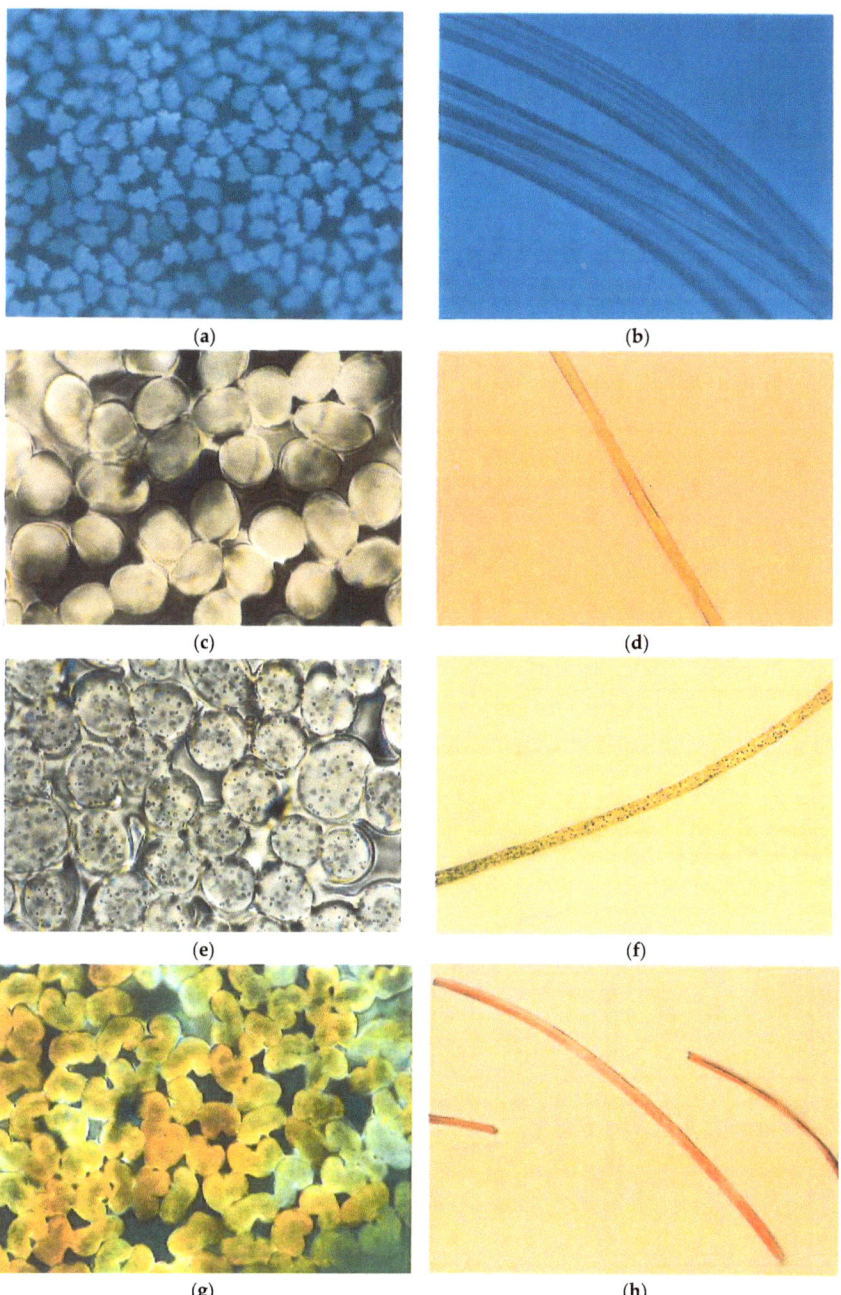

Figure 1. *Cont.*

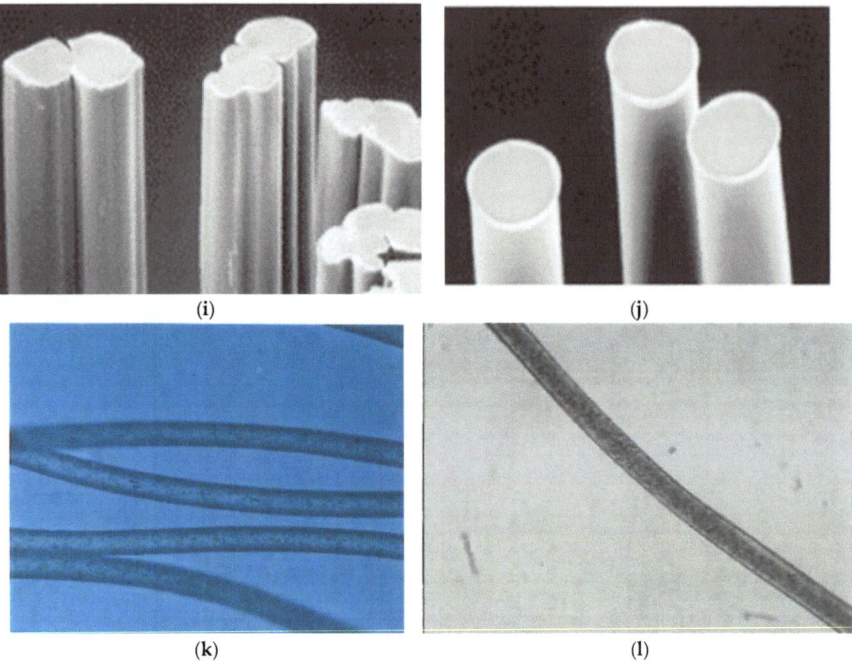

Figure 1. The microscopic views of regenerated cellulose fibers and some common synthetic fibers [24,25]: (**a**) microscopic view of viscose fiber cross-section; (**b**) microscopic longitudinal view of viscose fiber; (**c**) microscopic view of lyocell fiber cross-section; (**d**) microscopic longitudinal view of lyocell fiber; (**e**) microscopic view of cupro fiber cross-section; (**f**) microscopic longitudinal view of cupro fiber; (**g**) microscopic view of modal fiber cross-section; (**h**) microscopic longitudinal view of modal fiber; (**i**) microscopic view of acetate fiber; (**j**) microscopic view of typical melt spun synthetic fibers cross-section, i.e., polyester, polyamide, and olefin; (**k**) microscopic longitudinal view of polyester fiber; (**l**) microscopic longitudinal view of polyamide fiber; ((**a–h**,**k**,**l**) Reprinted with permission from Ref. [24]. Copyright 2012 Lithuanian Standards Board). ((**i**,**j**) Reprinted with permission from Ref. [25]. Copyright 2008 Elsevier).

1.2. The Classification of These Different Regenerated Cellulose Fibers Is Based on Fiber Production Method

- Viscose fibers

Viscose fibers are derived from natural cellulose, primarily sourced from wood pulp or cotton linters. The process of converting cellulose into viscose involves several chemical treatments, including steeping, shredding, and xanthation, resulting in fibers that closely resemble natural fibers like cotton and silk in texture and appearance. Viscose is appreciated for its versatility and biodegradability, making it a popular choice in various textile applications [26]. The development of viscose fibers dates back to the late 19th century when researchers sought to create a fiber that mimics the properties of natural silk. The viscose process was first patented by British chemists Charles Cross, Edward Bevan, and Clayton Beadle in 1892, marking the beginning of commercial production in the early 20th century [27].

Viscose rayon, the oldest and most widely used regenerated cellulose fiber, is produced by dissolving cellulose in a solution of sodium hydroxide and carbon disulfide to form cellulose xanthate. This solution is then extruded into fibers and regenerated in an acidic bath [28,29]. Viscose rayon is highly absorbent and soft, making it ideal for applications such as wound dressings, bandages, and surgical swabs. Its excellent sterilizability further enhances its suitability for medical applications [30,31].

Viscose rayon fibers typically have a round or slightly irregular cross-sectional shape. The cross-sectional geometry of viscose fibers is largely influenced by the coagulation conditions during the spinning process, which can lead to variations in shape and size. Studies have shown that these fibers often exhibit a circular or slightly oval cross-section depending on the specific conditions under which they are formed [32,33]. The internal structure often shows a central lumen (a hollow core) surrounded by a cellulosic wall with varying degrees of fibrillation. This lumen, a characteristic feature of regenerated cellulose fibers like viscose, is formed due to the rapid coagulation of the cellulose solution during spinning, which traps air or other gasses within the fiber structure [34]. The cellulosic wall surrounding the lumen is composed of aligned cellulose microfibrils, which can vary in their degree of crystallinity and orientation, affecting the fiber's mechanical properties [35]. The cross-section of viscose rayon appears with a smooth, even texture but can exhibit some degree of irregularity due to the manufacturing process. Variations in coagulation conditions, such as temperature and solvent concentration, can lead to irregularities in the fiber cross-section, contributing to a less uniform appearance [36,37]. This irregularity can also be attributed to fluctuations in the flow rate during extrusion, which affects the fiber's final shape. In the longitudinal view, viscose rayon fibers display a smooth surface with longitudinal striations or fibrils along their length. These striations result from the stretching and drawing processes during fiber production, which align the cellulose chains and create a characteristic fibrillar texture [32]. The surface of viscose fibers may also exhibit undulations or waviness, which are linked to the mechanical drawing and relaxation steps that the fibers undergo during spinning. The fibers are generally smooth but can show some periodic variations in diameter due to variations in spinning conditions. These variations in diameter are primarily caused by inconsistencies in the extrusion and coagulation phases, where factors like pressure, temperature, and draw ratio play significant roles [33]. The periodic changes in diameter can impact the fiber's overall mechanical performance and its suitability for various applications.

Recent advancements in viscose rayon technology have focused on improving fiber strength and reducing environmental impacts. For instance, innovations in the viscose process have led to the development of more sustainable production methods and enhanced fiber properties [38].

For medical textiles, viscose fibers may be engineered to have specific structural properties, such as enhanced porosity, higher purity, and tailored surface characteristics, to improve their performance in medical applications. Medical-grade viscose fibers require higher purity levels to avoid any adverse reactions. This involves more rigorous washing and bleaching processes to remove any residual chemicals or impurities [33]. The fiber structure can be adjusted to enhance porosity and surface area, which is critical for applications like wound dressings, where absorbency and moisture management are crucial. Viscose fibers for medical use may be coated with antimicrobial agents or other functional substances to provide additional benefits such as infection control [39,40].

- Lyocell

Lyocell fibers are a type of regenerated cellulose fiber known for their eco-friendly production process and superior performance characteristics. Lyocell is a form of rayon, primarily derived from wood pulp. Unlike other types of rayon, such as viscose, the production of lyocell involves an environmentally friendly process that uses a non-toxic solvent, N-methylmorpholine N-oxide (NMMO). Introduced in the 1990s, lyocell has quickly gained popularity due to its sustainability and high performance [41,42]. The development of lyocell fibers was driven by the need for a more sustainable alternative to traditional rayon fibers. The first commercial production of lyocell was by Courtaulds Fibers under the brand name Tencel in the early 1990s [33].

Lyocell fibers are renowned for their distinctive structural and mechanical properties, which set them apart from other regenerated cellulose fibers such as viscose. The cross-sectional profile of lyocell fibers is typically round and exhibits a smooth, uniform surface. Unlike viscose fibers, which often display a central lumen, lyocell fibers generally have a more compact and dense cellulosic structure with minimal or no lumen. This compact structure results from the solvent-spinning process used in lyocell production, where cellulose is dissolved in N-methylmorpholine N-oxide (NMMO) and then extruded and regenerated in a controlled manner [41,43]. The uniform and round cross-section of lyocell fibers contribute to their high tensile strength and smooth texture, making them particularly well suited for applications requiring durability and a fine hand feel. The absence of significant internal voids, such as a pronounced lumen, and the alignment of cellulose microfibrils during the spinning process result in fibers that are not only strong but also resistant to deformation. This is a crucial advantage in textile applications, where fiber consistency and performance are critical [33]. In terms of surface characteristics, lyocell fibers are characterized by their smooth and even surface, which is evident in both cross-sectional and longitudinal views. The spinning process ensures that the fibers have fewer surface defects compared to viscose fibers, which often show more pronounced striations and irregularities. The smooth surface of lyocell fibers contributes to their luxurious feel and low surface friction, making them ideal for high-end textiles, including apparel and home textiles. Moreover, this smoothness enhances the fibers' dyeing properties, allowing for vibrant and uniform coloration [44]. The longitudinal view of lyocell fibers further emphasizes their superior quality. The fibers are generally straight with a high degree of smoothness and uniformity. This straightness, combined with the fiber's inherent strength, reduces the likelihood of pilling and wear, which is a common issue with less uniform fibers. The enhanced mechanical properties of lyocell fibers, including their high tensile strength and durability, make them suitable for a wide range of applications, from fashion textiles to technical fabrics [45].

Lyocell fibers are produced using the N-methylmorpholine N-oxide (NMMO) process, which is considered more environmentally friendly compared to the viscose process [42,46]. The NMMO solvent effectively dissolves cellulose, which is then extruded into fibers and regenerated. Lyocell fibers exhibit high tensile strength, excellent moisture management, and biocompatibility, making them suitable for surgical gowns, drapes, and wound dressings [26,47].

Similarly to viscose, lyocell fibers intended for medical textiles undergo specific modifications and enhanced production methods to meet medical standards. Lyocell fibers for medical use are often engineered to have properties such as higher absorbency, enhanced biocompatibility, and tailored mechanical strength. Ensuring biocompatibility involves minimizing any potential cytotoxicity. This can be achieved by optimizing the fiber's chemical composition and ensuring the absence of harmful residual solvents [33]. Medical applications may require fibers with higher tensile strength and durability, especially for sutures or implants [48].

Recent studies have highlighted the advantages of lyocell fibers in medical textiles due to their enhanced mechanical properties and environmental sustainability. The lyocell process continues to evolve, with research focusing on improving fiber functionality and reducing environmental impacts [42]. The ongoing evolution of the lyocell process has focused on improving fiber functionality, such as controlling fibrillation and enhancing the durability of the fibers for demanding medical applications like wound dressings and surgical textiles. Advances in the lyocell process include the use of cross-linking agents and enzymatic treatments to minimize fibrillation, thus improving fiber performance in medical applications [49,50].

Moreover, the lyocell process continues to innovate with research aimed at further reducing its environmental impact, including the development of more efficient solvent recovery systems and the exploration of alternative raw materials for cellulose production [51,52].

- Modal

Modal fibers are a subtype of rayon specifically designed to improve upon the properties of regular viscose rayon. They are made from beech tree pulp and undergo a series of chemical processes to achieve their unique qualities. Modal fibers are often praised for their luxurious feel and durability, making them a popular choice in the textile industry.

Modal fibers were developed as a response to the need for a more robust and high-performing type of rayon. The name "modal" was coined by the Austrian company Lenzing AG, which remains one of the leading producers of these fibers [53,54]

Modal fibers, a type of regenerated cellulose fiber, are produced through a process similar to viscose but with modifications that enhance fiber strength and elasticity [52]. Modal fibers are known for their softness, durability, and resistance to shrinkage. These properties make them suitable for reusable medical textiles, such as hospital bed linens and patient clothing [47].

Modal fibers often have a slightly oval cross-section. The internal structure is similar to viscose but with improved structural integrity and uniformity due to modifications in the production process. The cross-sectional view shows a more consistent shape and density compared to standard viscose fibers. Modal fibers exhibit a smoother and more consistent surface compared to viscose. The longitudinal view shows a straight and uniform appearance with fewer surface irregularities. The fibers are typically smooth, with fewer longitudinal striations compared to viscose fibers. [55]

Modal fibers are increasingly being used in combination with other fibers to create high-performance textiles for medical applications. Research has shown that modal fibers can be blended with antimicrobial agents to enhance their functionality in medical textiles [56].

- Cupro (Cuprammonium Rayon)

Cupro fibers are derived from cellulose, specifically cotton linter, which is a byproduct of the cotton industry. The production of cupro fibers involves dissolving cellulose in a copper–ammonium solution, hence the name cuprammonium rayon [26,57].

Cupro fibers are produced using the cuprammonium process, where cellulose is dissolved in a copper–ammonium solution [58]. These fibers are smooth, lustrous, and hypoallergenic, making them suitable for applications requiring non-abrasive materials, such as advanced wound care products [59,60]

Recent advancements in the cuprammonium process have focused on improving fiber strength and exploring new applications in medical textiles. Cupro fibers are being studied for their potential use in high-performance medical textiles due to their unique properties.

Cupro fibers typically have a round or slightly irregular cross-sectional shape. The internal structure is dense and smooth, with minimal lumen. The fibers have a smooth and even cross-section, reflecting their fine, high-quality texture. In the longitudinal view, cupro fibers display a very smooth surface with minimal striations or surface defects. The fibers are straight and exhibit a high level of smoothness and sheen [60,61].

Cupro (cuprammonium rayon) fibers are characterized by their silk-like feel and high moisture absorption. For medical uses, the fibers are often engineered to enhance their biocompatibility and reduce potential allergic reactions.

- Acetate

Acetate fibers, also known as cellulose acetate fibers, are derived from cellulose. They are produced by acetylating the cellulose extracted from wood pulp or cotton linters. Acetate fibers are unique due to their blend of natural and synthetic properties, which provide distinct advantages in textile manufacturing [26,62].

Acetate fibers are produced through the acetylation of cellulose, where cellulose is reacted with acetic anhydride or acetyl chloride in the presence of a catalyst. This process yields a fiber with distinct properties compared to other regenerated cellulose fibers. Acetate fibers are characterized by their smooth texture, lustrous appearance, and relatively low moisture absorption [63].

Acetate fibers are generally biocompatible, which makes them suitable for certain medical applications. Their smooth surface can reduce irritation in contact with the skin. In addition, acetate fibers are known for their soft and silky feel, which can enhance comfort in medical textiles like patient clothing and surgical drapes. This type of fiber can be sterilized, making them suitable for medical textiles that require sterilization before use [64–66].

While acetate fibers are not as commonly used as some other regenerated cellulose fibers in medical textiles, they possess properties that make them suitable for specific applications. Their softness, biocompatibility, and potential for customization through chemical treatments can make them a viable option for certain medical textile products. Regenerated cellulose fibers (RCFs) offer a solution to the drawbacks of synthetic fibers while preserving the benefits of natural fibers. According to The Materials Market Report 2023 by Textile Exchange [67], the RCFs covered 6% (i.e., 7,3 mln. tons of fibers) of a global market in 2022; among that, viscose fiber production accounted for 80% of all the RCFs and a production volume of around 5.8 million tons in 2022. Acetate covered 13% of all the RCF market, and lyocell, modal, and cupro, respectively, 4%, 3%, and 0.2%. Along with that, 0.5% of all RCF market shares recycled regenerated cellulose fiber producers.

2. Production of RC

RCFs have attracted attention because they can be easily processed via fiber spinning to produce man-made fibers using bio-based feedstocks rather than the cases of natural fibers [68]. There are several methods for producing high-quality regenerated cellulose (RC) fibers through dissolution and regeneration processes [69,70]. Typically, these fibers are created using the spinning method. One of the most common techniques is wet spinning, where a cellulose solution is extruded through a spinneret into an acid bath for coagulation. Following this, the fibers are washed first with hot water and then with distilled water to remove salts formed during coagulation. The resulting RC fibers are then dried at room temperature [71].

The properties of regenerated cellulose (RC) fibers can be influenced by different solvent systems [68]. Each solvent system impacts the final characteristics of the RC fibers and offers distinct advantages. The solubility and spinnability of cellulose are influenced by multiple factors, including solvent properties (solvating power and viscosity), raw material characteristics (degree of polymerization, molecular structure, and chemical composition), process parameters (such as temperature, duration, and pressure), and the equipment used for dissolution and spinning [72,73].

The structural and mechanical properties of RC fibers can be optimized by adjusting parameters, such as draw ratio, spinning dopes, and the spinning process. Research [71,74] shows that at low draw ratios, regenerated cellulose (RC) fibers display an irregular morphology with pronounced surface grooves. Conversely, when fibers are produced at high draw ratios, they have a smoother surface and a more rounded structure.

High-purity cellulose pulp is essential for producing medical-grade viscose fibers. Advanced purification techniques, such as enzymatic treatment and bleaching, are employed to remove impurities and ensure biocompatibility [6,75,76].

2.1. Production Methods of Viscose Fibers

The production of viscose fibers involves several stages, including the preparation of cellulose pulp, dissolution, regeneration, and finishing. Specific production methods tailored for medical textiles focus on enhancing the purity and functionality of the fibers.

The first step in producing viscose fibers is extracting cellulose from wood pulp or cotton linters. The pulp undergoes purification to remove hemicellulose, lignin, and other impurities [77–79]. The purified cellulose is then treated with sodium hydroxide (NaOH) to form alkali cellulose. This process increases the cellulose's reactivity, preparing it for the subsequent chemical reactions [80]. The alkali cellulose is aged under controlled conditions, typically for 1–2 days. Aging reduces the degree of polymerization, which is necessary for achieving the desired viscosity in the final viscose solution [81]. Aging is followed by the xanthation process, where the alkali cellulose reacts with carbon disulfide (CS2) to form cellulose xanthate. This reaction imparts solubility to the cellulose in aqueous sodium hydroxide. The cellulose xanthate is dissolved in dilute sodium hydroxide to produce a viscous orange-yellow solution known as viscose. The solution is allowed to ripen for a specific period, enabling the cellulose chains to rearrange and achieve the required spinnability [29]. The ripened viscose solution is filtered to remove undissolved particles and degassed to eliminate air bubbles. These steps are crucial for ensuring the uniformity and quality of the fibers [82].

The viscose solution is extruded through a spinneret into an acid bath (usually sulfuric acid), where the cellulose is regenerated into solid fibers through a process of acid precipitation and coagulation. This step determines the final properties of the viscose fibers [83,84]. Viscose rayon fibers can be produced in two forms: continuous filament and staple fibers, but generally, viscose rayon is manufactured as staple fiber.

After regeneration, the fibers undergo various post-treatment processes, including washing, bleaching, and finishing. These treatments enhance the fibers' properties, such as whiteness, strength, and softness [70,85].

The production of viscose fibers involves the use of chemicals such as sodium hydroxide and carbon disulfide, which can pose environmental and health risks. However, advancements in production technologies aim to mitigate these impacts through improved chemical recovery and waste management systems [84,86].

The production methods of viscose fibers for medical textiles often incorporate additional steps to ensure the fibers meet medical standards. Medical-grade viscose fibers must be sterilizable. The production methods may include steps that ensure fibers can withstand sterilization processes, such as autoclaving, gamma irradiation, or ethylene oxide treatment. The entire production process is subjected to more rigorous quality control to ensure the fibers are free from contaminants and have consistent properties suitable for medical use [87,88].

The principal production scheme of viscose fibers is presented in Figure 2.

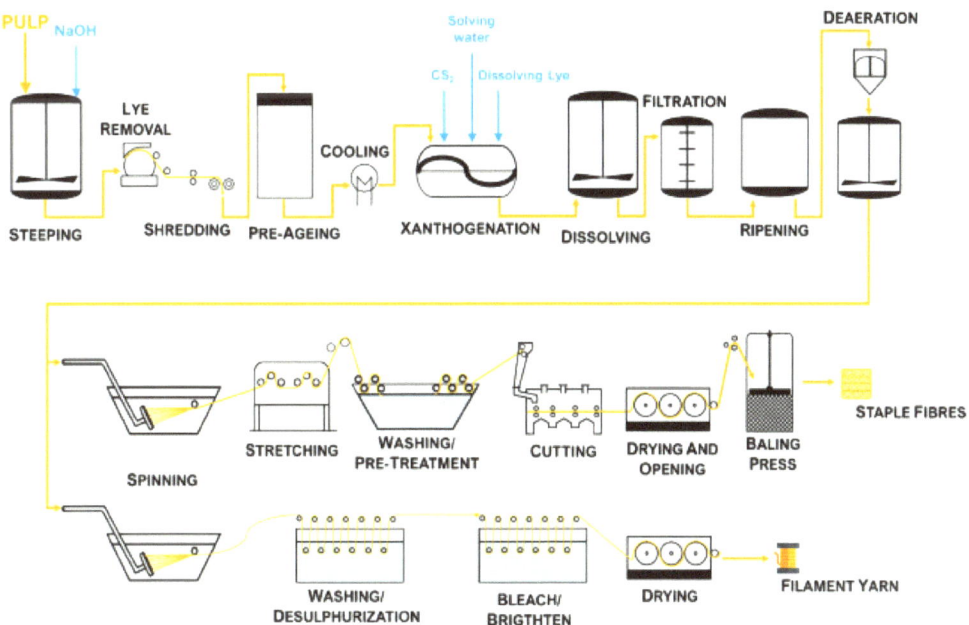

Figure 2. Production scheme of viscose fibers [89] (Reprinted with permission from Ref. [89]. Copyright 2021 Elsevier).

2.2. Production Methods of Lyocell Fibers

The production of lyocell fibers involves several key stages: cellulose extraction, dissolution, spinning, and post-treatment. The most distinguishing feature of lyocell production is the use of N-Methylmorpholine N-oxide (NMMO) as the solvent, which is non-toxic and recyclable.

The primary raw material for lyocell production is wood pulp from sustainably managed forests. The pulp is purified to remove lignin and hemicellulose, resulting in high-purity cellulose suitable for dissolution [90]. In the lyocell process, purified cellulose is dissolved directly in an aqueous solution of NMMO. This solvent system is highly effective in dissolving cellulose with no chemical derivatization, distinguishing lyocell from other types of rayon [46,91,92].

The cellulose-NMMO solution, known as the spinning dope, is filtered to remove undissolved particles and air bubbles. The filtered solution is then extruded through a spinneret into a coagulation bath, typically water or a dilute NMMO solution, where the cellulose fibers are regenerated [33,93]. After regeneration, the lyocell fibers undergo several post-treatment steps, including washing to remove residual NMMO, drying, and finishing. These steps are crucial for enhancing the mechanical properties and appearance of the fibers [91]. A significant advantage of the lyocell process is the efficient recovery and reuse of NMMO. Over 99% of the solvent is typically recovered and recycled, minimizing environmental impact and production costs [43,94].

The production of lyocell fibers is considered more environmentally friendly compared to other types of rayon due to the closed-loop solvent recovery system and the use of non-toxic chemicals. Additionally, lyocell fibers are biodegradable and derived from renewable resources, aligning with principles of sustainable development [15,70,95].

The production process for medical-grade lyocell fibers often includes additional purification and validation steps to meet stringent medical standards. The NMMO solvent recovery process is optimized to ensure that no toxic residues are present in the fibers. This is critical for medical applications where patient safety is paramount. Additional washing and bleaching steps are implemented to achieve the high purity required for medical use [96,97].

The principal production scheme of lyocell fibers is presented in Figure 3.

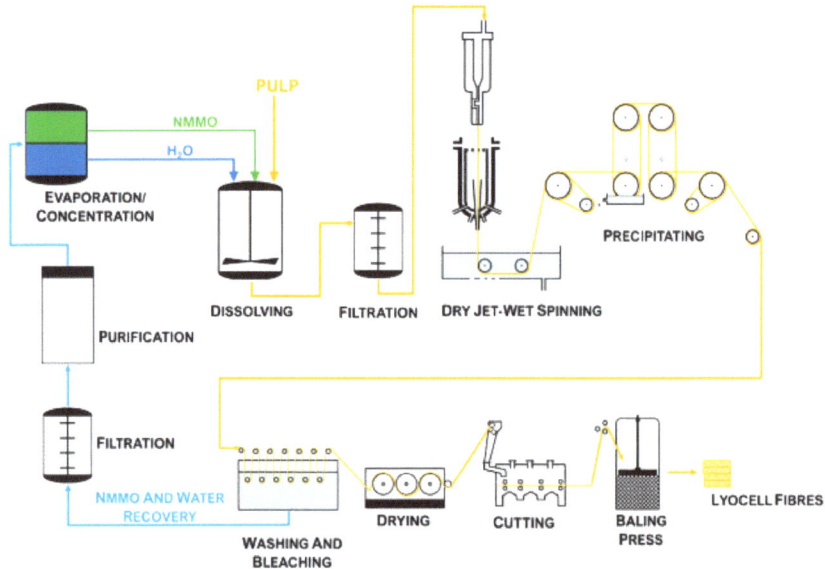

Figure 3. Production scheme of lyocell fibers [89] (reproduced with permission from Elsevier, *Carbohydrate Polymers*; published by Elsevier, 2021).

2.3. Production Methods of Modal Fibers

The primary raw material for modal fibers is beech wood pulp, which is processed to remove lignin and hemicellulose, resulting in high-purity cellulose [52,98].

The purified cellulose is treated with sodium hydroxide to form alkali cellulose, which is then reacted with carbon disulfide to form cellulose xanthate. This process is similar to the viscose process but is optimized for better fiber strength and quality [99–101].

The cellulose xanthate is dissolved in a dilute sodium hydroxide solution to form a viscous solution. This solution is further processed to ensure uniformity and the desired viscosity for spinning.

The spinning dope is filtered and degassed before being extruded through a spinneret into a coagulation bath, typically containing sulfuric acid. The coagulation process regenerates the cellulose, forming solid fibers [33,102].

After spinning, the fibers undergo several post-treatment processes, including washing to remove residual chemicals, bleaching to achieve desired whiteness, and finishing treatments to enhance fiber properties [50].

The production of modal fibers is considered more environmentally friendly compared to traditional viscose rayon. This is primarily due to the closed-loop process used by manufacturers like Lenzing, which recycles chemicals and reduces waste.

Beech wood, the primary source of cellulose for modal fibers, is typically sourced from sustainably managed forests, ensuring a renewable supply of raw materials [103,104].

The production process includes the efficient recovery and recycling of chemicals, such as sodium hydroxide and carbon disulfide, minimizing environmental impact [91].

The production of modal fibers for medical applications may involve additional purification steps to remove any residual chemicals and impurities that could cause adverse reactions. This can include more rigorous washing and bleaching processes. Special finishes or coatings may also be applied to improve antimicrobial properties [105].

2.4. Production Methods of Cupro Fibers

The primary raw material for cupro fibers is cotton linter, which is a short fiber left on the cottonseed after the longer cotton fibers have been removed. These linters are purified to remove impurities, resulting in high-purity cellulose [77].

The purified cellulose is dissolved in a cuprammonium solution, which is a mixture of copper(II) hydroxide and aqueous ammonia. This process creates a viscous solution that can be extruded to form fibers. The dissolution process is unique to cupro fibers and differs significantly from other cellulose regeneration methods, like the viscose process [55,106].

The cellulose solution is extruded through a spinneret into a coagulation bath, where the cellulose is regenerated, forming solid fibers. The bath typically contains a dilute sulfuric acid solution, which precipitates the cellulose from the solution.

Cupro fibers undergo post-treatment processes that involve washing to eliminate residual chemicals, bleaching for the desired whiteness, and further treatments to improve fiber properties. The fibers are also stretched to align the cellulose molecules, which improves their strength and uniformity.

The production of cupro fibers involves the use of copper and ammonia, which require careful handling and disposal to minimize environmental impact. Modern cupro production processes include systems for recovering and reusing these chemicals to improve sustainability [107].

To ensure safety and sterility, the manufacturing process for medical-grade cupro fibers includes additional steps to eliminate any residual copper ions and other potentially harmful substances. Enhanced purification and finishing processes are employed to achieve the required medical standards [108].

2.5. Production Methods of Acetate Fibers

The main raw materials for acetate fibers are wood pulp and cotton linters. These materials are purified to remove lignin, hemicellulose, and other impurities, yielding high-purity cellulose [109].

The purified cellulose is then subjected to acetylation, where it is reacted with acetic acid and acetic anhydride in the presence of a catalyst (usually sulfuric acid). This reaction converts the hydroxyl groups in the cellulose into acetyl groups, forming cellulose acetate [110].

The cellulose acetate is then dissolved in a suitable solvent, typically acetone, to form a viscous solution. This solution is prepared for the spinning process, where it will be extruded to form fibers [111].

The spinning dope (cellulose acetate solution) is extruded through a spinneret into a coagulation bath or directly into the air (dry spinning). In the coagulation bath, the solvent is removed, and the cellulose acetate fibers solidify. In dry spinning, the solvent evaporates as the fibers solidify [112].

After spinning, the fibers undergo post-treatment processes such as washing to remove residual solvents, stretching to orient the polymer chains, and finishing treatments to enhance properties like dyeability and luster. The production of acetate fibers involves the use of solvents and chemicals that require careful handling and disposal to minimize environmental impact. Advances in solvent recovery and recycling technologies have helped to improve the sustainability of acetate fiber production. Sustainability efforts include sourcing cellulose from sustainably managed forests and using eco-friendly solvents and catalysts during the acetylation and dissolution processes [6].

Modern production processes include systems for recovering and reusing solvents like acetone, which reduces waste and environmental impact. Additionally, the development of biodegradable acetate fibers contributes to their environmental sustainability [113,114].

The production of acetate fibers for medical textiles includes a rigorous purification process to remove any acetic acid and other residual chemicals. Additional treatments may be applied to impart antimicrobial properties, which are crucial for preventing infections in medical settings. This often involves coating the fibers with antimicrobial agents [9].

In recent years, researchers have investigated alternative solvents for cellulose dissolution, with substantial research and development concentrated on a category of solvents known as ionic liquids (ILs). These solvents are considered environmentally sustainable and have demonstrated suitability for spinning regenerated cellulose fibers [115].

Ionic liquids (ILs) are salts that melt below 100 °C and exhibit unique properties such as low vapor pressure, high thermal stability, and a strong ability to dissolve various organic and inorganic substances. These liquids dissolve cellulose by disrupting its hydrogen bonding network, a process that does not require derivatization. Commonly, the ILs used for this purpose feature cations like imidazolium, pyridinium, or ammonium, which are paired with anions such as chloride or acetate. Once dissolved, the cellulose can be regenerated into fibers through a coagulation bath. Often regarded as green solvents, ILs are non-volatile and recyclable, although their cost and biodegradability can vary widely [116].

Another promising process for cellulose dissolution is the urea/alkali process. The urea/alkali process, utilizing a mixture of urea and sodium hydroxide, offers a promising method for cellulose dissolution. In this process, urea serves as a hydrogen bond acceptor, effectively disrupting the hydrogen bonds within cellulose to facilitate dissolution, typically at temperatures below 0 °C, which enhances energy efficiency. This method not only stabilizes the cellulose solution but also is environmentally friendly due to the non-toxic nature of the solvents. However, to achieve commercial viability, this process still requires the optimization of conditions and parameters to enhance the mechanical properties of the fibers. Compared to ionic liquids, scaling this process for large-scale industrial applications presents challenges in terms of efficiency and scalability.

The CarbaCell process is a cellulose dissolution method that uses organic carbonates, such as dimethyl carbonate (DMC), as the primary solvent. This process is noted for its low toxicity and environmental friendliness, as DMC is a biodegradable and less hazardous solvent compared to traditional solvents like carbon disulfide (CS2) used in the viscose process. The CarbaCell process dissolves cellulose by activating the hydroxyl groups of the cellulose molecules with organic carbonates, which facilitate the breakdown of intermolecular hydrogen bonds within cellulose. The resulting cellulose solution can be regenerated into fibers through processes such as wet or dry-jet wet spinning [55].

Each of these processes offers distinct advantages for cellulose dissolution and regeneration into fibers. The CarbaCell process, in particular, stands out due to its low environmental impact and mild operating conditions, utilizing low-toxicity organic carbonates that are both biodegradable and easily recoverable. This makes it an attractive option for sustainable fiber production [117]. Ionic liquids (ILs), on the other hand, are highly effective solvents for cellulose dissolution, but they can be costly and have varying environmental profiles depending on the specific IL used. While ILs provide efficient dissolution and can be recycled, their biodegradability and overall environmental impact need careful consideration [118]. The urea/alkali process is also a promising method because of its low toxicity and cost-effectiveness, making it a sustainable option for cellulose processing. However, it may face challenges in terms of scalability and efficiency compared to ILs and CarbaCell [119].

3. Properties of Main Types Regenerated Cellulose Fibers

Cellulose fibers stand out among all the textile fibers for their vast diversity in structure and properties. Natural cellulose fibers feature highly crystalline fibrillar structures arranged in various helical patterns, while fibrils in regenerated cellulose are less in ordered arrangement, and exhibit a wide range of structural variations. Chemically altered and regenerated through different processes, regenerated cellulose fibers, such as viscose rayon, modal, lyocell (Tencel), acetate, and cupro (cuprammonium), are widely used in the textile industry and for medical applications due to their unique properties. The structural diversity of regenerated cellulose fibers results in a broad spectrum of properties (see Table 2) and applications [54].

Table 2. Properties of cotton and some regenerated cellulose fibers [50,70,77,120–122].

Characteristic	Cotton	Viscose Rayon	Modal	Lyocell	Tencel	Cupro	Acetate
Density, g/cm^3	1.52	1.46–1.54	1.53	1.5	1.5	1.5	1.29–1.33
Tenacity, cN/tex	19.5–35	8.8–24	34	30–45	36–44	9–28	12.8
Moisture regain, %	7–8	11–14	11.8	10–13	11	11–12.5	6–7
Elongation, %	7–14	17–30	12	12–18	16–18	6–25	24
Crystallinity, %	60–70	30–40	40–48	50–65	50–65	30–40	20–30
Circularity of fiber cross-section *	0.4–0.8	0.5–0.8	closer to 1	approaching 1	approaching 1	0.7–0.9	Lower circularity

Note: *—where 1 represents a perfect circle.

Regenerated cellulose fibers are composed of cellulose—a natural polymer of glucose. The chemical properties of regenerated cellulose fibers are influenced by the degree of substitution and the method of regeneration, e.g., viscose rayon fibers often feature a lower degree of polymerization and higher susceptibility to degradation compared to other regenerated cellulose types, and lyocell excels in improved mechanical and chemical properties, while modal exhibits enhanced strength and durability properties if compared to viscose [43].

The regenerated cellulose fibers are highly hydrophilic because of the hydroxyl groups (–OH) on the cellulose chain. This property enhances their ability to absorb and retain moisture and makes them suitable for medical applications like wound dressing and surgical sutures where moisture management is a key parameter [122,123].

Regenerated cellulose fibers exhibit moderate chemical stability. They can be affected by strong acids and alkalis, but they generally maintain their integrity in physiological conditions, which is advantageous for medical use [123]. Regenerated cellulose fibers are insoluble in water, but can be dissolved in certain solvents, e.g., lyocell dissolves in cuprammonium hydroxide and N-Methylmorpholine N-oxide (NMMO), and viscose dissolves in sodium hydroxide. The solubility characteristic is exploited in the fiber production process [15,124,125]. But these fibers exhibit limited solubility in biological fluids, which makes them suitable for medical applications, like wound dressing [126].

The fibers of regenerated cellulose decompose naturally under environment conditions due to the presence of cellulose. This property is particularly beneficial for medical applications as post-use disposal is a consideration [126]. The biocompatibility of regenerated cellulose fibers is a key property of medical applications [127].

Regenerated cellulose fibers decompose at temperatures typically ranging from 200 °C to 300 °C. The exact decomposition temperature can vary depending on the fiber type and degree of polymerization, e.g., lyocell fibers usually have a higher thermal stability compared to viscose [128]. The glass transition temperature (Tg) of regenerated cellulose fibers is influenced by their molecular weight and crystallinity. These fibers generally exhibit a Tg around 250 °C [129,130].

Regenerated cellulose fibers each have distinct properties, which make them versatile for various textile applications, from everyday clothing to high-end fashion. Influenced by their production processes and chemical treatments, the following properties can be observed [32,120,131]:

- Viscose rayon: Softness, good dyeability, and moderate strength.
- Modal: Stronger, more durable, and excellent moisture management.
- Lyocell (Tencel): High strength, environmentally friendly, and excellent moisture control.
- Acetate: Luxurious feel, moderate moisture absorption, and lower strength.
- Cuprammonium (Cupro): Silk-like texture, good drape, and moderate strength.

3.1. Viscose Rayon Fibers

Viscose rayon, a type of regenerated cellulose fiber, exhibits lower crystallinity compared to natural cellulose fibers like cotton. The crystallinity of viscose rayon depends on the specific manufacturing conditions and post-treatment processes. The process disrupts the highly ordered crystalline structure of the original cellulose, leading to a more amorphous arrangement in the resulting fibers. However, the lower crystallinity also means that viscose rayon fibers are generally less strong and durable compared to natural cellulose fibers like cotton, especially when wet, and they may exhibit lower resistance to mechanical and environmental stresses [77,132].

The properties imparted by this level of crystallinity in viscose rayon include the following [58,132]:

- Softness and drapability: The lower crystallinity contributes to a softer and more drapable fabric, making viscose rayon ideal for clothing and textiles that require a fluid drape.
- Enhanced dyeability: The increased amorphous regions in viscose rayon fibers allow for better dye penetration, resulting in vibrant and uniform colors.
- Moisture absorption: Viscose rayon has good moisture-wicking properties, similar to cotton, providing comfort in clothing.
- Versatility: Its blend of properties makes viscose rayon suitable for a wide range of applications, from fashion garments to home textiles.

3.2. Modal Fibers

Modal is a type of regenerated cellulose fiber similar to viscose rayon but produced through a modified process that results in higher crystallinity and improved properties.

The increased crystallinity of modal fibers imparts several advantageous properties [133,134]:
- Strength and durability: Higher crystallinity leads to increased tensile strength and durability, making modal fibers more robust and long-lasting. Modal fibers are stronger and more durable than viscose rayon, both wet and dry.
- Softness and smoothness: Despite the higher crystallinity, modal fibers retain a smooth, silky texture and soft hand feel, which is ideal for high-quality textiles and apparel, often used in premium fabrics.
- Dimensional stability: Modal fibers exhibit better dimensional stability, maintaining their shape and size after repeated washing and drying.
- Moisture absorption: Modal fibers have excellent moisture-wicking properties, providing superior comfort and breathability in clothing. Modal has excellent moisture-wicking properties, superior to cotton and viscose.
- Color retention: The enhanced structure allows modal fibers to hold dyes well, resulting in vibrant, long-lasting colors.

These properties make modal fibers a popular choice for a variety of textile applications, including underwear, activewear, and bed linens, where both comfort and durability are desired.

3.3. Lyocell (Tencel®) Fibers

Lyocell has similar strength as polyester and it is stronger than cotton or all other man-made staple cellulosic fibers. It also has a very high dry and wet modulus in both the dry and wet states. All man-made cellulosic fibers lose strength and modulus when wetted, but lyocell reduces by much less than the others. However, the fibers do fibrillate during wet abrasion and thus, specific finishing techniques are required to achieve the best results.

Tencel® can be processed via the established yarn manufacturing routes, using conventional machinery with few major changes to settings or procedures. Tencel® possesses a non-durable crimp; it has a high modulus and there is little fiber entanglement. Thus, Tencel® will open easily with little nep and yield yarns with high tensile strength and few imperfections. Tencel® blends well with other fibers, especially other cellulosic fibers. It adds strength to the final yarn and enhances the performance and aesthetic values of the final fabrics [54].

Lyocell is a type of regenerated cellulose fiber known for its high crystallinity and eco-friendly production process. The higher crystallinity is attributed to the solvent-spinning process used in its production, which involves dissolving cellulose in a non-toxic solvent (N-methylmorpholine N-oxide or NMMO) and then regenerating the cellulose fibers in a coagulation bath. The higher crystallinity of lyocell/Tencel® fibers results in several beneficial properties [93,135,136]:
- Strength and durability: The high degree of crystallinity provides lyocell fibers with exceptional tensile strength and durability, making them more resilient to wear and tear. The value of lyocell fiber tenacity is larger than for viscose and modal fibers, and is almost equal to polyester fiber. Lyocell is the only regenerated cellulose fiber with a wet tensile strength reaching the cotton wet strength. Lyocell has a significantly reduced elongation compared to viscose, but slightly above modal fibers.
- Softness and comfort: Despite their strength, lyocell fibers are also known for their smooth, soft texture, which enhances comfort in textiles and apparel. Lyocell fibers feature a fibrillar structure with microfibrils aligned parallel to the fiber axis because of a high degree of cellulose crystallinity, which allows lyocell fiber to easily develop a fibrillated surface under mechanical abrasion. Due to the high cellulose crystallinity produced via lyocell spinning, the moisture regain of lyocell fiber is slightly lower than for viscose.
- Moisture management: Lyocell fibers exhibit excellent moisture-wicking abilities, which help in maintaining dryness and comfort, making them suitable for activewear and intimate apparel.

- Biodegradability: The natural cellulose base and environmentally friendly production process contribute to the biodegradability of lyocell fibers, making them a sustainable choice.
- Sustainability: The solvent used in the production process is non-toxic and is recycled in a closed-loop system, making Tencel a more sustainable and environmentally friendly fiber.
- Versatility: The combination of strength, softness, and moisture management makes lyocell suitable for a wide range of applications, from fashion to home textiles.
- Environmental impact: The production process is more sustainable using a closed-loop system that recycles solvents.

These properties, driven by the high crystallinity of lyocell, contribute to its popularity as a premium, sustainable textile material.

3.4. Cuprammonium (Cupro) Fibers

Cuprammonium fibers, commonly known as cupro, are a type of regenerated cellulose fiber produced using the cuprammonium process. Regarding their crystallinity, cupro fibers generally exhibit lower crystallinity compared to natural cellulose fibers but higher crystallinity than some other types of regenerated cellulose fibers, such as viscose rayon. This intermediate level of crystallinity imparts cupro fibers with several desirable properties, including a silky texture, good drape, and a soft hand feel. Additionally, the lower crystallinity compared to natural fibers results in improved dyeability and a smoother surface, which enhances their suitability for high-quality textiles and apparel applications [58,137]:

- Softness and smoothness: Cupro fibers are known for their silk-like feel and smooth texture.
- Moisture absorption: Good moisture-wicking properties.
- Drape: Excellent drape and fluidity.
- Strength: Generally weaker compared to lyocell, but stronger than some types of viscose.

3.5. Cellulose Acetate Fiber

Acetate fibers, also known as cellulose acetate, typically exhibit lower crystallinity compared to other cellulose fibers like cotton, modal, or lyocell. It is no longer considered a regenerated cellulose fiber because the polymer formula to form acetate fiber is acetate (cellulose ester) instead of cellulose. The relatively low crystallinity is due to the chemical modification of cellulose during the production process. In this process, cellulose is acetylated by reacting it with acetic anhydride, resulting in cellulose acetate. This modification reduces the ability of the cellulose chains to align and form crystalline regions, leading to a more amorphous structure.

The properties imparted by this level of crystallinity in acetate fibers include the following [58]:

- Softness and drapability: The low crystallinity contributes a soft, smooth texture and excellent drapability, making acetate suitable for lightweight, flowy fabrics. Acetate fibers have a luxurious feel and a high sheen, and excellent drapability and fluidity.
- Sheen and luster: Acetate fibers have a natural sheen and luster, giving fabrics made from acetate an attractive, silky appearance.
- Moisture absorption: While not as absorbent as more crystalline cellulose fibers, acetate still has moderate moisture absorption, providing some level of comfort. It is moderate, though less effective compared to cotton and lyocell.
- Color retention: Acetate fibers take dyes well and retain color vibrantly, which is beneficial for fashion and decorative textiles.
- Resistance to shrinkage and wrinkling: The chemical structure of acetate fibers provides good resistance to shrinkage and wrinkling, enhancing their durability and ease of care.
- Strength: Lower tensile strength compared to other regenerated cellulose fibers.

However, the lower crystallinity also means that acetate fibers are generally less strong and durable compared to higher crystallinity cellulose fibers. They may also have lower resistance to heat and chemical damage.

4. Medical Applications of Main Regenerated Cellulose Fibers

Medical textiles are one of the fastest-growing sectors of technical textiles. The product range in this sector encompasses a wide array of items, spanning from disposable healthcare and hygiene products (e.g., napkins and diapers) to specialized products including operating room textiles, sutures, scaffolds, and sensors. The selection of fibers and fabrics for these applications varies significantly based on the specific property requirements of the medical textile raw materials. Medical textile products demand biocompatibility (biostability or biodegradability), non-toxicity and nonallergenic, absorbency and softness, elasticity and flexibility, and relative mechanical properties (tenacity, strength, durability, and sterilizability) [138].

Medical textile products are classified into four categories: (1) non-implantable materials, (2) implantable materials, (3) extracorporeal devices, and (4) healthcare and hygiene products [139–142].

Non-implantable materials are designed exclusively for external application in the human body. Their primary functions include protecting the skin from external infections, absorbing body fluids, or delivering medication to damaged skin. Non-implantable materials encompass wound dressings, plasters, bandages, gauze, and compression bandages, among others. These materials are essential for wound healing, providing protection against infection, and absorbing blood and exudates. These products are typically manufactured from various types of textile fibers, selected based on specific application requirements and end uses. [142–144]

The inherent comfort of cellulose-based materials like cotton and regenerated cellulosic fibers (viscose rayon) makes them ideal for non-implantable textile products (see Table 3). The porous structure of cellulose-based fibers provides excellent moisture-wicking capabilities, and it also has unique characteristics such as biodegradability and ease of dyeing, in addition to its being soft and comfortable [145,146]. Regenerated cellulose fibers are widely used in non-implantable materials due to their versatile properties, such as softness, absorbency, and environmental sustainability. These fibers are applied in various non-medical and industrial contexts.

Table 3. Regenerated cellulose fibers used for non-implantable materials [147,148].

Type of Regenerated Cellulose Fibers Used	Application Areas
Viscose and lyocell	Absorbent pad for wound care
Viscose, lyocell, and modal	Wound-contact layer
Viscose	Base material for wound care
Viscose and lyocell	Base material for pads and bandages
Viscose and lyocell	Simple bandages
Viscose and lyocell	High-support bandages
Viscose and lyocell	Compression bandages
Viscose and lyocell	Orthopedical bandages
Viscose	Plasters
Viscose and lyocell	Gauze dressing
Viscose and cotton linters	Wadding
Cotton linters	Virus removal filter

Regenerated cellulose fibers are versatile and used in a wide range of non-implantable materials:

- Viscose rayon: Common in textiles and nonwovens for its softness and absorbency. Viscose rayon fibers are used as wound dressings due to their absorbency and comfort and are employed in surgical drapes and gowns only in non-critical settings for their softness and cost-effectiveness.
- Lyocell (Tencel): Used in apparel and home textiles for its strength and moisture management. Lyocell fibers are advanced wound dressings due to biocompatibility and moisture management. As materials for surgical gowns and drapes, lyocell fibers are used in high-performance medical textiles.
- Modal: Employed in underwear and towels for its softness and absorbency. Modal fibers are used in high-performance medical textiles such as patients' gowns and beddings, and for wound dressings due to softness and moisture management.
- Acetate: Used in fashion fabrics and linings for its luster and drape. Used in some medical linens for its softness.
- Cuprammonium fibers (cupro): Applied in luxury textiles and some technical fabrics for their softness and breathability. Used in high-end medical linens due to their softness—for luxury medical linens. Emerging applications due to their sustainability and performance for medical textiles.

Implantable material textiles are materials designed for use within the human body or beneath the skin. The primary requirement for these applications is biocompatibility, versatility, and suitability for medical applications. These materials are generally used for repairing damaged internal organs, skin, or wounds, particularly during surgery. They can be categorized into soft tissue implants, orthopedic implants, and cardiovascular implants. Soft tissue implants are used in areas such as ligaments and cartilage. Orthopedic implants function as artificial bones, and cardiovascular implants are employed in heart valves and vascular grafts (see Table 4). Surgical sutures, which are commonly used to close skin wounds, also fall under this category. The fibers used in these applications range from natural materials like silk and biodegradable polymers such as chitin, collagen, and chitosan, to synthetic polymers including polyester, polyamide, polyethylene, and polytetrafluoroethylene [143,144,147].

Table 4. Regenerated cellulose fibers used for implantable materials [53,143,144,147].

Type of Regenerated Cellulose Fibers Used	Application Areas
Viscose	sutures
Viscose, lyocell, and Ioncell	surgical meshes, e.g., for hernia
Lyocell, cupro, and Ioncell	scaffolds for tissue engineering
Modal	surgical dressings
Acetate	controlled drug release systems
Acetate	biodegradable meshes
Cupro	surgical textiles

Regenerated cellulose fibers are employed in various implantable materials due to their properties:

- Viscose rayon: Used in sutures and surgical meshes for its absorbability and biocompatibility, but in general, the use of viscose is limited and usually not used for implantable materials due to lower durability and potential for degradation.
- Lyocell (Tencel): Ideal for tissue engineering scaffolds and surgical meshes due to its strength and biocompatibility. Lyocell fibers are utilized in scaffolds for tissue regeneration.
- Modal: Applied in implantable textiles and surgical dressings for its softness and flexibility. Generally not used for implantable materials or is limited in use.

- Acetate: Utilized in controlled release systems and biodegradable meshes for its biodegradability and film-forming capabilities. Limited used, not typically used for implantable materials due to lower durability.
- Cuprammonium fibers (cupro): Used in tissue engineering and surgical textiles for its softness and smoothness. Rarely used for implantable materials—limited use.
- Ioncell: Investigated for advanced applications in tissue engineering and surgical meshes due to its high strength and biocompatibility. Potential for implantable materials because of biocompatibility.

These fibers provide various benefits in implantable materials, enhancing performance and patient outcomes.

Analogous to implantable textiles, extracorporeal devices are implantable artificial organs composed of textile materials. These artificial organs are utilized within the human body to support vital organ functions, including artificial kidneys, livers, and lungs. These devices are used to replace or support the function of diseased organs. The primary function of these artificial organs is to purify the blood during bodily functions through processes such as dialysis, filtration, and adsorption [93]. Mostly synthetic fibers are used for such products, but regenerated fibers are employed as well (see Table 5).

Table 5. Regenerated cellulose fibers used for extracorporeal devices [54,144,147,149].

Type of Regenerated Cellulose Fibers Used	Application Areas	Function
Hollow viscose	artificial kidney	remove waste products from patients' blood
Hollow viscose	artificial liver	separate and dispose of patients' plasma, and supply fresh plasma
Viscose, lyocell, and modal	hemodialysis membranes	the selective filtration of waste products from the blood
Viscose, lyocell, modal, and acetate	peritoneal dialysis	to facilitate the exchange of waste products and electrolytes through the peritoneal membrane
Viscose, lyocell, and modal	plasma filters	for removing proteins, toxins, and other unwanted substances from the blood
Viscose, lyocell, and modal	dialyzer units	for both hemodialysis and hemofiltration

Regenerated cellulose fibers have been employed in extracorporeal devices due to their favorable properties, such as biocompatibility, high absorbency, and structural integrity. Extracorporeal devices are medical devices that perform functions outside the body, such as in dialysis machines and blood filtration systems.

Regenerated cellulose fibers are crucial in extracorporeal devices due to their properties:

- Viscose rayon: Used in dialysis membranes and blood filtration due to high absorbency and biocompatibility, but general viscose fibers are not commonly used in extracorporeal devices.
- Lyocell (Tencel): Preferred for its strength and excellent biocompatibility in dialysis and blood filtration. The protectional applications of lyocell fibers for emerging uses in extracorporeal devices due to biocompatibility.
- Modal: Applied in specific blood filtration systems with good moisture absorption, but in general, not commonly used in extracorporeal devices or in limited use.
- Acetate: Utilized in niche applications for its softness and moderate moisture management. Not commonly used or in limited application.

- Cuprammonium fibers (cupro): Employed in high-performance dialysis membranes and specialized blood filtration systems due to their smooth texture. Limited application in extracorporeal devices, not commonly used.
- Ioncell: Emerging uses—potential applications in extracorporeal devices.

These fibers offer different advantages depending on the specific requirements of the extracorporeal devices, enhancing performance and patient safety.

Healthcare and hygiene products are designed to protect both medical personnel and patients from infections. They can be either washable or disposable after use. This category includes items ranging from personal protective clothing to disposable masks. Additionally, personal hygiene products such as napkins, tissues, hospital bed linens, uniforms, and diapers fall under this category. Using regenerated cellulose fibers in healthcare and hygiene products includes face masks, personal protective equipment (PPE), and sanitary products. The COVID-19 pandemic significantly increased the demand for these products. The primary requirements for these textiles are that they must be nonallergenic, non-toxic, and non-carcinogenic. Nonwoven materials, often made from regenerated cellulose fibers, have been widely used due to their breathability, strength, and bacterial resistance [144,148,150,151].

Cellulose-based materials like cotton and regenerated cellulosic fibers (viscose rayon) are predominantly used in medical applications due to their high moisture absorbency and comfort properties, which makes them ideal for hygienic and healthcare clothing (see Table 6). The porous structure of cellulose-based fibers provides excellent moisture-wicking capabilities, along with a soft handle and drape, which identifies it as a skin-friendly material [145,146].

Table 6. Regenerated cellulose fibers used for healthcare/hygiene products [54,144,147].

Type of Regenerated Cellulose Fibers Used	Application Areas	Structure of the Fabric
Viscose	Surgical caps	Nonwoven
Viscose	Surgical masks	Nonwoven
Superabsorbent fibers and wood fluff, modal, lyocell, and acetate	Absorbent layer for incontinence diaper/sheet	Nonwoven
Viscose and lyocell, modal, and cupro	Surgical swabs, drapes, and cloths/wipes,	Nonwoven

Regenerated cellulose fibers are extensively used in healthcare and hygiene products due to their excellent absorbency, softness, and biocompatibility. These fibers are commonly found in items such as wound dressings, surgical drapes, hygiene products, and more.

A summary of the medical application of conventional regenerated cellulose fibers is presented in Table 7.

Table 7. Medical application of conventional regenerated cellulose fibers.

Fiber Type	Primary Medical Application	Advantages	Challenges	References
Viscose rayon	Wound care and surgical sponges	High absorbency and cost-effective	Residual chemicals from production	[21]
Cupro	Bandages and medical textiles	Soft and hypoallergenic	Copper residue may require further purification	[152]
Modal	Patient garments and hospital bedding	Durable and moisture-wicking	Moderate cost	[20]
Lyocell	Advanced wound dressings and tissue scaffolds	Biodegradable and high absorbency	Limited production capacity	[153]
Acetate	Medical packaging and drug delivery systems	Good barrier properties and biocompatible	Low absorbency	[154]

Regenerated cellulose fibers offer numerous benefits for healthcare and hygiene products:
- Viscose rayon: Used in wound dressings, surgical drapes, and hygiene products for its absorbency and softness, and because of its comfort and absorbency, viscose fibers are utilized in hygiene products, such as sanitary pads and medical linens.
- Lyocell (Tencel): Applied in advanced wound care and hygiene products for its moisture management and biocompatibility. These fibers are used for hospital bedding and garments because of their comfort and performance.
- Modal: Used in hygiene products and medical textiles for its softness and absorbency; for sanitary pads and medical linens, these fibers are used for their high absorbency and comfort.
- Acetate: Utilized in wound dressings and some hygiene products for its softness and moisture management. Occasionally used in nonwoven hygiene products (sanitary products).
- Cuprammonium fibers (cupro): Employed in healthcare textiles and premium hygiene products for their softness and biocompatibility. Employed in certain high-end healthcare textiles.

These fibers provide various advantages in healthcare and hygiene applications, enhancing the performance and comfort of these products.

5. The Newest Regenerated Cellulose Fibers Used for Medical Applications

The newest regenerated cellulose fibers used for medical applications are typically advanced variations in traditional cellulose fibers, incorporating innovations in production techniques and functional properties to meet specific medical needs. Some latest developed regenerated cellulose fibers used for medical applications are as follows:
- Nanocellulose fibers: Fibers are produced by breaking down cellulose into nanoscale dimensions. This can be performed through chemical, mechanical, or enzymatic methods. These fibers exhibit extraordinary mechanical properties, high surface area, and biocompatibility. These properties contribute to their suitability for various biomedical applications, such as wound dressings, tissue engineering scaffolds, and drug delivery systems. Applications: used in wound dressings, drug delivery systems, and tissue engineering scaffolds due to their high strength and ability to support cell growth [155,156].
- Regenerated cellulose nanofibers (RCNFs): RCNFs are produced using advanced methods to extract and refine cellulose fibers to the nanometer scale. They offer enhanced mechanical properties, high surface area, and improved interactions with biological tissues. These properties enhance their suitability for various medical applications, including wound care, tissue engineering, and drug delivery systems. Application: utilized in advanced wound care products, tissue engineering, and as carriers for drug delivery systems [157,158].
- Bioactive cellulose fibers: Bioactive cellulose fibers are engineered to incorporate active agents such as antimicrobial or anti-inflammatory agents within the cellulose matrix. These fibers provide additional therapeutic benefits beyond the structural support of traditional cellulose. Bioactive cellulose fibers are designed to have specific properties that enhance their performance in medical applications. These fibers are often functionalized with bioactive agents to provide additional therapeutic benefits, such as antimicrobial properties or enhanced healing. These properties contribute to their effectiveness in various medical applications, such as wound care, drug delivery systems, and implants. Applications: employed in wound dressings, surgical sutures, and implants to enhance healing and reduce infection [159,160].
- Electrospun cellulose nanofibers: Electrospinning techniques are used to produce ultrafine cellulose fibers with diameters in the nanometer range. These fibers have high surface area–volume ratios, which are beneficial for medical applications. Electrospun cellulose nanofibers, produced using electrospinning techniques, possess specific properties that make them highly suitable for medical applications. These properties

are crucial for their performance in areas such as tissue engineering, wound healing, and drug delivery. These properties make electrospun cellulose nanofibers highly effective for applications in tissue engineering, wound care, and drug delivery systems. Applications: used in creating scaffolds for tissue engineering, drug delivery systems, and wound care products [161,162].

- Lyocell-like fibers with enhanced functionalization: Recent advancements in lyocell-like fibers involve functionalizing the fibers with additional properties such as enhanced biocompatibility, controlled drug release, or specific mechanical attributes tailored for medical use. Applications: employed in a range of medical textiles, including wound dressings, surgical gowns, and drug delivery systems [163].
- Ioncell is a relatively new type of regenerated cellulose fiber produced using an ionic liquid process. The circularity of Ioncell fibers is generally high, reflecting their more regular cross-sectional shape compared to other regenerated cellulose fibers. Ioncell is a type of regenerated cellulose fiber produced using an ionic liquid-based process. This innovative process offers several advantages over traditional methods. The crystallinity of Ioncell fibers is typically high, often ranging from 50% to 60%. This high crystallinity is achieved through the controlled dissolution and regeneration process, which promotes the formation of well-ordered crystalline regions within the fiber. Applications: Applied in advanced hygiene products and medical textiles for their strength and durability. Used in certain medical textiles (healthcare textiles) for their properties [164,165].

The main properties of some new regenerated cellulose fibers, described above, are presented in Table 8.

Table 8. Properties of some new regenerated cellulose fibers [155–165].

Characteristic	Nanocellulose Fiber	Regenerated Cellulose Nanofibers (RCNFs)	Bioactive Cellulose Fibers	Electrospun Cellulose Nanofibers	Lyocell-like Fibers	Ioncell
Density, g/cm^3	1.5–1.6	1.5–1.6	1.5–1.6	1.5–1.6	1.48–1.52	1.5
Tenacity, cN/tex	1×10^8–2×10^8	1×10^8–2×10^8	0.8×10^8–1.5×10^8	0.5×10^8–1.5×10^8	40–60	20.5–36.5
Moisture regain, %	10–15	10–15	8–15	8–15	10–15	10–12
Elongation, %	5–10	5–10	5–10	5–10	10–20	7–13
Crystallinity, %	70–90	70–90	60–90	70–90	40–55	50–60
Circularity of fiber cross-section *	Close to 1	Close to 1	Close to 1	Close to 1	0.8–1	0.7–0.9.

Note: *—where 1 represents a perfect circle.

The newest regenerated cellulose fibers for medical applications represent significant advancements in fiber technology, focusing on enhancing properties such as strength, biocompatibility, and functionality.

6. Recycling and Challenges of Regenerated Cellulose Fibers for Medical Applications

The industry of regenerated cellulose fibers, including materials like viscose and lyocell, has shown significant growth in various medical applications from 2020 to 2024. The medical textile market, including the segments mentioned above, was valued at approximately USD 24.7 billion in 2020 and is expected to grow at a compound annual growth rate (CAGR) of around 3% from 2021 to 2028. This growth is attributed to the rising demand for medical textiles in various applications, increasing healthcare awareness, and advancements in textile technology [144,150,151]

Overall, the industry for regenerated cellulose fibers in medical applications has seen robust growth driven by technological advancements, increasing healthcare needs, and a heightened focus on hygiene and safety, particularly due to the COVID-19 pandemic.

In general, all regenerated cellulosic fibers offer many advantages for medical applications, including biocompatibility, absorbency, and comfort. However, their use in medical settings comes with several challenges (see Table 9).

Table 9. Challenges of usage of regenerated cellulose fibers for medical applications [71,84,166–170].

Challenge	Description
Biocompatibility	Allergic reactions: Despite being generally biocompatible, some individuals may experience allergic reactions or sensitivities to regenerated cellulose fibers.
	Infection risk: Medical-grade regenerated cellulose fibers must be carefully processed to minimize the risk of infections or inflammatory responses.
Contamination and sterilization	Sterilization: Regenerated cellulose fibers used in medical applications must withstand sterilization processes (such as autoclaving or gamma irradiation) without degrading. Some fibers may degrade or lose their properties during these processes.
	Contamination: Ensuring the fibers are free from contaminants and pathogens is crucial, especially for applications like wound dressings and surgical materials.
Mechanical properties	Strength and durability: Some regenerated cellulose fibers may lack the mechanical strength and durability required for certain medical applications, such as surgical sutures and implants.
	Wear and tear: Medical textiles made from regenerated cellulose can be prone to wear and tear, which can impact their performance and safety.
Environmental impact	Environmental sustainability: The production of regenerated cellulose fibers involves the use of chemicals and processes that can have significant environmental impacts. Addressing the sustainability of these processes is crucial.
	Recycling: Recycling regenerated cellulose fibers, particularly those used in medical applications, poses challenges due to contamination and the need for specialized recycling systems.
Cost and economic viability	Cost: Regenerated cellulose fibers can be more expensive than other materials, particularly when high purity and specific properties are required for medical use.
	Economic viability: Balancing cost with the need for high-performance medical textiles can be challenging, especially in low-resource settings.
Process and quality control	Consistency and quality control: Ensuring consistency in fiber quality and performance is critical for medical applications, where variations can impact safety and efficacy.
	Manufacturing process: The complexity of manufacturing processes for regenerated cellulose fibers can affect their quality and performance.

The recycling of regenerated cellulose fibers used in medical applications is a significant area of research due to the increasing focus on sustainability and reducing environmental impact [71,166–168]. Regenerated cellulose fibers are valuable in medical textiles, but their disposal and recycling present challenges. RCF-based products have the potential to be recycled into new fibers or repurposed into other products, contributing to a circular economy in healthcare. This can help minimize the need for virgin resources and reduce overall waste.

Regulations on recycling regenerated cellulose fibers for medical purposes emphasize safety, environmental care, and the overall impact. These laws ensure that recycled fibers comply with rigorous health and safety standards for medical use, reducing risks to patients and medical staff. In most jurisdictions, regenerated cellulose fibers intended for direct or indirect use in or on the human body (e.g., wound dressings and surgical sutures) are classified as medical devices. This classification places them under stringent regulatory oversight. The classification levels are as follows:

- Class I (low risk): Basic dressings made from regenerated cellulose may fall under this classification.
- Class II (moderate risk): Absorbable regenerated cellulose sutures and hemostatic agents may fall under this classification.
- Class III (high risk): Regenerated cellulose materials used in critical applications, like internal implants, may require this classification.

Regulations also support sustainable recycling practices that cut waste and save resources while limiting environmental damage from the chemicals or byproducts of fiber processing. Ultimately, these regulations strive to balance advances in fiber recycling with strict standards for health, safety, and environmental responsibility. When used in medical products (like wound dressings or sutures), regenerated cellulose fibers must meet health standards set by authorities, such as the FDA in the U.S. and the EMA in Europe. These fibers also need to comply with strict biocompatibility and safety standards (e.g., ISO 10993) to ensure safe use in or on the human body. Manufacturers must implement quality control systems according to standards like ISO 13485 to meet medical device production standards [171].

For recycling and waste management, the EU's Waste Framework Directive mandates recycling and waste management practices for textiles, including cellulose fibers, to reduce landfill waste. In some regions, Extended Producer Responsibility laws require companies to manage their products' post-consumer waste, encouraging recycling and reuse programs for cellulose-based textiles. Regulations like the EU's REACH program restrict hazardous chemicals in fiber production to protect the environment and ensure chemical safety [172].

Countries like those in the EU promote sustainable textile policies, like a circular economy, including cellulose fibers, by supporting recycling and reuse, encouraging fiber innovation, and investing in recycling infrastructure. Regions such as the EU have established textile recycling targets and mandatory collection systems (beginning in 2025) to ensure cellulose-based textiles are recycled instead of landfilled [172].

These regulations guide the safe and sustainable production, use, and disposal of regenerated cellulose fibers, supporting both medical and environmental goals.

6.1. Recycling of Viscose Rayon Fiber

Viscose rayon can be recycled via chemical or mechanical processes. The chemical recycling process involves dissolving viscose rayon in a solvent and then reconstituting the cellulose into new fibers or products. This process is complex because it involves breaking down the fibers into their chemical components and then reprocessing them again into fibers. This method can be more effective in maintaining fiber quality but is costly. The mechanical recycling of viscose involves shredding the fibers and re-spinning them into new products, though this is less common for medical-grade viscose due to contamination issues. However, this process often results in quality degradation (loss of such properties as strength, flexibility, and comfort), which can be problematic for applications requiring high performance, such as medical textiles [89,173–175].

A specific challenge of viscose rayon recycling is that medical-grade viscose often comes into contact with bodily fluids and contaminants, complicating recycling processes as fibers need to meet strict hygiene and sterilization requirements. The effective sterilization of viscose rayon fibers is a significant challenge.

Innovations in the recycling technologies of viscose rayon fibers, such as enzyme-based recycling methods, might potentially improve the efficiency and effectiveness of recycling viscose fibers. Developing closed-loop recycling systems where viscose fibers are continuously recycled within the same supply chain could enhance sustainability and reduce waste [173,174].

6.2. Recycling of Lyocell (Tencel®) Fibers

Lyocell fibers are well suited to closed-loop recycling systems where the fibers are dissolved and re-spun into new lyocell fibers. This process is relatively efficient and retains the fiber's original properties. Similarly to closed-loop systems, lyocell can be chemically recycled by dissolving and regenerating cellulose. Lyocell fibers can be mechanically recycled as well; that is, fibers are ground into smaller pieces and then re-spined. But, the recycling of lyocell fibers faces some challenges. The closed-loop process can be more expensive compared to chemical and mechanical recycling methods. Ensuring that chemically recycled lyocell maintains the same quality as virgin fibers is a key challenge. However, this method is more complex and expensive compared to mechanical recycling, but the mechanical recycling method can cause degradation in fiber quality, which may be problematic for medical applications where fiber integrity is crucial [15,50,91,175].

Specific challenges for medical applications for recycled lyocell and Tencel® fibers are that they must meet stringent hygiene and sterilization standards for medical textiles and the recycling process needs to ensure that fibers are free from contaminants and can be properly sterilized.

As for lyocell fibers, an innovation in the recycling process, such as enzyme-based recycling and more efficient chemical processes, could improve the effectiveness and efficiency of recycling lyocell and Tencel® fibers. Implementing closed-loop recycling systems, where fibers are continuously recycled within the same supply chain, could enhance sustainability and reduce waste [175,176].

6.3. Recycling of Modal Fibers

Modal fibers can be recycled chemically or mechanically. Chemical recycling is similar to lyocell, where they are dissolved and reconstituted. While this method can preserve the quality of fibers better than mechanical recycling, it is complex and often more expensive. Modal fibers can be mechanically recycled into nonwoven fabrics or other products, but contamination from medical use can limit this method. As with other medical regenerated cellulose fibers, the presence of contaminants can complicate the recycling process. But the primary challenge is that mechanical recycling (shredding the fibers and re-spinning them into new yarns) can degrade the fiber quality, which is critical for high-performance applications such as medical textiles. Recycled modal fibers need to maintain their performance characteristics, such as strength, elasticity, and moisture management, which are important for medical textiles that must perform reliably in various conditions [71,175,177].

Advanced recycling technologies, such as enzyme-based recycling methods or closed-loop systems, can enhance the efficiency and effectiveness of recycling modal fibers, potentially addressing some challenges associated with traditional methods. Developing circular economy models for modal fibers, where fibers are continuously recycled within the supply chain, can improve sustainability and reduce waste [167].

6.4. Recycling of Cupro Fibers

As other regenerated cellulose fibers, described above, cupro fibers can be recycled in mechanical or chemical way. Mechanical recycling may lead to degradation in fiber quality due to shredding and re-spinning cupro fibers. The chemical recycling process can be complex and may have environmental implications due to the chemicals used. Specific challenges for the medical applications of recycled cupro fibers are the removal of contaminants and ensuring effective sterilization [71,168].

Advanced recycling technologies, such as solvent-based recycling and closed-loop systems, may improve the efficiency and sustainability of recycling cupro fibers. Developing circular economy models where cupro fibers are continuously recycled within the supply chain could enhance sustainability and reduce waste [178].

6.5. Recycling of Acetate Fibers

The mechanical recycling process of acetate fibers involves grinding the fibers and re-spinning them. This process is the same as for the regenerated cellulose fibers and can lead to a loss of fiber quality, which can be problematic in medical textiles. Also, it leads to challenges in removing contaminants, allowing the effective sterilization of the fibers, and ensuring that recycled acetate fibers maintain their initial properties, such as strength, flexibility, and moisture management [175,179,180].

As for cupro fibers, solvent-based recycling and improved chemical processes could enhance the efficiency and sustainability of recycling acetate fibers, and closed-loop systems where acetate fibers are continuously recycled within the supply chain could improve sustainability and reduce waste [179,180].

Regenerated cellulose fibers have become an exciting area of research and application in healthcare, driven by their biocompatibility, sustainability, and versatility. These fibers include varieties like lyocell, viscose, modal, cupro, and acetate, each bringing unique characteristics to medical textiles and biomaterials. Their potential to transform medical technologies is significant, particularly in areas such as tissue engineering, wound care, drug delivery, and sustainable healthcare solutions. A summary of the benefits and emerging technologies of regenerated cellulose fibers for medical applications is presented in Table 10.

Table 10. Challenges of usage of regenerated cellulose fibers for medical applications.

Category	Potential/Benefit	Emerging Technologies	Example	References
Tissue engineering and regenerative medicine	RCFs like lyocell and bacterial nanocellulose (BNC) serve as scaffolds due to their porosity, strength, and biocompatibility.	Three-dimensional bioprinting enables customizable tissue scaffolds for skin regeneration and bone tissue.	Lyocell is explored for bone scaffolds, reducing recovery times in orthopedic surgeries.	[181]
Smart wound dressings and wearable biosensors	Smart textiles with biosensors can monitor wound environment (pH, temperature, and moisture) in real time.	Wearable devices using RCFs with embedded electronics reduce the need for frequent inspections.	Lyocell-based smart wound dressings transmit real-time data to healthcare providers, reducing hospital visits.	[182]
Drug delivery systems	RCFs (e.g., cupro and lyocell) allow for localized, controlled drug release, improving patient compliance.	Drug-eluting fibers release therapeutic agents over time, aiding in chronic wound care.	Cupro fibers in curcumin-loaded dressings for diabetic ulcers promoting healing and offering sustained drug release.	[183]
Biodegradability and waste reduction	RCFs are biodegradable, breaking down without leaving harmful residues, and reducing medical waste.	RCFs in disposable medical textiles (e.g., gowns and bandages) help reduce plastic waste.	Lyocell gowns and drapes degrade faster than synthetic alternatives.	[184]
Sustainable production processes	RCFs like lyocell use closed-loop processes where chemicals are recovered and recycled, reducing environmental impact.	Closed-loop production recycles over 99% of the solvent used, making lyocell eco-friendly.	Lyocell fiber production is a model for sustainable practices, aligning with global healthcare eco-initiatives.	[185]

7. Conclusions

Regenerated cellulose fibers are increasingly being explored for medical applications due to their desirable properties, such as biocompatibility, biodegradability, and comfort. The future of these fibers in medical textiles is shaped by several emerging trends and technological advancements. Future advancements are likely to focus on enhancing closed-loop production processes to make them more efficient and cost-effective. This includes better solvent recovery systems and reduced environmental impact. The integration of nanotechnology can be carried out to create nanofibers with enhanced properties, such as increased surface area, antimicrobial activity, and improved mechanical strength.

While RCFs are known to be biocompatible, there is a lack of long-term clinical data on how these fibers perform over extended periods, especially in complex medical applications such as implants or tissue scaffolds. Research should focus on the long-term effects of RCFs on tissue regeneration, their degradation rates, and their mechanical stability in demanding environments such as bone scaffolds or vascular implants. Developing reinforced RCF composites that can withstand higher mechanical stress while maintaining biodegradability could open up new applications in orthopedics and cardiovascular surgery. The functionalization of RCFs (e.g., with antimicrobial agents or drug-loaded fibers) has shown promise, but cost-effective manufacturing processes for large-scale production remain a challenge. The high cost of incorporating agents like silver nanoparticles can limit the use of these advanced materials in lower-resource settings. Creating affordable antimicrobial RCF-based dressings that can be deployed in global healthcare markets, especially in areas with limited access to expensive wound care technologies, presents a significant opportunity. Smart textiles using RCFs with embedded biosensors have shown great potential, but their scalability and commercial viability are still in question. The integration of electronics into cellulose-based textiles poses challenges related to durability, data transmission, and cost. There is a need for clearer regulatory frameworks and standards for the use of RCF-based medical products, especially in emerging applications like tissue engineering scaffolds and drug delivery systems. Navigating the regulatory approval process can be slow and complex, delaying the commercialization of innovative RCF products.

Author Contributions: Conceptualization, S.V.-Ž. and J.B.-G.; formal analysis, S.V.-Ž. and J.B.-G.; investigation, S.V.-Ž. and J.B.-G.; resources, J.B.-G.; data curation, S.V.-Ž. and J.B.-G.; writing—original draft preparation, S.V.-Ž. and J.B.-G.; writing—review and editing, S.V.-Ž. and J.B.-G.; visualization, S.V.-Ž. and J.B.-G.; supervision, J.B.-G.; project administration, S.V.-Ž.; funding acquisition, J.B.-G. All authors have read and agreed to the published version of the manuscript.

Funding: This research received no external funding.

Data Availability Statement: The original contributions presented in the study are included in the article, further inquiries can be directed to the corresponding author.

Conflicts of Interest: The authors declare no conflicts of interest.

References

1. Molla, S.; Abedin, M.M.; Siddique, I.M. Exploring the versatility of medical textiles: Applications in implantable and non-implantable medical textiles. *World J. Adv. Res. Rev.* **2024**, *21*, 603–615. [CrossRef]
2. Schwarz, I.; Kovačević, S. Textile application: From need to imagination. In *Textiles for Advanced Applications*; IntechOpen: London, UK, 2017.
3. Hossain, M.T.; Shahid, M.A.; Limon, M.G.M.; Hossain, I.; Mahmud, N. Techniques, applications, and challenges in textiles for sustainable future. *J. Open Innov. Technol. Mark. Complex.* **2024**, *10*, 100230. [CrossRef]
4. Sherif, F.S. A New Prospects to Enhance the Commercial and Economical Status in Textile Industry. *Int. Des. J.* **2016**, *6*, 141–148. [CrossRef]
5. Bao, H.; Hong, Y.; Yan, T.; Xie, X.; Zeng, X. A systematic review of biodegradable materials in the textile and apparel industry. *J. Text. Inst.* **2024**, *115*, 1173–1192. [CrossRef]
6. Felgueiras, C.; Azoia, N.G.; Gonçalves, C.; Gama, M.; Dourado, F. Trends on the cellulose-based textiles: Raw materials and technologies. *Front. Bioeng. Biotechnol.* **2021**, *9*, 608826. [CrossRef]
7. Shirvan, A.R.; Nouri, A. Medical textiles. *Adv. Funct. Prot. Text.* **2020**, 291–333. [CrossRef]

8. Srinivasan, J.; Kalyana Kumar, M. Polypropylene for high performance medical applications. *Asian Text. J.* **2005**, *2*, 73–81.
9. Chang, C.; Ginn, B.; Livingston, N.K.; Yao, Z.; Slavin, B.; King, M.W.; Chung, S.; Mao, H.Q. Medical fibers and biotextiles. In *Biomaterials Science*; Academic Press: Cambridge, MA, USA, 2020; pp 575–600.
10. Sundarraj, A.A.; Ranganathan, T.V. A review on cellulose and its utilization from agro-industrial waste. *Drug Invent. Today* **2018**, *10*, 89–94.
11. Tofanica, B.M.; Belosinschi, D.; Volf, I. Gels, Aerogels and Hydrogels: A Challenge for the Cellulose-Based Product Industries. *Gels* **2022**, *8*, 497. [CrossRef]
12. Abdul Khalil, H.P.S.; Adnan, A.S.; Yahya, E.B.; Olaiya, N.G.; Safrida, S.; Hossain, M.S.; Balakrishnan, V.; Gopakumar, D.A.; Abdullah, C.K.; Oyekanmi, A.A.; et al. A review on plant cellulose nanofibre-based aerogels for biomedical applications. *Polymers* **2020**, *12*, 1759. [CrossRef]
13. Yue, Y.; Han, G.; Wu, Q. Transitional properties of cotton fibers from cellulose I to cellulose II structure. *BioResources* **2013**, *8*, 6460–6471. [CrossRef]
14. Zainul Armir, N.A.; Zulkifli, A.; Gunaseelan, S.; Palanivelu, S.D.; Salleh, K.M.; Che Othman, M.H.; Zakaria, S. Regenerated cellulose products for agricultural and their potential: A review. *Polymers* **2021**, *13*, 3586. [CrossRef] [PubMed]
15. Periyasamy, A.P.; Militky, J. Sustainability in regenerated textile fibers. In *Sustainability in the Textile and Apparel Industries: Sourcing Synthetic and Novel Alternative Raw Materials*; Springer: Berlin/Heidelberg, Germany, 2020; pp 63–95.
16. Ramamoorthy, S.K.; Skrifvars, M.; Persson, A. A review of natural fibers used in biocomposites: Plant, animal and regenerated cellulose fibers. *Polym. Rev.* **2015**, *55*, 107–162. [CrossRef]
17. Morris, H.; Murray, R. Medical textiles. *Text. Prog.* **2020**, *52*, 1e127. [CrossRef]
18. Tu, H.; Zhu, M.; Duan, B.; Zhang, L. Recent progress in high-strength and robust regenerated cellulose materials. *Adv. Mater.* **2021**, *33*, 2000682. [CrossRef]
19. Rostamitabar, M.; Seide, G.; Jockenhoevel, S.; Ghazanfari, S. Effect of Cellulose Characteristics on the Properties of the Wet-Spun Aerogel Fibers. *Appl. Sci.* **2021**, *11*, 1525. [CrossRef]
20. Pinho, E.; Soares, G. Functionalization of cotton cellulose for improved wound healing. *J. Mater. Chem. B* **2018**, *6*, 1887–1898. [CrossRef]
21. Ahmed, J.; Gultekinoglu, M.; Edirisinghe, M. Bacterial cellulose micro-nano fibres for wound healing applications. *Biotechnol. Adv.* **2020**, *41*, 107549. [CrossRef] [PubMed]
22. Miao, J.; Pangule, R.C.; Paskaleva, E.E.; Hwang, E.E.; Kane, R.S.; Linhardt, R.J.; Dordick, J.S. Lysostaphin-functionalized cellulose fibers with antistaphylococcal activity for wound healing applications. *Biomaterials* **2011**, *32*, 9557–9567. [CrossRef]
23. Ninan, N.; Muthiah, M.; Park, I.K.; Kalarikkal, N.; Elain, A.; Wong, T.W.; Thomas, S.; Grohens, Y. Wound healing analysis of pectin/carboxymethyl cellulose/microfibrillated cellulose based composite scaffolds. *Mater. Lett.* **2014**, *132*, 34–37. [CrossRef]
24. *ISO/TR 11827:2012*; Textiles—Composition Testing—Identification of Fibres. International Organization for Standardization: Geneva, Switzerland, 2012.
25. Gupta, B.S. Textile fiber morphology, structure and properties in relation to friction. In *Friction in Textile Material*; Woodhead Publishing: Cambridge, UK, 2008; pp 3–36.
26. Klemm, D.; Heublein, B.; Fink, H.P.; Bohn, A. Cellulose: Fascinating biopolymer and sustainable raw material. *Angew. Chem. Int. Ed.* **2005**, *44*, 3358–3393. [CrossRef] [PubMed]
27. Mark, R.E. Viscose fiber production and its historical development. *J. Text. Inst.* **2001**, *92*, 68–75.
28. Hämmerle, F.M. The cellulose gap (the future of cellulose fibers). *Lenzing. Berichte* **2011**, *89*, 12–21.
29. Shaikh, T.; Chaudhari, S.; Varma, A. Viscose rayon: A legendary development in the manmade textile. *Int. J. Eng. Res.* **2012**, *2*, 675–680.
30. Vartiainen, J.; Vähä-Nissi, M.; Harlin, A. Biopolymer films and coatings in packaging applications—A review of recent developments. *J. Appl. Polym. Sci.* **2017**, *134*, 44367. [CrossRef]
31. Kuen, T.H.; Lau, T.Y.; Ng, L.T. A Study on the Sterilization of Regenerated Cellulose Fiber for Medical Textiles. *J. Ind. Text.* **2018**, *47*, 1803–1820.
32. Jiang, G.; Huang, W.; Li, L.; Wang, X.; Pang, F.; Zhang, Y.; Wang, H. Structure and properties of regenerated cellulose fibers from different technology processes. *Carbohydr. Polym.* **2012**, *87*, 2012–2018. [CrossRef]
33. Fink, H.-P.; Weigel, P.; Purz, H.J.; Ganster, J. Structure formation of regenerated cellulose materials from NMMO-solutions. *Prog. Polym. Sci.* **2001**, *26*, 1473–1524. [CrossRef]
34. Kreze, T.; Malej, S. Structural characteristics of new and conventional regenerated cellulosic fibers. *Text. Res. J.* **2003**, *73*, 675–684. [CrossRef]
35. Wang, H.; Sixta, H.; Lorenz, M. Properties and structure of viscose fibers spun from NMMO-cellulose solution. *Cellulose* **2004**, *11*, 163–172.
36. Kim, H.G.; Bai, B.C.; In, S.J.; Lee, Y.S. Effects of an inorganic ammonium salt treatment on the flame-retardant performance of Lyocell fibers. *Carbon Lett.* **2016**, *17*, 74–78. [CrossRef]
37. Liu, X.-H.; Zhang, Y.-G.; Cheng, B.-W.; Ren, Y.-L.; Zhang, Q.-Y.; Ding, C.; Peng, B. Preparation of durable and flame retardant Lyocell fibers by a one-pot chemical treatment. *Cellulose* **2018**, *25*, 6745–6758. [CrossRef]
38. Sixta, H.; Michud, A.; Hummel, M.; Anghelescu-Hakala, A.; Haslinger, S.; Maunu, S. Ioncell-F: A High-strength regenerated cellulose fiber. *Nord. Pulp Pap. Res. J.* **2015**, *30*, 43–57. [CrossRef]

39. Mahltig, B.; Textor, T. Silver-containing antimicrobial coatings for textiles. *J. Mater. Chem.* **2011**, *21*, 19265–19273.
40. Shankar, S.; Rhim, J.W. Preparation of nanocellulose from micro-crystalline cellulose: The effect on antimicrobial activity. *Carbohydr. Polym.* **2018**, *200*, 163–172.
41. Sixta, H. *Handbook of Pulp*; Wiley-VCH: Hoboken, NJ, USA, 2006.
42. Shen, L.; Worrell, E.; Patel, M.K. Environmental impact assessment of man-made cellulose fibers. *Resour. Conserv. Recycl.* **2010**, *55*, 260–274. [CrossRef]
43. Zhang, S.; Chen, C.; Duan, C.; Hu, H.; Li, H.; Li, J.; Liu, Y.; Ma, X.; Stavik, J.; Ni, Y. Regenerated cellulose by the lyocell process, a brief review of the process and properties. *BioResources* **2018**, *13*, 4577–4592. [CrossRef]
44. Kim, H.S.; Yun, C.; Lee, D.S. Characterization of the mechanical properties of lyocell and viscose fibers. *J. Ind. Text.* **2016**, *45*, 875–890.
45. Müssig, J.; Haag, K. The tensile properties of lyocell made from wood pulp and from bamboo pulp fibers. *Compos. Appl. Sci. Manuf.* **2011**, *42*, 1185–1192.
46. Rosenau, T.; Potthast, A.; Sixta, H.; Kosma, P. The chemistry of side reactions and byproduct formation in the system NMMO/cellulose (Lyocell process). *Prog. Polym. Sci.* **2001**, *26*, 1763–1837. [CrossRef]
47. Eichhorn, S.J.; Baillie, C.A.; Zafeiropoulos, N.; Mwaikambo, L.Y.; Ansell, M.P.; Dufresne, A.; Entwistle, K.M. Review: Current international research into cellulose nanofibres and nanocomposites. *J. Mater. Sci.* **2001**, *36*, 2107–2131. [CrossRef]
48. Siro, I.; Plackett, D. Microfibrillated cellulose and new nanocomposite materials: A review. *Cellulose* **2021**, *17*, 459–494. [CrossRef]
49. Carrillo, F.; Colom, X.; Valldeperas, J.; Evans, D.; Huson, M.; Church, J. Structural characterization and properties of lyocell fibers after fibrillation and enzymatic defibrillation finishing treatments. *Text. Res. J.* **2003**, *73*, 1024–1030. [CrossRef]
50. Edgar, K.J.; Zhang, H. Antibacterial modification of Lyocell fiber: A review. *Carbohydr. Polym.* **2020**, *250*, 116932. [CrossRef]
51. Wang, X.; Wang, S.; Li, Y.; Jin, X.; Dong, C. Preparation and properties of low fibrillated antibacterial Lyocell fiber. *Arab. J. Chem.* **2024**, *17*, 105658. [CrossRef]
52. Parajuli, P.; Acharya, S.; Rumi, S.S.; Hossain, M.T.; Abidi, N. Regenerated cellulose in textiles: Rayon, lyocell, modal and other fibres. In *Fundamentals of Natural Fibres and Textiles*; Woodhead Publishing: Cambridge, UK, 2021; pp 87–110.
53. Lenzing, A.G. *Lenzing Modal®—The Fiber with a History of Innovation*; Lenzing AG: Lenzing, Austria, 2020; Available online: https://www.lenzing.com/ (accessed on 21 July 2024).
54. Woodings, C. *Regenerated Cellulose Fibres*; Woodhead Publishing: Cambridge, UK, 2001.
55. Shen, H.; Sun, T.; Zhou, J. Recent progress in regenerated cellulose fibers by wet spinning. *Macromo. Mater. Eng.* **2023**, *308*, 2300089. [CrossRef]
56. Zemljic, L.F.; Sauperl, O.; Kreze, T.; Strnad, S. Characterization of regenerated cellulose fibers antimicrobial functionalized by chitosan. *Text. Res. J.* **2013**, *83*, 185–196.e2. [CrossRef]
57. Kamide, K.; Nishiyama, K. Cuprammonium processes. *Regen. Cellul. Fibers* **2001**, *88*–155. [CrossRef]
58. Hearle, J.W. Physical structure and fibre properties. *Regen. Cellul. Fibres* **2001**, *18*, 199.
59. Seisl, S.; Hengstmann, R. Manmade Cellulosic Fibers (MMCF)—A Historical Introduction and Existing Solutions to a More Sustainable Production. In *Sustainable Textile and Fashion Value Chains*; Matthes, A., Beyer, K., Cebulla, H., Arnold, M.G., Schumann, A., Eds.; Springer: Cham, Switzerland, 2021. [CrossRef]
60. Das, M. Man-made cellulose fibre reinforcements (MMCFR). In *Biocomposites for High-Performance Applications*; Woodhead Publishing: Cambridge, UK, 2017; pp 23–55.
61. Oral, O.; Dirgar, E.; İlleez, A.A. Cupro fibers, properties and usage areas. In Proceedings of the 3rd International Congress of Innovative Textiles ICONTEX 2022, Izmir, Turkey, 18–19 May 2022; p. 26.
62. Chen, J.; Xu, J.; Wang, K.; Cao, X.; Sun, R. Cellulose acetate fibers prepared from different raw materials with rapid synthesis method. *Carbohydr. Polym.* **2016**, *137*, 685–692. [CrossRef]
63. Fischer, S.; Thümmler, K.; Volkert, B.; Hettrich, K.; Schmidt, I.; Fischer, K. Properties and applications of cellulose acetate. In *Macromolecular Symposia*; WILEY-VCH Verlag: Weinheim, Germany, 2008; Volume 262, pp 89–96.
64. Wsoo, M.A.; Shahir, S.; Bohari, S.P.M.; Nayan, N.H.M.; Abd Razak, S.I. A review on the properties of electrospun cellulose acetate and its application in drug delivery systems: A new perspective. *Carbohyd. Res.* **2020**, *491*, 107978. [CrossRef] [PubMed]
65. Gouda, M.; Hebeish, A.A.; Aljafari, A.I. Synthesis and characterization of novel drug delivery system based on cellulose acetate electrospun nanofiber mats. *J. Ind. Text.* **2014**, *43*, 319–329. [CrossRef]
66. Hu, J.; Li, R.; Zhu, S.; Zhang, G.; Zhu, P. Facile preparation and performance study of antibacterial regenerated cellulose carbamate fiber based on N-halamine. *Cellulose* **2021**, *28*, 4991–5003. [CrossRef] [PubMed]
67. Materials Market Report. Available online: https://textileexchange.org/app/uploads/2023/11/Materials-Market-Report-2023.pdf (accessed on 21 July 2024).
68. Kim, T.; Kim, D.; Park, Y. Recent progress in regenerated fibers for "green" textile products. *J. Clean. Prod.* **2022**, *376*, 134226. [CrossRef]
69. Wang, S.; Lu, A.; Zhang, L. Recent advances in regenerated cellulose materials. *Prog. Polym. Sci.* **2016**, *53*, 169–206. [CrossRef]
70. Shabbir, M.; Mohammad, F. Sustainable production of regenerated cellulosic fibres. In *Sustainable Fibres and Textiles*; Woodhead Publishing: Cambridge, UK, 2017; pp 171–189.
71. Ma, Y.; Nasri-Nasrabadi, B.; You, X.; Wang, X.; Rainey, T.J.; Byrne, N. Regenerated cellulose fibers wetspun from different waste cellulose types. *J. Nat. Fibers* **2021**, *18*, 2338–2350. [CrossRef]

72. De Silva, R.; Vongsanga, K.; Wang, X.; Byrne, N. Understanding key wet spinning parameters in an ionic liquid spun regenerated cellulosic fibre. *Cellulose* **2016**, *23*, 2741–2751. [CrossRef]
73. Hauru, L.K.; Hummel, M.; King, A.W.; Kilpeläinen, I.; Sixta, H. Role of solvent parameters in the regeneration of cellulose from ionic liquid solutions. *Biomacromolecules* **2012**, *13*, 2896–2905. [CrossRef]
74. Ma, Y.; Rosson, L.; Wang, X.; Byrne, N. Upcycling of waste textiles into regenerated cellulose fibres: Impact of pretreatments. *J. Text. Inst.* **2020**, *111*, 630–638. [CrossRef]
75. Gorade, V.; Chaudhary, B.; Parmaj, O.; Kale, R. Preparation and characterization of chitosan/viscose rayon filament biocomposite. *J. Nat. Fibers* **2022**, *19*, 1189–1200. [CrossRef]
76. Eid, B.M.; Ibrahim, N.A. Recent developments in sustainable finishing of cellulosic textiles employing biotechnology. *J. Clean. Prod.* **2021**, *284*, 124701. [CrossRef]
77. Chen, J. Synthetic textile fibers: Regenerated cellulose fibers. In *Textiles and Fashion*; Woodhead Publishing: Cambridge, UK, 2015; pp 79–95.
78. Lavanya, D.K.P.K.; Kulkarni, P.K.; Dixit, M.; Raavi, P.K.; Krishna, L.N.V. Sources of cellulose and their applications—A review. *Int. J. Drug Formul. Res.* **2011**, *2*, 19–38.
79. Wilkes, A.G. The viscose process. *Regen. Cellul. Fibres* **2001**, 37–61. [CrossRef]
80. Fechter, C.; Fischer, S.; Reimann, F.; Brelid, H.; Heinze, T. Influence of pulp characteristics on the properties of alkali cellulose. *Cellulose* **2020**, *27*, 7227–7241. [CrossRef]
81. Hon, D.N.S. Cellulose and its derivatives: Structures, reactions, and medical uses. In *Polysaccharides in Medicinal Applications*; Routledge: New York, NY, USA, 2017; pp 87–105.
82. Dyer, J.; Smith, F.R. New Methods for Studying Particles in Viscose and Their Effects on Viscose Rayon Production. In *Textile and Paper Chemistry and Technology*; Published Online 2009; ACS Symposium Series; American Chemical Society: Washington DC, USA, 1977; Volume 49, Chapter 1; pp 3–19.
83. Ozipek, B.; Karakas, H. Wet spinning of synthetic polymer fibers. In *Advances in Filament Yarn Spinning of Textiles and Polymers*; Woodhead Publishing: Cambridge, UK, 2014; pp 174–186.
84. El Seoud, O.A.; Kostag, M.; Jedvert, K.; Malek, N.I. Cellulose regeneration and chemical recycling: Closing the "cellulose gap" using environmentally benign solvents. *Macromol. Mater. Eng.* **2020**, *305*, 1900832. [CrossRef]
85. Mondal, M.I.H.; Ahmed, F.; Rahman, M.H. Surface Modification of Viscose Fabric with Silane Coupling Agents-An Aspect of Comfort Properties. *ChemistrySelect* **2024**, *9*, e202400727. [CrossRef]
86. Emam, H.E.; El-Hawary, N.S.; Ahmed, H.B. Green technology for durable finishing of viscose fibers via self-formation of AuNPs. *Int. J. Biol. Macromol.* **2017**, *96*, 697–705. [CrossRef] [PubMed]
87. Gogoi, N.; Bhuyan, S. Medical textiles: It's present and prospects. *J. Pharm. Innov.* **2020**, *9*. [CrossRef]
88. Durand, H.; Smyth, M.; Bras, J. Nanocellulose: A new biopolymer for biomedical application. *Biopoly. Biomed. Biotechnol. Appl.* **2021**, 129–179. [CrossRef]
89. Mendes, I.S.; Prates, A.; Evtuguin, D.V. Production of rayon fibres from cellulosic pulps: State of the art and current developments. *Carbohyd. Polym.* **2021**, *273*, 118466. [CrossRef]
90. Sixta, H.; Iakovlev, M.; Testova, L.; Roselli, A.; Hummel, M.; Borrega, M.; van Heiningen, A.; Carmen, F.; Schottenberger, H. Novel concepts of dissolving pulp production. *Cellulose* **2013**, *20*, 1547–1561. [CrossRef]
91. Jiang, X.; Bai, Y.; Chen, X.; Liu, W. A review on raw materials, commercial production and properties of lyocell fiber. *J. Bioresour. Bioprod.* **2020**, *5*, 16–25. [CrossRef]
92. Perepelkin, K.E.E. Lyocell fibres based on direct dissolution of cellulose in N-methylmorpholine N-oxide: Development and prospects. *Fibre Chem.* **2007**, *39*, 163–172. [CrossRef]
93. Sayyed, A.J.; Gupta, D.; Deshmukh, N.A.; Mohite, L.V.; Pinjari, D.V. Influence of intensified cellulose dissolution process on spinning and properties of lyocell fibres. *Chem. Eng. Process.-Process Intensif.* **2020**, *155*, 108063. [CrossRef]
94. Jadhav, S.; Lidhure, A.; Thakre, S.; Ganvir, V. Modified Lyocell process to improve dissolution of cellulosic pulp and pulp blends in NMMO solvent. *Cellulose* **2021**, *28*, 973–990. [CrossRef]
95. Foroughi, F.; Rezvani Ghomi, E.; Morshedi Dehaghi, F.; Borayek, R.; Ramakrishna, S. A review on the life cycle assessment of cellulose: From properties to the potential of making it a low carbon material. *Materials* **2021**, *14*, 714. [CrossRef]
96. Rac-Rumijowska, O.; Fiedot, M.; Karbownik, I.; Suchorska-Woźniak, P.; Teterycz, H. Synthesis of silver nanoparticles in NMMO and their in situ doping into cellulose fibers. *Cellulose* **2017**, *24*, 1355–1370. [CrossRef]
97. Janjic, S.; Kostic, M.; Vucinic, V.; Dimitrijevic, S.; Popovic, K.; Ristic, M.; Skundric, P. Biologically active fibers based on chitosan-coated lyocell fibers. *Carbohyd. Polym.* **2009**, *78*, 240–246. [CrossRef]
98. Röder, T.; Moosbauer, J.; Wöss, K.; Schlader, S.; Kraft, G. Man-made cellulose fibres–a comparison based on morphology and mechanical properties. *Lenzing. Berichte* **2013**, *91*, 7–12.
99. Bali, G.; Meng, X.; Deneff, J.I.; Sun, Q.; Ragauskas, A.J. The effect of alkaline pretreatment methods on cellulose structure and accessibility. *ChemSusChem* **2015**, *8*, 275–279. [CrossRef]
100. Colom, X.; Carrillo, F. Crystallinity changes in lyocell and viscose-type fibres by caustic treatment. *Eur. Polym. J.* **2002**, *38*, 2225–2230. [CrossRef]
101. Singh, S.C.; Murthy, Z.V.P. Study of cellulosic fibres morphological features and their modifications using hemicelluloses. *Cellulose* **2017**, *24*, 3119–3130. [CrossRef]

102. Wang, B.; Nie, Y.; Kang, Z.; Liu, X. Effects of coagulating conditions on the crystallinity, orientation and mechanical properties of regenerated cellulose fibers. *Int. J. Biol. Macromol.* **2023**, *225*, 1374–1383. [CrossRef]
103. Isogai, A.; Bergström, L. Preparation of cellulose nanofibers using green and sustainable chemistry. *Curr. Opin. Green Sust. Chem.* **2018**, *12*, 15–21. [CrossRef]
104. Hasan, K.F.; Champramary, S.; Al Hasan, K.N.; Indic, B.; Ahmed, T.; Pervez, M.N.; Horvath, P.G.; Bak, M.; Sandor, B.; Hofmann, T.; et al. Eco-friendly production of cellulosic fibers from Scots pine wood and sustainable nanosilver modification: A path toward sustainability. *Res. Eng.* **2023**, *19*, 101244. [CrossRef]
105. Strnad, S.; Šauperl, O.; Fras-Zemljič, L. Cellulose fibres functionalised by chitosan: Characterization and application. *Biopolymers* **2010**, 181–200. [CrossRef]
106. Miyamoto, I.; Matsuoka, Y.; Matsui, T.; Saito, M.; Okajima, K. Studies on structure of cuprammonium cellulose III. Structure of regenerated cellulose treated by cuprammonium solution. *Polym. J.* **1996**, *28*, 276–281. [CrossRef]
107. Karthik, T.; Gopalakrishnan, D. Environmental analysis of textile value chain: An overview. In *Roadmap to Sustainable Textiles and Clothing: Environmental and Social Aspects of Textiles and Clothing Supply Chain*; Springer: Berlin/Heidelberg, Germany, 2014; pp 153–188.
108. Ide, S. Filter made of cuprammonium regenerated cellulose for virus removal: A mini-review. *Cellulose* **2022**, *29*, 2779–2793. [CrossRef]
109. Cheng, H.N.; Dowd, M.K.; Selling, G.W.; Biswas, A. Synthesis of cellulose acetate from cotton byproducts. *Carbohyd. Polym.* **2010**, *80*, 449–452. [CrossRef]
110. Homem, N.C.; Amorim, M.T.P. Synthesis of cellulose acetate using as raw material textile wastes. *Mater. Today Proc.* **2020**, *31* (Suppl. S2), S315–S317. [CrossRef]
111. Nawaz, H.; He, A.; Wu, Z.; Wang, X.; Jiang, Y.; Ullah, A.; Xu, F.; Xie, F. Revisiting various mechanistic approaches for cellulose dissolution in different solvent systems: A comprehensive review. *Int. J. Biolog. Macromol.* **2024**, *273*, 133012. [CrossRef] [PubMed]
112. Swapnil, S.I.; Datta, N.; Mahmud, M.M.; Jahan, R.A.; Arafat, M.T. Morphology, mechanical, and physical properties of wet-spun cellulose acetate fiber in different solvent-coagulant systems and in-situ crosslinked environment. *J. Appl Polym. Sci.* **2021**, *138*, 50358. [CrossRef]
113. Slejko, E.A.; Tuan, A.; Scuor, N. From waste to value: Characterization of recycled cellulose acetate for sustainable waste management. *Waste Manag. Bull.* **2024**, *1*, 67–73. [CrossRef]
114. Wolfs, J.; Meier, M.A. A more sustainable synthesis approach for cellulose acetate using the DBU/CO_2 switchable solvent system. *Green Chem.* **2021**, *23*, 4410–4420. [CrossRef]
115. Kosan, B.; Michels, C.; Meister, F. Dissolution and forming of cellulose with ionic liquids. *Cellulose* **2008**, *15*, 59–66. [CrossRef]
116. Zhou, L.; Kang, Z.; Nie, Y.; Li, L. Fabrication of regenerated cellulose fibers with good strength and biocompatibility from green spinning process of ionic liquid. *Macromol. Mater. Eng.* **2021**, *306*, 2000741. [CrossRef]
117. Wang, C.G.; Li, N.; Wu, G.; Lin, T.T.; Lee, A.M.X.; Yang, S.W.; Li, Z.; Luo, H.K. Carbon Dioxide Mediated Cellulose Dissolution and Derivatization to Cellulose Carbonates in a Low-pressure System. *Carbohydr. Polym. Technol. Appl.* **2022**, *3*, 100186. [CrossRef]
118. Zhang, J.; Wu, J.; Yu, J.; Zhang, X.; He, J.; Zhang, J. Application of ionic liquids for dissolving cellulose and fabricating cellulose-based materials: State of the art and future trends. *Mater. Chem. Front.* **2017**, *1*, 1273–1290. [CrossRef]
119. Xiong, B.; Zhao, P.; Hu, K.; Zhang, L.; Cheng, G. Dissolution of cellulose in aqueous NaOH/urea solution: Role of urea. *Cellulose* **2014**, *21*, 1183–1192. [CrossRef]
120. Kayseri, G.Ö.; Bozdoğan, F.; Hes, L. Performance properties of regenerated cellulose fibers. *Text. Appar.* **2010**, *20*, 208–212.
121. Maher, R.R.; Warsman, R.H. *The Chemistry of Textile Fibers*, 2nd ed.; Printing; Royal Society Chem.: London, UK, 2015; pp 127–128. ISBN 978-1-78262-023-5.
122. Abu-Rous, M.; Cafuta, D.; Schuster, K.C. The role of the enhanced moisture absorption of regenerated cellulose fibers (MMCF) in textile comfort and the thermoregulation of the body. *Lenzing. Berichte* **2023**, *98*, 5–13.
123. Oprea, M.; Voicu, S.I. Recent advances in composites based on cellulose derivatives for biomedical applications. *Carbohyd. Polym.* **2020**, *247*, 116683. [CrossRef] [PubMed]
124. Kreze, T.; Jeler, S.; Strnad, S. Correlation between structure characteristics and adsorption properties of regenerated cellulose fibers. *Mater. Res. Innov.* **2002**, *5*, 277–283. [CrossRef]
125. Gizaw, M.; Thompson, J.; Faglie, A.; Lee, S.Y.; Neuenschwander, P.; Chou, S.F. Electrospun fibers as a dressing material for drug and biological agent delivery in wound healing applications. *Bioengineering* **2018**, *5*, 9. [CrossRef]
126. Zemljič, L.F.; Kreže, T.; Strnad, S.; Šauperl, O.; Vesel, A. Functional Vellulose Fibers for Hygienic and Medical Applications. In *Textiles: Types, Uses and Production Methods*; Nova Science Publishers, Inc.: Hauppauge, NY, USA, 2012; pp 467–487.
127. de Araújo, A.M., Jr.; Braido, G.; Saska, S.; Barud, H.S.; Franchi, L.P.; Assunção, R.M.; Scarel-Caminaga, R.M.; Capote, T.S.O.; Messaddeq, Y.; Ribeiro, S.J. Regenerated cellulose scaffolds: Preparation, characterization and toxicological evaluation. *Carbohyd. Polym.* **2016**, *136*, 892–898. [CrossRef] [PubMed]
128. Suñol, J.J.; Miralpeix, D.; Saurina, J.; Carrillo, F.; Colom, X. Thermal behavior of cellulose fibers with enzymatic or Na_2CO_3 treatment. *J. Therm. Anal. Calorim.* **2005**, *80*, 117–121. [CrossRef]
129. Nam, S.; Hillyer, M.B.; Condon, B.D. Method for identifying the triple transition (glass transition-dehydration-crystallization) of amorphous cellulose in cotton. *Carbohyd. Polym.* **2020**, *228*, 115374. [CrossRef]
130. Huson, M.; Denham, C. The Glass Transition of Cotton. *Cotton Fibres* **2017**, *43*, 354.

131. Roseveare, W.E.; Waller, R.C.; Wilson, J.N. Structure and Properties of Regenerated Cellulose. *Text. Res. J.* **1948**, *18*, 114–123. [CrossRef]
132. Rebenfeld, L. Fibers and fibrous materials. In *Textile Science and Technology*; Elsevier: Amsterdam, The Netherlands, 2002; Volume 3, pp 199–232.
133. Latif, W.; Basit, A.; Rehman, A.; Ashraf, M.; Iqbal, K.; Jabbar, A.; Baig, S.A.; Maqsood, S. Study of mechanical and comfort properties of modal with cotton and regenerated fibers blended woven fabrics. *J. Nat. Fibers* **2019**, *16*, 836–845. [CrossRef]
134. Kıvrak, N.M.; Ozdıl, N.; Mengüç, G.S. Characteristics of the yarns spun from regenerated cellulosic fibers. *Text. Appar.* **2018**, *28*, 107–117.
135. Zhang, H.; Shen, Y.; Edgar, K.J.; Yang, G.; Shao, H. Influence of cross-section shape on structure and properties of Lyocell fibers. *Cellulose* **2021**, *28*, 1191–1201. [CrossRef]
136. Udomkichdecha, W.; Chiarakorn, S.; Potiyaraj, P. Relationships between fibrillation behavior of lyocell fibers and their physical properties. *Text. Res. J.* **2002**, *72*, 939–943. [CrossRef]
137. Iqbal, D.; Zhao, R.; Sarwar, M.I.; Ning, X. Cellulose-Based Nanofibers Electrospun from Cuprammonium Solutions. *Preprint*, 2023. [CrossRef]
138. Yimin, Q. *Medical Textile Materials*, 1st ed.; Elsevier, Textile Institute: Amsterdam, The Netherlands, 2019.
139. Rajendran, S.; Anand, S.C. Developments in Medical Textiles. *Text. Prog.* **2002**, *32*, 1–42. [CrossRef]
140. Adanur, S. *Wellington Sears Handbook of Industrial Textiles*; Routledge: New York, NY, USA, 1995.
141. Anand, S.C. Medical Textile. In Proceedings of the 2nd international Conference, Bolton, UK, 24–25 August 1999; Anand, S.C., Ed.; Bolton Institute: Bolton, UK, 1999; pp 1–256.
142. Zhezhova, S.; Jordeva, S.; Golomeova Longurova, S.G.; Jovanov, S. Application of Technical Textile in Medicine. *Tekst. Ind.* **2021**, *2*, 21–29. [CrossRef]
143. Guru, R.; Kumar, A.; Grewal, D.; Kumar, R. Study of the Implantable and Non-Implantable Application in Medical Textile. In *Next-Generation Textiles*; IntechOpen: London, UK, 2022.
144. Medival Textile by Orgiline. Available online: https://leartex.com/medical-textile/ (accessed on 21 July 2024).
145. Tu, H.; Li, X.; Liu, Y.; Luo, L.; Duan, B.; Zhang, R. Recent progress in regenerated cellulose-based fibers from alkali/urea system via spinning process. *Carbohyd. Polym.* **2022**, *296*, 119942. [CrossRef]
146. Kim, H.C.; Kim, D.; Lee, J.Y.; Zhai, L.; Kim, J. Effect of Wet Spinning and Stretching to Enhance Mechanical Properties of Cellulose Nanofiber Filament. *Int. J. Precis. Eng. Manuf.-Green Technol.* **2019**, *6*, 567–575. [CrossRef]
147. Alistair, J.R.; Subhash, C.A. Medical textiles. In *Handbook of Technical Textiles*; Bolton Institute: Bolton, UK, 2001; pp 407–424.
148. Parvin, F.; Islam, I.; Urmy, Z.; Ahmed, S. A study on the textile materials applied in human medical treatment. *Eur. J. Physiother. Rehabil. Stud.* **2020**, *1*, 1–25.
149. Chaudary, S.N.; Borkar, S.P. Textiles for extracorporeal devices. *Indian Text. J.* **2009**, *65*, 79–84.
150. Medical Textiles Market Report by Product Type (Non-Woven, Knitted, Woven, and Others), Application (Implantable Goods, Non-Implantable Goods, Healthcare & Hygiene Products, and Others), and Region 2024–2032. Available online: https://www.researchandmarkets.com/reports/5946655/medical-textiles-market-report-product-type-non (accessed on 21 July 2024).
151. Medical Textiles Market Report by Product Type (Non-Woven, Knitted, Woven, and Others), Application (Implantable Goods, Non-Implantable Goods, Healthcare & Hygiene Products, and Others), and Region 2024–2032. Available online: https://www.marketresearch.com/IMARC-v3797/Medical-Textiles-Product-Type-Non-36504686/ (accessed on 21 July 2024).
152. Cullen, B.; Watt, P.W.; Lundqvist, C.; Silcock, D.; Schmidt, R.J.; Bogan, D.; Light, N.D. The role of oxidised regenerated cellulose/collagen in chronic wound repair and its potential mechanism of action. *Int. J. Biochem. Cell Biol.* **2002**, *34*, 1544–1556. [CrossRef]
153. Pei, Y.; Ye, D.; Zhao, Q.; Wang, X.; Zhang, C.; Huang, W.; Zhang, N.; Liu, S.; Zhang, L. Effectively promoting wound healing with cellulose/gelatin sponges constructed directly from a cellulose solution. *J. Mater. Chem. B* **2015**, *3*, 7518–7528. [CrossRef] [PubMed]
154. Chowdhury, N.A.; Al-Jumaily, A.M. Regenerated cellulose/polypyrrole/silver nanoparticles/ionic liquid composite films for potential wound healing applications. *Wound Med.* **2016**, *14*, 16–18. [CrossRef]
155. Abitbol, T.; Rivkin, A.; Cao, Y.; Nevo, Y.; Abraham, E.; Ben-Shalom, T.; Lapidot, S.; Shoseyov, O. Nanocellulose, a tiny fiber with huge applications. *Curr. Opin. Biotechnol.* **2016**, *39*, 76–88. [CrossRef] [PubMed]
156. Randhawa, A.; Dutta, S.D.; Ganguly, K.; Patil, T.V.; Patel, D.K.; Lim, K.T. A review of properties of nanocellulose, its synthesis, and potential in biomedical applications. *Appl. Sci.* **2022**, *12*, 7090. [CrossRef]
157. Maharjan, B.; Park, J.; Kaliannagounder, V.K.; Awasthi, G.P.; Joshi, M.K.; Park, C.H.; Kim, C.S. Regenerated cellulose nanofiber reinforced chitosan hydrogel scaffolds for bone tissue engineering. *Carbohyd. Polym.* **2021**, *251*, 117023. [CrossRef]
158. Zamel, D.; Khan, A.U.; Khan, A.N.; Waris, A.; Ilyas, M.; Ali, A.; Baset, A. Regenerated Cellulose and Composites for Biomedical Applications. In *Regenerated Cellulose and Composites: Morphology-Property Relationship*; Springer Nature: Singapore, 2023; pp 265–311.
159. Credou, J.; Berthelot, T. Cellulose: From biocompatible to bioactive material. *J. Mater. Chem. B* **2014**, *2*, 4767–4788. [CrossRef] [PubMed]
160. Kamel, S.A.; Khattab, T. Recent advances in cellulose-based biosensors for medical diagnosis. *Biosensors* **2020**, *10*, 67. [CrossRef]

161. Anusiya, G.; Jaiganesh, R. A review on fabrication methods of nanofibers and a special focus on application of cellulose nanofibers. *Carbohydr. Polym. Technol. Appl.* **2022**, *4*, 100262. [CrossRef]
162. Meftahi, A.; Momeni Heravi, M.E.; Barhoum, A.; Samyn, P.; Najarzadeh, H.; Alibakhshi, S. Cellulose nanofibers: Synthesis, unique properties, and emerging applications. In *Handbook of Nanocelluloses: Classification, Properties, Fabrication, and Emerging Applications*; Springer: Berlin/Heidelberg, Germany, 2022; pp 233–262.
163. Baye, B.; Tesfaye, T. The new generation fibers: A review of high performance and specialty fibers. *Polym. Bull.* **2022**, *79*, 9221–9235. [CrossRef]
164. Michud, A.; Tanttu, M.; Asaadi, S.; Ma, Y.; Netti, E.; Kääriainen, P.; Persson, A.; Berntsson, A.; Hummel, M.; Sixta, H. Ioncell-F: Ionic liquid-based cellulosic textile fibers as an alternative to viscose and Lyocell. *Text. Res. J.* **2016**, *86*, 543–552. [CrossRef]
165. Hart, E. Ioncell Fibres from Alternative Cellulose Sources for Nonwovens. Master's Thesis, Kemian Tekniikan Korkeakoulu, Espoo, Finland, 2024.
166. Elsayed, S.; Hellsten, S.; Guizani, C.; Witos, J.; Rissanen, M.; Rantamäki, A.H.; Varis, P.; Wiedmer, S.K.; Sixta, H. Recycling of superbase-based ionic liquid solvents for the production of textile-grade regenerated cellulose fibers in the lyocell process. *ACS Sustain. Chem. Eng.* **2020**, *8*, 14217–14227. [CrossRef]
167. Liu, H.; Fan, W.; Miao, Y.; Dou, H.; Shi, Y.; Wang, S.; Zhang, X.; Hou, L.; Yu, X.; Lam, S.S.; et al. Closed-loop recycling of colored regenerated cellulose fibers from the dyed cotton textile waste. *Cellulose* **2023**, *30*, 2597–2610. [CrossRef]
168. Liu, X.; Tian, Y.; Wang, L.; Chen, L.; Jin, Z.; Zhang, Q. A Cost-Effective and Chemical-Recycling Approach for Facile Preparation of Regenerated Cellulose Materials. *Nano Lett.* **2024**, *24*, 9074–9081. [CrossRef]
169. Liu, W.; Liu, S.; Liu, T.; Liu, T.; Zhang, J.; Liu, H. Eco-friendly post-consumer cotton waste recycling for regenerated cellulose fibers. *Carbohyd. Polym.* **2019**, *206*, 141–148. [CrossRef]
170. Zeng, C.; Wang, H.; Bu, F.; Cui, Q.; Li, J.; Liang, Z.; Zhao, L.; Yi, C. A regenerated cellulose fiber with high mechanical properties for temperature-adaptive thermal management. *Int. J. Biol. Macromol.* **2024**, *274*, 133550. [CrossRef]
171. Available online: https://www.enicbcmed.eu/sites/default/files/2023-01/A%20Study%20on%20technologies%20for%20recycling%20and%20re-use%20of%20textile%20scraps%20-%20TMA%20WP6.pdf (accessed on 28 October 2024).
172. Forsberg, D.C.R.; Bengtsson, J.; Hollinger, N.; Kaldéus, T. Towards Sustainable Viscose-to-Viscose Production: Strategies for Recycling of Viscose Fibres. *Sustainability* **2024**, *16*, 4127. [CrossRef]
173. Li, Y.; Peng, J.; Liu, X.; Song, D.; Xu, W.; Zhu, K. Dissolving waste viscose to spin cellulose fibers. *Polymers* **2021**, *237*, 124349. [CrossRef]
174. dos Santos, R.F.; Oliveira, F.R.; Rocha, M.R.D.; Velez, R.A.; Steffens, F. Reinforced cementitious composite using viscose rayon fiber from textile industry waste. *J. Eng. Fibers Fabr.* **2022**, *17*, 15589250221115722. [CrossRef]
175. Heikkilä, P.; Määttänen, M.; Jetsu, P.; Kamppuri, T.; Paunonen, S. *Nonwovens from Mechanically Recycled Fibres for Medical Applications*; VTT Research Report No. VTT-R-00923-20; VTT Technical Research Centre of Finland: Espoo, Finland, 2020.
176. Lakshmanan, S.O.; Raghavendran, G. Regenerated Sustainable Fibres. In *Sustainable Innovations in Textile Fibres. Textile Science and Clothing Technology*; Muthu, S., Ed.; Springer: Singapore, 2018.
177. Yang, X.; Fan, W.; Wang, H.; Shi, Y.; Wang, S.; Liew, R.K.; Ge, S. Recycling of bast textile wastes into high value-added products: A review. *Environ. Chem. Lett.* **2022**, *20*, 3747–3763. [CrossRef]
178. Wojciechowska, P. Fibres and textiles in the circular economy. In *Fundamentals of Natural Fibres and Textiles*; Woodhead Publishing: Cambridge, UK, 2021; pp 691–717.
179. Bracciale, M.P.; de Caprariis, B.; Musivand, S.; Damizia, M.; De Filippis, P. Chemical Recycling of Cellulose Acetate Eyewear Industry Waste by Hydrothermal Treatment. *Ind. Eng. Chem. Res.* **2024**, *63*, 5078–5088. [CrossRef]
180. Rodrigues Filho, G.; Monteiro, D.S.; da Silva Meireles, C.; de Assunção, R.M.N.; Cerqueira, D.A.; Barud, H.S.; Ribeiro, S.J.L.; Messadeq, Y. Synthesis and characterization of cellulose acetate produced from recycled newspaper. *Carbohyd. Polym.* **2008**, *73*, 74–82. [CrossRef]
181. Jiji, S.; Maharajan, K.; Kadirvelu, K. Recent developments of bacterial nanocellulose porous scaffolds in biomedical applications. *Nanocellulose Mater.* **2022**, 83–104.
182. Farahani, M.; Shafiee, A. Wound healing: From passive to smart dressings. *Adv. Healthc. Mater.* **2021**, *10*, 2100477. [CrossRef] [PubMed]
183. Boateng, J.S.; Matthews, K.H.; Stevens, H.N.; Eccleston, G.M. Wound healing dressings and drug delivery systems: A review. *J. Pharm. Sci.* **2008**, *97*, 2892–2923. [CrossRef]
184. Das, S.K.; Chinnappan, A.; Jayathilaka, W.A.D.M.; Gosh, R.; Baskar, C.; Ramakrishna, S. Challenges and potential solutions for 100% recycling of medical textiles. *Mater. Circ. Econ.* **2021**, *3*, 13. [CrossRef]
185. Prete, S.; Dattilo, M.; Patitucci, F.; Pezzi, G.; Parisi, O.I.; Puoci, F. Natural and synthetic polymeric biomaterials for application in wound management. *J. Funct. Biomater.* **2023**, *14*, 455. [CrossRef]

Disclaimer/Publisher's Note: The statements, opinions and data contained in all publications are solely those of the individual author(s) and contributor(s) and not of MDPI and/or the editor(s). MDPI and/or the editor(s) disclaim responsibility for any injury to people or property resulting from any ideas, methods, instructions or products referred to in the content.

MDPI AG
Grosspeteranlage 5
4052 Basel
Switzerland
Tel.: +41 61 683 77 34

Journal of Functional Biomaterials Editorial Office
E-mail: jfb@mdpi.com
www.mdpi.com/journal/jfb

Disclaimer/Publisher's Note: The title and front matter of this reprint are at the discretion of the Guest Editors. The publisher is not responsible for their content or any associated concerns. The statements, opinions and data contained in all individual articles are solely those of the individual Editors and contributors and not of MDPI. MDPI disclaims responsibility for any injury to people or property resulting from any ideas, methods, instructions or products referred to in the content.

www.ingramcontent.com/pod-product-compliance
Lightning Source LLC
LaVergne TN
LVHW072331090526
838202LV00019B/2393